PROGRESS IN CLINICAL AND BIOLOGICAL RESEARCH

RECENT TITLES

Vol 137: **PCB Poisoning in Japan and Taiwan,** Masanori Kuratsune, Raymond E. Shapiro, *Editors*

Vol 138: **Chemical Taxonomy, Molecular Biology, and Function of Plant Lectins,** Irwin J. Goldstein, Marilynn E. Etzler, *Editors*

Vol 139: **Difficult Decisions in Medical Ethics: The Fourth Volume in a Series on Ethics, Humanism, and Medicine,** Doreen Ganos, Rachel E. Lipson, Gwynedd Warren, Barbara J. Weil, *Editors*

Vol 140: **Abnormal Functional Development of the Heart, Lungs, and Kidneys: Approaches to Functional Teratology,** Casimer T. Grabowski, Robert J. Kavlock, *Editors*

Vol 141: **Industrial Hazards of Plastics and Synthetic Elastomers,** Jorma Järvisalo, Pirkko Pfäffli, Harri Vainio, *Editors*

Vol 142: **Hormones and Cancer,** Erlio Gurpide, Ricardo Calandra, Carlos Levy, Roberto J. Soto, *Editors*

Vol 143: **Viral Hepatitis and Delta Infection,** Giorgio Verme, Ferruccio Bonino, Mario Rizzetto, *Editors*

Vol 144: **Chemical and Biological Aspects of Vitamin B_6** *Catalysis,* A.E. Evangelopoulos, *Editor*. Published in 2 volumes. Part A: *Metabolism, Structure, and Function of Phosphorylases, Decarboxylases, and Other Enzymes*. Part B: *Metabolism, Structure, and Function of Transaminases*

Vol 145: **New Approaches to the Study of Benign Prostatic Hyperplasia,** Frances A. Kimball, Allen E. Buhl, Donald B. Carter, *Editors*

Vol 146: **Experimental Allergic Encephalomyelitis: A Useful Model for Multiple Sclerosis,** Ellsworth C. Alvord, Jr., Marian W. Kies, Anthony J. Suckling, *Editors*

Vol 147: **Genetic Epidemiology of Coronary Heart Disease: Past, Present, and Future,** D.C. Rao, Robert C. Elston, Lewis H. Kuller, Manning Feinleib, Christine Carter, Richard Havlik, *Editors*

Vol 148: **Aplastic Anemia: Stem Cell Biology and Advances in Treatment,** Neal S. Young, Alan S. Levine, R. Keith Humphries, *Editors*

Vol 149: **Advances in Immunobiology: Blood Cell Antigens and Bone Marrow Transplantation,** Jeffrey McCullough, S. Gerald Sandler, *Editors*

Vol 150: **Factor VIII Inhibitors,** Leon W. Hoyer, *Editor*

Vol. 151: **Matrices and Cell Differentiation,** R.B. Kemp, J.R. Hinchliffe, *Editors*

Vol 152: **Intrahepatic Calculi,** Kunio Okuda, Fumio Nakayama, John Wong, *Editors*

Vol 153: **Progress and Controversies in Oncological Urology,** Karl H. Kurth, Frans M.J. Debruyne, Fritz H. Schroeder, Ted A.W. Splinter, Theo D.J. Wagener, *Editors*

Vol 154: **Myelofibrosis and the Biology of Connective Tissue,** Paul D. Berk, Hugo Castro-Malaspina, Louis R. Wasserman, *Editors*

Vol 155: **Malaria and the Red Cell,** John W. Eaton, George J. Brewer, *Editors*

Vol 156: **Advances in Cancer Control: Epidemiology and Research,** Paul F. Engstrom, Paul N. Anderson, Lee E. Mortenson, *Editors*

Vol 157: **Recognition Proteins, Receptors, and Probes: Invertebrates,** Elias Cohen, *Editor*

Vol 158: **The Methylxanthine Beverages and Foods: Chemistry, Consumption, and Health Effects,** Gene A. Spiller, *Editor*

Vol 159: **Erythrocyte Membranes 3: Recent Clinical and Experimental Advances,** Walter C. Kruckeberg, John W. Eaton, Jon Aster, George J. Brewer, *Editors*

Vol 160: **Reproduction: The New Frontier in Occupational and Environmental Health Research,** James E. Lockey, Grace K. Lemasters, William R. Keye, Jr., *Editors*

Vol 161: **Chemical Regulation of Immunity in Veterinary Medicine,** Meir Kende, Joseph Gainer, Michael Chirigos, *Editors*

Vol 162: **Bladder Cancer**, René Küss, Saad Khoury, Louis J. Denis, Gerald P. Murphy, James P. Karr, *Editors*. Published in 2 volumes: Part A: *Pathology, Diagnosis, and Surgery*. Part B: *Radiation, Local and Systemic Chemotherapy, and New Treatment Modalities*

Vol 163: **Prevention of Physical and Mental Congenital Defects**, Maurice Marois, *Editor.* Published in 3 volumes: Part A: *The Scope of the Problem.* Part B: *Epidemiology, Early Detection and Therapy, and Environmental Factors.* Part C: *Basic and Medical Science, Education, and Future Strategies*

Vol 164: **Information and Energy Transduction in Biological Membranes,** C. Liana Bolis, Ernst J.M. Helmreich, Hermann Passow, *Editors*

Vol 165: **The Red Cell: Sixth Ann Arbor Conference**, George J. Brewer, *Editor*

Vol 166: **Human Alkaline Phosphatases,** Torgny Stigbrand, William H. Fishman, *Editors*

Vol 167: **Ethopharmacological Aggression Research**, Klaus A. Miczek, Menno R. Kruk, Berend Olivier, *Editors*

Vol 168: **Epithelial Calcium and Phosphate Transport: Molecular and Cellular Aspects,** Felix Bronner, Meinrad Peterlik, *Editors*

Vol 169: **Biological Perspectives on Aggression,** Kevin J. Flannelly, Robert J. Blanchard, D. Caroline Blanchard, *Editors*

Vol 170: **Porphyrin Localization and Treatment of Tumors,** Daniel R. Doiron, Charles J. Gomer, *Editors*

Vol 171: **Developmental Mechanisms: Normal and Abnormal,** James W. Lash, Lauri Saxén, *Editors*

Vol 172: **Molecular Basis of Cancer,** Robert Rein, *Editor*. Published in 2 volumes: Part A: *Macromolecular Structure, Carcinogens, and Oncogenes.* Part B: *Macromolecular Recognition, Chemotherapy, and Immunology*

Vol 173: **Pepsinogens in Man: Clinical and Genetic Advances,** Johanna Kreuning, I. Michael Samloff, Jerome I. Rotter, Aldur Eriksson, *Editors*

Vol 174: **Enzymology of Carbonyl Metabolism 2: Aldehyde Dehydrogenase, Aldo-Keto Reductase, and Alcohol Dehydrogenase,** T. Geoffrey Flynn, Henry Weiner, *Editors*

Vol 175: **Advances in Neuroblastoma Research,** Audrey E. Evans, Giulio J. D'Angio, Robert C. Seeger, *Editors*

Vol 176: **Contemporary Sensory Neurobiology,** Manning J. Correia, Adrian A. Perachio, *Editors*

Vol 177: **Medical Genetics: Past, Present, Future,** Kåre Berg, *Editor*

Vol 178: **Bluetongue and Related Orbiviruses,** T. Lynwood Barber, Michael M. Jochim, *Editors*

Vol 179: **Taurine: Biological Actions and Clinical Perspectives,** Simo S. Oja, Liisa Ahtee, Pirjo Kontro, Matti K. Paasonen, *Editors*

Vol 180: **Intracellular Protein Catabolism,** Edward A. Khairallah, Judith S. Bond, John W.C. Bird, *Editors*

Vol. 181: **Germfree Research: Microflora Control and Its Application to the Biomedical Sciences,** Bernard S. Wostmann, *Editor*, Julian R. Pleasants, Morris Pollard, Bernard A. Teah, Morris Wagner, *Co-Editors*

Please contact the publisher for information about previous titles in this series.

GERMFREE RESEARCH
MICROFLORA CONTROL AND ITS APPLICATION TO THE BIOMEDICAL SCIENCES

GERMFREE RESEARCH
MICROFLORA CONTROL AND ITS APPLICATION TO THE BIOMEDICAL SCIENCES

Proceedings of the VIIIth International Symposium on Germfree Research, held at Notre Dame, Indiana
June 17–22, 1984

Editor

Bernard S. Wostmann
Lobund Laboratory
University of Notre Dame
Notre Dame, Indiana

Co-Editors

Julian R. Pleasants **Bernard A. Teah**
Morris Pollard **Morris Wagner**
Lobund Laboratory
University of Notre Dame
Notre Dame, Indiana

ALAN R. LISS, INC. • NEW YORK

Address all Inquiries to the Publisher
Alan R. Liss, Inc., 41 East 11th Street, New York, NY 10003

Printed in the United States of America.

Library of Congress Cataloging in Publication Data

International Symposium on Germfree Research (8th : 1984 :
Notre Dame, Ind.)
Germfree research.

Includes index.
1. Germfree life—Congresses. I. Wostmann, Bernard S.
II. Title.
QH324.4.I48 1984 619 85-4306
ISBN 0-8451-5031-6

Contents

Contributors

Mark Achtman, Max-Planck-Institut für Molekulare Genetik, Berlin, West Germany **[403]**

Carl M. Allen, Section of Oral Pathology and Diagnosis, The Ohio State University, College of Dentistry, Columbus, OH **[283]**

Antoine O. Andremont, Department of Medical Microbiology, Institut Gustave-Roussy, Villejuif, France **[175,255,275,279]**

Patrice Ardouin, Service d'Expérimentation Animale, Institut Gustave-Roussy, Villejuif, France **[43,55,175]**

Renate Arnold, Zentrum Innere Medizin, Universität Ulm, Ulm, West Germany **[433]**

Junpei Asai, Laboratory of Germfree Life, Research Institute for Disease Mechanism and Control, Nagoya University School of Medicine, Nagoya, Japan **[377]**

Karen A. Baker, Division of Hematology, Brigham and Women's Hospital, Boston, MA **[443]**

Edward Balish, Departments of Medical Microbiology and Surgery, University of Wisconsin Medical School, Madison, WI **[199,333,407]**

Kenneth F. Bartizal, Jr., Department of Medical Microbiology and Immunology, Kirksville College of Osteopathic Medicine, Kirksville, MO **[333,407]**

Patricia M. Bealmear, Department of Dermatology and Viral Leukemia, Roswell Park Memorial Institute, Buffalo, NY **[455,461,475]**

Bernard B. Beaulieu, Jr., School of Public Health, Harvard University, Boston, MA **[127]**

Margaret Beaver-Johnson, Lobund Laboratory, University of Notre Dame, Notre Dame, IN **[107,111]**

E.X. Beck, Department of Oral Biology, College of Dentistry, Ohio State University, Columbus, OH **[211]**

Frank M. Beck, Department of Oral Pathology and Diagnosis, The Ohio State University, College of Dentistry, Columbus, OH **[211,283]**

R. Benner, Department of Cell Biology and Genetics, Erasmus University, Rotterdam, The Netherlands **[355,359]**

Rodney D. Berg, Department of Microbiology and Immunology, Louisiana State University Medical Center, Shreveport, LA **[195]**

Hilaire Bewa, Laboratoire d'Ecologie Microbienne, INRA–CNRZ, Jouy-en-Josas, France **[163]**

J. Biemond, Departments of Infectious Diseases and Hematology, Gastroenterology, and Human Genetics, University Hospital of Leiden, Leiden, The Netherlands **[429]**

W.M. Bitter, Department of Cell Biology and Genetics, Erasmus University, Rotterdam, The Netherlands **[359]**

Arvid Bjørneklett, National Hospital, Oslo, Norway **[241]**

Hollis Bodman, Department of Microbiology, Brigham and Women's Hospital, Boston, MA **[443]**

The number in brackets is the opening page number of the contributor's article.

N. Bos, Department of Cell Biology and Genetics, Erasmus University, Rotterdam, The Netherlands **[355]**

Chantal Bridonneau, Laboratoire d'Ecologie Microbienne, INRA–CNRZ, Jouy-en-Josas, France **[163]**

C.D. Buckner, Department of Medicine, Division of Oncology, The Fred Hutchinson Cancer Research Center, Seattle, WA **[423]**

Gerd D. Burchard, Clinical Department, Bernhard-Nocht-Institut für Schiffs- und Tropenkrankheiten, Hamburg, Federal Republic of Germany **[231]**

Bodil Carlstedt-Duke, Karolinska Institutet, Stockholm, Sweden **[241]**

Carl E. Cerniglia, Department of Microbiology, National Center for Toxicological Research, Food and Drug Administration, Jefferson, AR **[133]**

M. Chaigneau, Université René Descartes, Paris, France **[119]**

Otar V. Chakhava, Laboratory of Gnotobiology, The Gamaleya Institute for Epidemiology and Microbiology, Moscow, USSR **[159]**

Arjun D. Chanana, Medical Research Center, Brookhaven National Laboratory, Upton, NY **[367]**

J.P. Chappuis, Laboratoire d'Ecologie Microbienne, INRA–CNRZ, Jouy-en-Josas, France **[187]**

Seto M. Chice, Department of Pathology, State University of New York, Downstate Medical Center, Brooklyn, NY **[387]**

Chu Guangda, Department of Gnotobiology, Medical Laboratory Animal Center, Chinese Academy of Medical Sciences, Beijing, People's Republic of China **[191]**

Tae J. Chung, Department of Internal Medicine, Hanyang University Hospital, Seoul, Korea **[329]**

Marie E. Coates, Robens Institute, University of Surrey, Guildford, Surrey, United Kingdom **[123]**

G. Corthier, Laboratoire d'Ecologie Microbienne, INRA–CNRZ, Jouy-en-Josas, France **[179,183,207]**

A. Dekker, Department of Hematology, University Hospital of Utrecht, Utrecht, The Netherlands **[251]**

E.D. Dennis, Department of Animal Supply, National Animal Disease Center, Ames, IA **[69]**

G. De Pauw, Department of Microbiology, The Rega Institute, Leuven, Belgium **[103]**

H.G. de Vries-Hospers, Department of Medical Microbiology, University of Groningen, Groningen, The Netherlands **[155,251,259]**

N. Dierick, Laboratorium voor Voeding en Hygiene, Melle, Belgium **[119]**

Manfred Dietrich, Clinical Department, Bernhard-Nocht-Institut für Schiffs - und Tropenkrankheiten, Hamburg, Federal Republic of Germany **[231]**

Susan B. Dillon, Department of Microbiology, Thomas Jefferson Medical College, Philadelphia, PA **[381]**

F. Dubos, Laboratoire d'Ecologie Microbienne, INRA–CNRZ, Jouy-en-Josas, France **[179,207]**

Robert Ducluzeau, Laboratoire d'Ecologie Microbienne, INRA–CNRZ, Jouy-en-Josas, France **[171,391]**

Helen G. Durkin, Deparment of Pathology, State University of New York, Downstate Medical Center, Brooklyn, NY **[387]**

Yvonne Duval-Iflah, Laboratoire d'Ecologie Microbienne, INRA–CNRZ, Jouy-en-Josas, France **[187]**

Rudolf Edenharder, Institute of Hygiene, University of Mainz, Mainz, Federal Republic of Germany **[305]**

Karam El-Bayoumy, Division of Nutrition and Endocrinology, American Health Foundation, Valhalla, NY **[293]**

Stefan Engelhorn, Clinical Department, Bernhard-Nocht-Institut für Schiffs - und Tropenkrankheiten, Hamburg, Federal Republic of Germany **[231]**

Joshua Epstein, Department of Medical Oncology, Roswell Park Memorial Institute, Buffalo, NY **[455,461]**

E.D. Exum, Departments of Experimental Hematology and Biochemistry, Armed Forces Radiobiology Research Institute, Bethesda, MD **[139]**

H. Eyssen, Department of Microbiology, The Rega Institute, Leuven, Belgium **[103]**

Walter R. Farkas, Department of Environmental Practice, The University of Tennessee College of Veterinary Medicine, Knoxville, TN **[91]**

Elisabeth Faure, Service d'Expérimentation Animale, Institut Gustave-Roussy, Villejuif, France **[175]**

D. Fiere, Department of Hematology, Hôpital Edouard Herriot, Lyon, France **[263]**

Audrey Fjelde, Department of Biological Resources, Roswell Park Memorial Institute, Buffalo, NY **[319]**

Katrina G. Fuller, Department of Microbiology and Immunology, Louisiana State University Medical Center, Shreveport, LA **[195]**

W. Gaus, Department of Documentation and Statistics, University of Ulm, Ulm, Federal Republic of Germany **[251]**

H.M. Gelston, Jr., Department of Experimental Hematology, National Naval Medical Center, Bethesda, MD **[139]**

Akira Ghoda, Kitasato University, Tokyo, Japan **[325]**

Peter Goldman, Department of Nutrition, Food and Drug Laboratory, Boston, MA **[127]**

Charles Gosse, Service d'Expérimentation Animale, Institut Gustave-Roussy, Villejuif, France **[43,55,175]**

Gerda Groen, Department of Medical Microbiology, University of Groningen, Groningen, The Netherlands **[151]**

Gu Fangzhou, Chinese Academy of Medical Sciences, Beijing, People's Republic of China **[47]**

H.F.L. Guiot, Department of Infectious Diseases, University Hospital of Leiden; J.A. Cohen Institute, Leiden, The Netherlands **[267,429]**

Bengt E. Gustafsson, Department of Germfree Research, Karolinska Institutet, Stockholm, Sweden **[17,99,241]**

D. Guyotat, Department of Hematology, Hôpital Edouard Herriot, Lyon, France **[263]**

Frank E. Halleck, American Sterilizer Company, Erie, PA **[61]**

Rainer Hammann, Institute for Medical Microbiology, University of Bonn, Bonn, Federal Republic of Germany **[305]**

Elsa Haralambie, Institut für Medizinische Mikrobiologie, Universitätsklinikum, Essen, Federal Republic of Germany **[155,225]**

Suzanne Harbin, Department of Surgery, Louisiana State University School of Medicine, New Orleans, LA **[467]**

Kazuo Hashimoto, Department of Microbiology, Tokai University School of Medicine, Isehara, Kanagawa, Japan **[347,351,363]**

P.J. Heidt, Department of Microbiology and Gnotobiology, Radiobiological Institute TNO, Rijswijk, The Netherlands **[437]**

H. Heimpel, Zentrum Innere Medizin, Universität Ulm, Ulm, West Germany **[433]**

Willi P.O. Heine, Zentralinstitut für Versuchstierzucht, Hannover, Federal Republic of Germany **[35]**

W. Heit, Zentrum Innere Medizin, Universität Ulm, Ulm, West Germany **[433]**

G. Helmus, Department of Medical Microbiology, University of Groningen, Groningen, The Netherlands **[155]**

James B. Heneghan, Departments of Surgery and Physiology, Louisiana State University School of Medicine, New Orleans, LA **[315,467]**

Michael J. Herrmann, Department of Dermatology and Viral Leukemia, Roswell Park Memorial Institute, Buffalo, NY **[455]**

Ole A. Holtermann, Department of Dermatology, Roswell Park Memorial Institute, Buffalo, NY **[319]**

H. Hooijkaas, Department of Cell Biology and Genetics, Erasmus University, Rotterdam, The Netherlands **[355,359]**

Torgeir Høverstad, National Hospital, Oslo, Norway **[241]**

Michel Huard, Usine d'Alimentation Rationelle, Villemoisson Sur Orge, France **[55]**

Sylvie Hudault, Laboratoire d'Ecologie Microbienne, INRA–CNRZ, Jouy-en-Josas, France **[163]**

Kikuji Itoh, Animal Psychology Laboratory, The Institute of Physical and Chemical Research, Bunkyo-ku, Tokyo, Japan **[217]**

W.E. Jackson, III, Departments of Experimental Hematology and Biochemistry, Armed Forces Radiobiology Research Institute, Bethesda, MD **[139]**

Mark T. Jensen, Department of Environmental Health and Safety, National Animal Disease Center, Ames, IA **[69]**

Jing Shaoliang, Department of Gnotobiology, Medical Laboratory Animal Center, Chinese Academy of Medical Sciences, Beijing, People's Republic of China **[191]**

Darrel D. Joel, Medical Research Center, Brookhaven National Laboratory, Upton, NY **[367]**

Kang Bai, Laboratory for Normal Microbial Flora, Dalian Medical College, Dalian, People's Republic of China **[169]**

Shizuo Katagiri, The Kitasato Institute Hospital, Tokyo, Japan **[325]**

Eiji Kawamura, The Kitasato Institute Hospital, Tokyo, Japan **[325]**

T. Kawashima, Biomedical Research Laboratory, Morinaga Milk Industry Co. Ltd., Tokyo, Japan **[399]**

M.J. Kieskamp, Department of Microbiology and Gnotobiology, Radiobiological Institute TNO, Rijswijk, The Netherlands **[437]**

Yoon B. Kim, Department of Microbiology and Immunology, University of Health Science and The Chicago Medical School, North Chicago, IL **[329,337,371]**

T.R. Kingsley, South Bend Center for Medical Education, Indiana University, and Lobund Laboratory, University of Notre Dame, Notre Dame, IN **[77]**

E.C. Klasen, Departments of Infectious Diseases and Hematology, Gastroenterology, and Human Genetics, University of Leiden, Leiden, The Netherlands **[429]**

Ronald L. Koch, School of Public Health, Harvard University, Boston, MA **[127]**

Takashi Koshikawa, Laboratory of Germfree Life, Research Institute for Disease Mechanism and Control, Nagoya University School of Medicine, Nagoya, Japan **[377]**

D. Krieger, Zentrum Innere Medizin, Universität Ulm, Ulm, West Germany **[433]**

B. Kubanek, Abteilung Transfusionsmedizin, Zentrum Innere Medizin, Universität Ulm, Ulm, West Germany **[433]**

Yoshiyuki Kubota, Kitasato University, Tokyo, Japan **[325]**

Eberhard Kurrle, Zentrum Inner Medizin, Universität Ulm, Ulm, West Germany **[251,433]**

Monique Ladire, Laboratoire d'Ecologie Microbienne, INRA–CNRZ, Jouy-en-Josas, France **[171]**

G.D. Ledney, Department of Experimental Hematology, National Naval Medical Center, Bethesda, MD **[139]**

Florence C. Léonard, Department of Medical Microbiology, Institut Gustave-Roussy, Villejuif, France **[279]**

Li Zhongyuan, Laboratory of Microbiology, Institute of Epidemiology and Microbiology, Chinese Center for Preventive Medicine, Beijing, People's Republic of China **[191]**

Yen S. Liao, Laboratory of Microbiology, Jiangxi Academy of Sciences, Nanchang, Jiangxi, People's Republic of China **[221]**

Louise I. Lindholm, Lobund Laboratory, University of Notre Dame, Notre Dame, IN **[145]**

Egil Lingaas, National Hospital, Oslo, Norway **[241]**

G. Linzenmeier, Institut für Medizinische Mikrobiologie, Universitätsklinikum, Essen, Federal Republic of Germany **[155,225]**

Yves M. Louis, American Health Foundation, Valhalla, NY **[293]**

Alistair G. Low, Department of Nutrition, National Institute for Research in Dairying, Reading, Berks, England **[115]**

Lu Longmu, Department of Laboratory Animal Science, Wuhan Institute for Biological Products, Wuchang, Hubei, China **[311]**

Phyllis H. Luckert, Lobund Laboratory, University of Notre Dame, Notre Dame, IN **[289,447]**

Fredrick A. Lurie, Section of Oral Pathology and Diagnosis, The Ohio State University, College of Dentistry, Columbus, OH **[283]**

Thomas T. MacDonald, Department of Microbiology, Thomas Jefferson Medical College, Philadelphia, PA **[381]**

Emilio A. Machado, Department of Medical Biology, The University of Tennessee, Knoxville, TN **[91]**

K. Machii, Department of Veterinary Public Health, The Institute of Public Health, Tokyo, Japan **[399]**

Noriyuki Maruo, The Kitasato Institute Hospital, Tokyo, Japan **[325]**

Peter J. Matthews, Department of Animal Supply, National Animal Disease Center, Ames, IA **[69]**

Carol McGarigle, Division of Nursing, Bone Marrow Transplant Unit, Brigham and Women's Hospital, Boston, MA **[443]**

Martha A. McLafferty, School of Public Health, Harvard University, Boston, MA **[127]**

A. Meshorer, Lobund Laboratory, University of Notre Dame, Notre Dame, IN **[447]**

Tore Midtvedt, Department of Germfree Research, Karolinska Institutet, Stockholm, Sweden **[99,241]**

Olgerts P. Miniats, Department of Clinical Studies, Ontario Veterinary College, University of Guelph, Guelph, Ontario, Canada **[411]**

Tomotari Mitsuoka, Department of Biomedical Science, Faculty of Agriculture, The University of Tokyo, Bunkyo-Ku, Tokyo, Japan **[217,297]**

Masasumi Miyakawa, Laboratory of Germfree Life, Research Institute of Disease Mechanism and Control, University School of Medicine, Nagoya, Japan **[301]**

Takeo Mizutani, Animal Physiology Laboratory, The Institute of Physical and Chemical Reserach, Wako-shi, Japan **[297]**

H. Momose, Biomedical Research Laboratory, Morinaga Milk Industry Co. Ltd., Tokyo, Japan **[399]**

Tasso Moraes-Santos, Bioquimica-Immunologia, Instituto Ciencias Biologicas, Belo Horizonte, Brasil **[235]**

C. Moreau, Laboratoire d'Ecologie Microbienne, INRA–CNRZ, Jouy-en-Josas, France **[179,183]**

M.-Ch. Moreau, Laboratoire d'Ecologie Microbienne, INRA–CNRZ, Jouy-en-Josas, France **[391]**

Makio Mukai, Department of Pathology, Keio University, School of Medicine, Tokyo, Japan **[203]**

C.J.J. Mulder, Departments of Infectious Diseases and Hematology, Gastroenterology, and Human Genetics, University Hospital of Leiden, Leiden, The Netherlands **[429]**

M.C. Muller, Laboratoire d'Ecologie Microbienne, INRA–CNRZ, Jouy-en-Josas, France **[179,183,391]**

Hiroshi Nagura, Laboratory of Germfree Life, Research Institute for Disease Mechanism and Control, Nagoya University School of Medicine, Nagoya, Japan **[377]**

Paul K. Nakane, Department of Cell Biology, Tokai University School of Medicine, Isehara, Kanagawa, Japan **[363]**

Kikuo Nomoto, Kyushu University, Fukuoka, Japan **[325]**

Karin E. Norin, Karolinska Institutet, Stockholm, Sweden **[241]**

Lionelle Nugon-Baudon, Laboratoire d'Ecologie Microbienne, INRA–CNRZ, Jouy-en-Josas, France **[119]**

Nobuhiko Ohnishi, Department of Microbiology, Tokai University School of Medicine, Isehara, Kanagawa, Japan **[363]**

Tetsuro Okano, Kitasato University, Tokyo, Japan **[325]**

Atsushi Ozawa, Department of Microbiology, Tokai University School of Medicine, Isehara, Kanagawa, Japan **[363]**

Sureshkant Patel, Department of Dermatology, Roswell Park Memorial Institute, Buffalo, NY **[319]**

Sophie S. Pecquet, Department of Microbiology, Institut Gustave-Roussy, Villejuif, France **[255]**

Barbara A. Peri, Department of Pediatrics, The University of Chicago, Chicago, IL **[395]**

Alan G. Peters, Production Department, OLAC 1976 Ltd., Bicester, Oxon, England **[39]**

Arthur W. Phillips, Department of Biology, Biological Research Laboratories, Syracuse University, Syracuse, NY **[81]**

H.M. Pinsky, Section of Oral Pathology and Diagnosis, The Ohio State University, College of Dentistry, Columbus, OH **[283]**

Julian R. Pleasants, Lobund Laboratory, University of Notre Dame, Notre Dame, IN **[87,355,359]**

C. Plonton, Department of Hematology, Hôpital Edouard Herriot, Lyon, France **[263]**

Gerd Pluschke, Basel Institute for Immunology, Basel, Switzerland **[403]**

Morris Pollard, Lobund Laboratory, University of Notre Dame, Notre Dame, IN **[75, 77, 289, 343, 447]**

Zdzislow W. Przyjalkowski, Institute of Parasitology, Polish Academy of Sciences, Warsaw, Poland **[415]**

Pierre Raibaud, Laboratoire d'Ecologie Microbienne, INRA–CNRZ, Jouy-en-Josas, France **[119, 163, 171, 207, 391]**

P. Rapine, Laboratoire d'Ecologie Microbienne, INRA–CNRZ, Jouy-en-Josas, France **[207]**

Joel M. Rappeport, Department of Hematology, Brigham and Women's Hospital, Boston, MA **[443]**

Brian Ratcliffe, Department of Nutrition, National Institute for Research in Dairying, Reading, Berks, England **[115]**

Bandaru S. Reddy, Division of Nutrition and Endocrinology, American Health Foundation, Valhalla, NY **[293]**

Jon P. Reyniers, Department of Biology, Texas A and I University, Kingsville, TX **[91]**

Edward F. Rose, Department of Radiation Medicine, Massachusetts General Hospital, Boston, MA **[65]**

Sam Rosen, Department of Oral Biology, College of Dentistry, Ohio State University, Columbus, OH **[211]**

Soren Rosendal, Department of Veterinary Microbiology and Immunology, University of Guelph, Guelph, Ontario, Canada **[411]**

Sol R. Rosenthal, Department of Preventive Medicine, University of Illinois, Chicago, IL **[145]**

Robert Rothlein, Department of Membrane Immunochemistry, Dana-Farber Cancer Institute, Boston, MA **[371]**

S.Z. Ruban, Laboratory of Gnotobiology, The Gamaleya Institute for Epidemiology and Microbiology, Moscow, USSR **[159]**

Bernard Saint Martin, La Calhene Isotechnie Division, Bezons, France **[31]**

Cindy A. Salkowski, Departments of Medical Microbiology and Surgery, University of Wisconsin Medical School, Madison, WI **[333, 407]**

Shogo Sasaki, Department of Microbiology, Tokai University School of Medicine, Isehara, Kanagawa, Japan **[9, 347, 351, 363]**

Henri Saxerholt, Department of Germfree Research, Karolinska Institutet, Stockholm, Sweden **[99, 241]**

Thomas Schmeiser, Zentrum Innere Medizin, Universität Ulm, Ulm, West Germany **[433]**

Robert S. Sedlacek, Department of Radiation Medicine, Massachusetts General Hospital, Boston, MA **[65]**

Shi Shun-di, Medical Laboratory Animal Center, Chinese Academy of Medical Sciences, Beijing, People's Republic of China **[111]**

Charles C-Y. Shih, Winter Research Laboratory, University of Wisconsin Medical School, Mount Sinai Medical Center, Milwaukee, WI **[451]**

Makoto Shimizu, Keio University, Tokyo, Japan **[325]**

Irving L. Shklair, Naval Dental Research Institute, Great Lakes, IL **[211]**

N.M. Shustrova, Laboratory of Gnotobiology, The Gamaleya Institute for Epidemiology and Microbiology, Moscow, USSR **[159]**

David L. Snyder, Lobund Laboratory, University of Notre Dame, Notre Dame, IN **[95, 111]**

Mary A. South, National Institute for Neurological and Communicative Disorders and Stroke, Bethesda, MD **[475]**

L.K. Steel, Department of Biochemistry, National Naval Medical Center, Bethesda, MD **[139]**

Martin Steinbakk, National Hospital, Oslo, Norway **[241]**

Takiko Sugiyama, Department of Microbiology, Keio University, School of Medicine, Tokyo, Japan **[203]**

Yukiko Sumi, Laboratory of Germfree Life, Research Institute of Disease Mechanism and Control, University School of Medicine, Nagoya, Japan **[301]**

Keisuke Sunakawa, Department of Pediatrics, Keio University, School of Medicine, Tokyo, Japan **[203]**

Tatsuo Suzuki, Biomedical Laboratory, The Kitasato Institute Hospital, Tokyo, Japan **[325]**

Yumiko Suzuki, The Kitasato Institute Hospital, Tokyo, Japan **[325]**

Odette Szylit, Laboratoire d'Ecologie Microbienne, INRA–CNRZ, Jouy-en-Josas, France **[119]**

Cyrille H. Tancrède, Department of Medical Microbiology, Institut Gustave-Roussy, Villejuif, France **[175, 255, 275, 279]**

Seiki Tazume, Department of Microbiology, Tokai University School of Medicine, Isehara, Japan; Lobund Laboratory, University of Notre Dame, Notre Dame, IN **[343,347,351]**

Bernard A. Teah, Lobund Laboratory, University of Notre Dame, Notre Dame, IN **[59]**

C.P.J. Timmermans, Department of Microbiology and Gnotobiology, Radiobiological Institute TNO, Rijswijk, The Netherlands **[437]**

R.H.J. Tonk, Department of Microbiology, University Groningen, Groningen, The Netherlands **[155]**

A.G. Toorop-Bouma, Laboratory for Medical Microbiology, University Hospital, Groningen, The Netherlands **[271]**

E.F. Tordoir-van der Bent, Department of Infectious Diseases, University Hospital of Leiden; J.A. Cohen Institute, Leiden, The Netherlands **[267]**

Philip C. Trexler, Lobund Laboratory, University of Notre Dame, Notre Dame, IN; Robens Institute, University of Surrey, Guilford, Surrey, United Kingdom **[3,27]**

Robert L. Truitt, Winter Research Laboratory, University of Wisconsin Medical School, Mount Sinai Medical Center, Milwaukee, WI **[451]**

S. Tsuyuki, Biomedical Research Laboratory, Morinaga Milk Industry Co. Ltd., Tokyo, Japan **[399]**

K. Ueda, Department of Veterinary Public Health, The Institute of Public Health, Tokyo, Japan **[399]**

Keiko Umehara, Department of Microbiology, Tokai University School of Medicine, Isehara, Japan **[347,351]**

A.A. van der Linde-Preesman, Department of Cell Biology and Genetics, Erasmus University, Rotterdam, The Netherlands **[359]**

J.W.M. van der Meer, Department of Infectious Diseases, University Hospital of Leiden, Leiden, The Netherlands **[429]**

D. van der Waaij, Department of Medical Microbiology, University of Groningen, Groningen, The Netherlands **[155,245,251,259,271]**

J.M. van der Waaij, Department of Medical Microbiology, University of Groningen, Groningen, The Netherlands **[245]**

J. Van Eldere, Department of Microbiology, The Rega Institute, Leuven, Belgium **[103]**

Enio C. Vieira, Bioquimica-Immunologia, Instituto Ciencias Biologicas, Belo Horizonte, Brasil **[235]**

Leda Q. Vieira, Bioquimica-Immunologia, Instituto Ciencias Biologicas, Belo Horizonte, Brasil **[235]**

A. Wade, Lobund Laboratory, University of Notre Dame, Notre Dame, IN **[343]**

Morris Wagner, Lobund Laboratory, University of Notre Dame, Notre Dame, IN **[51,107]**

Ron Walker, Robens Institute, University of Surrey, Guildford, Surrey, United Kingdom **[123]**

Wang Anjing, Department of Gnotobiology, Medical Laboratory Animal Center, Chinese Academy of Medical Genetics, Beijing, People's Republic of China **[191]**

Frank W. Ward, Robens Institute, University of Surrey, Guildford, Surrey, United Kingdom **[123]**

Keiichi Watanabe, Department of Pathology, Tokai University School of Medicine, Isehara, Kanagawa, Japan **[363]**

Rachelle A. Weiss, Department of Pathology, State University of New York at Stony Brook, Stony Brook, NY **[367]**

Gjalt W. Welling Department of Medical Microbiology, University of Groningen, Groningen, The Netherlands **[151,155,259]**

F. Wendt, Department of Hematology and Oncology, Evangelisches Krankenhaus Essen-Werden, Essen, Federal Republic of Germany **[251]**

Sheila C. Wiedl, Department of Viral Leukemia, Roswell Park Memorial Institute, Buffalo, NY **[461]**

Raphael Wilson, Department of Biology, King's College, Wilkes Barre, PA **[475]**

Bernard S. Wostmann, Lobund Laboratory, University of Notre Dame, Notre Dame, IN **[xxv,75,87,95,107,111, 355, 359]**

Wu Juyuan, Department of Gnotobiology, Medical Laboratory Animal Center, Chinese Academy of Medical Sciences, Beijing, People's Republic of China **[191]**

Takaya Yamada, The Kitasato Institute Hospital, Tokyo, Japan **[325]**

Ryoko Yamashita, Department of Pediatrics, Keio University, School of Medicine, Tokyo, Japan **[203,325]**

S. Yamazaki, Veterinary Public Health, The Institute of Public Health, Tokyo, Japan **[399]**

Tin-Chuen Yeung, School of Public Health, Harvard University, Boston, MA **[127]**

John W. Yrios, Department of Medical Microbiology, University of Wisconsin Medical School, Madison, WI **[199]**

Nevzat Yurdusev, Laboratoire d'Ecologie Microbienne, INRA–CNRZ, Jouy-en-Josas, France **[171]**

Zhang Xufan, Laboratory of Microbiology, Institute of Epidemiology and Microbiology, Chinese Center for Preventive Medicine, Beijing, People's Republic of China **[191]**

F.E. Zwaan, Department of Infectious Diseases, University Hospital of Leiden, Leiden, The Netherlands **[429]**

Introduction

When we started to organize the VIIIth International Symposium on Germfree Research, it was decided not to give special emphasis to any specific field of application, but rather to try to present a more general picture of gnotobiotic research in the broadest sense of the word. As a result, a wide spectrum of papers was presented, the majority from the European countries, a sizable number from the United States and from Japan, but also from the People's Republic of China and from Latin America.

It would seem, therefore, that these Proceedings truly represent ongoing work in host-microflora relationships and its pertinent applications to the fields of immunology, transplantation, cancer, metabolism and other related fields of the biomedical sciences. Three introductory papers by Drs. Philip Trexler, Shogo Sasaki and Bengt Gustafsson present a view of past history and present achievements, and provide an encouraging glimpse of the future. Dr. Patricia Bealmear depicts for us the scientific, social and emotional aspects of the life of David, "the Boy in the Bubble," who died early in 1984 after 12 years of isolation in gnotobiotic systems.

More than one hundred scientific papers have eventually reached us for publication. As editor, I am especially grateful for the assistance of my colleagues Drs. Julian Pleasants, Morris Pollard, David Snyder and Morris Wagner, and Mr. Bernard Teah for their help in editing the various manuscripts. And it would have been impossible to have all manuscripts properly processed without the able help of my assistant, Mrs. Marge Johnson, the computer knowledge of Mrs. Arlene Snyder, and the skillful typing of Ms. Vicky Rodebush.

Bernard S. Wostmann

I GERMFREE RESEARCH: PAST, PRESENT, AND FUTURE

Germfree Research: Microflora Control and Its Application to the Biomedical Sciences, pages 3–7

THE EVOLUTION OF GNOTOBIOTIC TECHNOLOGY

Philip C. Trexler

Lobund Laboratory, University of Notre Dame

Notre Dame, Indiana 46556 U.S.A.

Gnotobiotics originated towards the close of the past century with the speculation of Pasteur regarding the importance of the microbiota for the higher animals and the first experimental results of Nuttall and Thierfelder. Several investigators continued the work but it was not until 1914 that apparatus having the main features of the present day isolators was developed by Küster and used in the successful rearing of a germfree goat. The early history of gnotobiotics has been related on many occasions (Luckey 1963). There were three centers in which germfree animal research developed under Gustafsson in Sweden, Miyakawa in Japan and Reyniers here at Notre Dame. The apparatus used in Japan and here was similar in that it held the pressure used in steam sterilization while that in Sweden was sterilized within a large autoclave.

This appears to be an appropriate occasion to recount in greater detail the development of Gnotobiotics, particularly the technology, as it occurred here. In 1930, during my freshman year, Father Wenninger, Dean of the College of Science, announced that the night before he had assisted the young Instructor in Bacteriology, Reyniers, to obtain the first ever germfree animals, a litter of guinea pigs. This was later reported in the Proceedings of the Indiana Academy of Science (1932-33). Reyniers was not proud of this report of his work and seldom if ever referred to it. I consider this work to be a notable achievement considering that Reyniers had no formal training in research. He left the University of Chicago shortly after enrolling as a graduate student to come back here to teach Bacteriology upon the death of Father Albertson. Reyniers had no knowledge of the early work with germfree animals since the biological research activities at the University of Notre Dame at that time were confined to plant taxonomy. The apparatus he used, which I did not see, and the results obtained did not differ greatly from that of thirty years before. The rearing of germfree animals was but one project in a research program Reyniers devised for investigating the problem of bacterial variation. At that time the mechanisms of inheritance were generally considered to be quite different between the bacteria and higher forms: opinions differed greatly as to the extent and type of variation in bacteria. Reyniers' research program involved (1) the development of single-cell isolation techniques to make certain that cultures started from a single cell, (2) mechanical methods for increasing the accuracy and reducing the labor required to make viable counts and (3) germfree animals to make certain that no contamination occurred during the study of bacteria associated with a host.

During 1934 Reyniers again started to work with apparatus for the rearing of germfree animals and I was persuaded to drop my graduate studies in Cellular Biology to assist in the design and construction of apparatus. Interestingly, the apparatus we designed and built resembled that of Kuster's though we used glass wool air filters and had provisions for resterilizing airfilters and changing gloves. Reyniers obtained funds for constructing a series of isolators which were cylindrical pressure vessels capable of withstanding sterilizing steam pressure and high vacuum. Through the publication of Glimstedt (1936), we became acquainted with the earlier work and we were pleased to learn that we had obtained better results than previous investigators. Reyniers persuaded the University to put up a building for Biology with an entire floor devoted to Microbiology. In this building, apparatus was installed which enabled us to remove the fetus directly from the uterus into an isolator through a plastic film barrier. Chickens and monkeys were reared in this apparatus as well as guinea pigs. The first germfree rats were weaned in 1944: reproduction occurred during the war. At that time, half of our isolators had been taken to Fort Detrick and most of our effort was devoted to war related research. Apparatus was developed here for spray-drying toxic and pathogenic agents so as to produce fine particles of predetermined sizes and shapes, thereby demonstrating that isolators were useful for containment as well as exclusion.

Since we had demonstrated that it was possible to obtain reproduction in a germfree rat, an attempt was made to reduce the cost of producing germfree animals by using a large tank isolator 2.4m dia. x 4.5 m long into which a person could enter in a ventilated sealed protective garment through a germicidal bath. While we did not succeed in reducing the costs of animals with this isolator, we did develop flexible film isolators as a result of studies of methods for sterilizing the surfaces of protective garments.

Reyniers was undoubtedly a genius as a promoter; he succeeded in raising over three million dollars for germfree research. When the Germfree Life Building, now the Reyniers Germfree Life Building, was dedicated in 1950, he made plans for a research center on the site comprising two dozen buildings. By that time the facility was usually referred to with the acronym Lobund. Reyniers was of the opinion that the Federal Government should support such a center in much the same way as the large astronomic observatories and centers for nuclear physics. I held that the production and use of these animals should be made simple and relatively inexpensive so that research could be carried on in any laboratory in which there were the appropriate animals. Unfortunately, we never did succeed in reconciling these differing viewpoints. During the early 50s Reyniers became the Chief Scientific Advisor of the Damon Runyon Fund and developed an interest in cancer research which superceded his interest in developing Lobund. As a result, in 1959 he left to set up the Germfree Life Research Center in Tampa, Florida.

A Symposium on Gnotobiotic Technology was held at Lobund in 1959. An introductory question was posed: "Can we expect the gnotobiote to be of more general use or even the common laboratory animal of the future?: and this question is still relevant. The advantages and disadvantages of using SPF and defined flora animals were examined. In an attempt to focus attention on problems beyond those of immediate importance, man's relationship to the internal microbiota was compared with that to the external macrobiota. Only in very primitive societies is man satisfied to control the predators and then harvest and hunt for food as it naturally occurs; the next step is to clear away the weeds and engage in agriculture; to do this internally is an obvious goal.

The following year a workshop was held for Commercial Breeders of Laboratory Animals. All ten breeders succeeded in rearing some germfree animals, used them in rederiving their stocks and a few breeders offered germfree animals for sale. Since then, large flexible film isolators have been developed and all of the common farm and domestic animals have been reared as gnotobiotes. The small laboratory animals can be maintained gnotobiotically in filter-top cages and serviced within a Laminar Airflow Workbench. There is now a renewed interest in the use of ventilated protective garments to work either in a sterile or hazardous environment. Isolators have been used in hospitals for surgery, the protection of wounds and the treatment and care of patients in a truly sterile environment. Patients suffering from infectious diseases that present a hazard to others can be treated in isolators for as long as required and moved by ambulance or air while remaining within isolators. Gnotobiotic Technology has been developed to accomodate almost any type of work within a protected environment. The main thrust of future developments seems likely to be directed towards more cost effective methods and those that are simpler to use. Support for further development may continue to be difficult to obtain unless a sufficient market develops to attract industrial interests. Fortunately, the increasing concern for protection from toxic and pathogenic agents in the workplace has created considerable interest in containment isolation. Since much of the isolation apparatus can serve for either exclusion or containment isolation, the size of the potential market has been considerably increased. In addition, there is a growing demand for laboratory animals that are virus-antigen-free and for those strains that are particularly susceptible to infections. The technology required to produce and use these animals approaches that required for gnotobiotes.

At the Meeting a quarter of a century ago, it was suggested that it seems likely that the microbiota with which man is associated can be improved since diet and environment is now so different from that in which the species evolved. Furthermore, at present there is concern about helping those aflicted with chronic diseases and the elderly; any evolutionary forces which helped to form the composition of, and responses to, our microbiota would have acted against this section of the population since evolution is survival of the fittest. During the past twenty-five years, there has been a substantial increase in the average lifespan in this country but relatively little increase in the span of healthy old age. The role of the microbiota in these conditions is not known but the increased lifespan of germfree animals is suggestive and the advantages of antibiotics in the food of farm animals is well known. It would not be surprising if future generations take a microbiota pill along with their multivitamins, perhaps in the form of an improved yogurt.

Gnotobiotics as a science resembles Genetics and in many respects the gnotobiote is anälogous to an inbred animal. The incorporation of viral DNA into the genome of a host reinforces this resemblance. The biological importance of both sciences is enhanced in combination with other disciplines, e.g. cytogenetics, molecular biology. The technology required to make the defined flora animal as readily available as inbred animals is now a realistic goal. When it is reached the use of gnotobiotes will not be considered to be a special biology, i.e. Gnotobiology, anymore than the use of an inbred animal is considered to be a special "Inbred-biology". In the abstracts I suggested that Gnotobiotics has reached a stage in its development comparable to that of Genetics a few years after Mendel was rediscovered; but, after reading through the Abstracts for this Meeting it is obvious that we are now considerably farther along the road.

Glimstedt G (1936). Der Stoffwechsel bakterienfreier Tiere. Skand Arch Physiol 74:48.

Luckey TD (1963). "Germfree Life and Gnotobiology." New York: Academic Press.

Germfree Research: Microflora Control and Its Application to the Biomedical Sciences, pages 9–15

THE PRESENT STATUS OF CLINICAL APPLICATIONS OF GERMFREE TECHNOLOGY

Shogo Sasaki, M.D.

Dean, Tokai University School of Medicine
Bohseidai, Isehara 259-11, Japan

Never before in the history of medical bacteriology has the normal flora received so much attention. This is probably due to the resolution of many of the problems concerning infectious diseases due to advances in medical technology, bringing forth the problem of the compromised host which had been lurking in the background. Some refer to it by the name opportunistic infection, but whatever it is called, it points to the fact that attention is now focused on infections by bacteria with low pathogenicity under normal circumstances - for instance, those comprising normal flora.

On another front, the development in organ transplantations has brought about the demand for germfree technology, as has development in the field of immunology, where clarification of the many aspects of immune deficiencies have highlighted the need for germfree technology in their therapy.

Furthermore, the development of this technology has not only opened new fields in medical research, but it is also becoming an integral part in the fast-developing field of biotechnology.

Now, let me touch on some of the topics of interest in turn, starting at the beginning.

The use of negative pressure isolators for quarantine of disease and the use of germfree animals for determination of pathogens is still far from wide spread. However, once the antigen-free diet, nude mice and germfree technology can be worked out in appropriate combinations, and can create animals suitable for mass production, a phenomenal increase in their application is expected.

And then, there is the issue of the application of germfree technology to clinical medicine. There exists two basic directions to this application, one of which is the prevention of infection by creating a germfree state, and the other is the active administration and promotion of bacteria, that is implantation. In understanding these concepts, one needs to first of all reflect upon the mechanism of natural defenses. Or more specifically, the nonspecific defense mechanisms which are activated before establishment of the immunological defenses. Included in this category are the effects of peristalsis of the intestines, various secretions, ciliary movement, normal indigenous bacterial flora, pH, and the physical barriers of the epidermis or mucous membranes. And past these defenses are the factors such as interferon, lactoferin, lysozyme, complement, macrophages, neutrophilic leukocytes, Kupffer's cells and natural killer cells. At this stage, immunological resistance in the form of specific response is put into motion.

Given the present state of germfree technology, the extent to which it can be applied to clinical medicine is the promotion of the host's non-specific defenses, i.e. isolation for protection from exogenous infections and selective or complete decontamination for protection against endogenous infections. And a dysfunction of the neutrophilic leukocytes would indicate the need for bone marrow transplantation.

And then, although it may be possible to isolate and maintain an organism in a germfree state, there still remains the problem of the capacities and reactivity carried over from the previous generations. In other words, it needs to be kept in mind that although placing an organism in a germfree environment is effective for prevention of direct invasion by external bacteria, internally, it has the effect of lowering the capabilities of the host's cellular defenses.

Next, let me trace the stages involved in the establishment of infection.

First, there is the mechanism whereby the bacteria become attached to the host cells. This phenomenon is called adhesion or simply attachment, and immunoglobulin and glycocalyx are examples of substances known to play a role in its occurence.

Next is the stage wherein the bacteria multiply until they reach a certain number around the initial point of adhesion. This is called multiplication, which together with the concept of antagonism is termed colonization, and their status as pathogens is recognized upon triggering an abnormal response in the host. Adhesion, multiplication and antagonism are entirely different mechanisms which are frequently confused as one, but this is a point of crucial importance requiring immediate clarification, and to do so, the use of germfree animals becomes a necessity.

Now, I would like to briefly reflect upon the significance of microflora in relation to the living organism.

Upon being born into this world, organisms encounter infections which normally result in the establishment of normal flora. The effect of this flora on the host organism is almost unaccountably extensive, and their influence on such manifestations of life as the urea cycle, bile acid metabolism, digestion, antibody production and hormonal regulation are considerable. Of special significance is the effect of flora on nutrition, and faced as we are with the contrasting problems stemming from over-nutrition in the advanced nations of the world, and those arising from insufficient nutrition in the developing countries, the study of flora as the one factor the two conditions have in common in relation to food is one of the problems demanding immediate attention.

The other form of infection known is the opportunistic infection. It is brought about by a decline in the host's defense capabilities, and leukemia and the immunodeficiency syndrome are triggers of this form of infection. Specifically, such infections are caused by bacteria of low pathogenicity. But it is a fact that part of the normal flora invade the blood stream with relatively high frequency. This phenomenon is called translocation, and under normal circumstances, nothing would come of the invasion as the alien organisms would be removed by phagocytosis of the macrophages. However, they are fully capable of being the causative agents of infection given changes in the host's line of defense.

However, it can be seen that little is known about the underlying mechanisms of these phenomena when taken individually and in detail, although they have been recognized for quite some time.

Now let me turn to the present state of medical therapy taken against the backdrop of the conditions giving rise to infection.

The antibiotics, anti-cancer agents, steroid hormones and radiation therapy in current use incur the lowering of the patient's natural defenses. Hence, potential pathogens in normal flora and similar exogenous bacteria need to be controlled. From experience with germfree animals, it would seem that for protective isolation of the patient, complete removal of all bacteria from the patient would be the logical method, but experience tells us that this is for all practical purposes, impossible. Therefore, the total decontamination we speak of is by no means absoulte. On the other hand, there is the process of selective decontamination promoted by Dr. van der Waaij and others, which is the selective removal of facultative anaerobic bacteria leaving the obligate anaerobes as the majority population. The basis of this process is the view that the facultative anaerobes have great potential as possible pathogens. However, when taken in light of the fact that normal flora is established as the result of equilibrium maintained between the many species present, the judgement on which form of decontamination is better, to which I will return to later on, should be determined by the disorder in question and is not one to be made lightly.

As for implantation, the representative species presently employed are E. coli, Lactobacillus and Streptococcus, but the results from their use are not always constant, and for the moment, one can only hope for further development in this area in the future.

Now, a few words on some aspects pertaining to immunology.

More than a few studies have been conducted on macrophages and lymphocytes derived from germfree animals, and they are frequently employed in research on HLA, interferon and xenografts. However, their greatest impact in relation to clinical medicine has been their role in studies surrounding bone marrow transplantation.

The technique of transplantation as well as transfusion is performed disregarding the complex immunological responses of the subject, thus provoking responses which cannot be seen under ordinary circumstances. Bone marrow transplantation is in effect, the transplantation of immunologically competent cells. As such, the immunological defense mechanisms of the host must first be suppressed, and an environment in which the newly introduced cells can exercise their immunological defenses must be created. However, such priming can in itself bring on disturbance or impairment of the immunological defenses, and affect the viability of the graft or of the host itself.

The control of rejection is a problem also encountered in the transplantation of organs, but in bone marrow transplantations for treatment of malignant tumors, there exists the problem of GvH disease in addition. Complete suppression of the GvH reaction, which is a manifestation of the host's defenses, could in overall terms be detrimental to the therapy, thus adding to the complexity of the problem. Moreover, although rejection of graft marrow can be solved by repeated transplants, the onset of GvH reactions harmful to the host must be avoided. However, cases of aplastic anemia and immunodeficiency require transplantation even in HLA incompatible cases, so that all we can do at present is to do our best to prevent GvH disease at all costs.

Then there is the method of allogeneic bone marrow transplantations which require various steps to purge the T-cells which bring about rejection, so that needless to say, autogenic transplants are the ideal. However, in such cases, the cancerous cells must be isolated and totally removed from the marrow to be transplanted, and radiation and monoclonal antibody have been used routinely for this purpose. And recently, dyes such as Merocyanin 540 and other agents such as the cyclophosphamide-derivatives have come to be employed. The dye selectively combines with the membrane lipids of leukemia cells whereas monoclonal antibody reacts with protein or carbohydrate so that the combined use of these agents are given a high rating.

In the field of internal medicine, germfree technology has been applied to problems pertaining to cardio-vascular disorders and endocrinology, and in surgical medicine, it has found applications in orthopedics in such processes as the total replacement of the hip joint, in neurosurgery, cardiosurgery - especially in the insertion of cardiac pacemakers - and in the treatment of burns, or specifically, local isolation. In the latter, recent attention has turned to flora translocation in relation to burns. And developments in central venous nutrition have evolved at a striking rate, permitting life and return to society as an active member to patients without any oral ingestion of food. This advancement has been so amazing, as to warrant re-evaluation of the significance of the digestive tract itself. Anyhow, the problems encountered here calling for germfree technology is the method of disinfection of the entry point of the catheter, and this problem is also inherent in the use of such devices as the hemodialyzer and artificial anus for which the development of an efficient method of local isolation is eagerly awaited.

In the field of dentistry, starting with the problem of dental caries, interest has turned to oral ecology, and on another front, to vaccines, and continued development in these and in new directions are expected.

Nevertheless, the point of greatest interest is undoubtedly cancer.

Research in this area is numerous, and while the relationships between cancer of the digestive organs and flora, diet, chemicals and other factors are being pursued on one front, other topics of diverse nature abound, such as the anti-tumorigenic effect exhibited by certain strains of E. coli, Lactobacillus and Bifidobacteria. And the problem being given renewed attention is the prevention of exogenous infection in the everyday care of patients, as well as how the rate of endogenous infections can be effectively reduced.

What has become a definite turning point in cancer research was the discovery of the oncogene. However, there are several unclarified points remaining, such as:

1) Are the so-called C-oncogenes all of uniform structure.
2) What is the function of the proto-oncogene present in the healthy cell under normal circumstances? And what is the mechanism whereby their function as a phosphokinase, or as a growth factor or a receptor of the growth factor is related to transformation? And,
3) What is the trigger for the transformation?

These questions await resolution, and high expectations are held for the results from Dr. Pollard's long-term study on spontaneous tumors.

Upon incorporating the isolator system into clinical practice, maintaining the psychological health of the patient has always been cited as a problem. However, it is unfortunately true that not much has been accomplished in this area.

According to a recent report, a mother attending to a child with infantile cancer placed under protective isolation presented strong signs of prison psychosis and demanded immediate withdrawal from the room. It would seem that encountering such psychologically unstable conditions in patients placed under isolation would not be unusual, but then, it may be that they have lost the capacity to give such strong reactions. Whatever the case, there is definitely a need for reassessment of the situation.

Lastly, let me touch upon the situation in Japan. Recently, a research team of the Ministry of Health and Welfare conducted a survey on the primary immune deficiency syndrome, and 556 cases were registered. Of the many forms of the disorder found, those encountered with high frequency were common variable immunodeficiency, selective IgA deficiency, chronic granulomatosis and juvenile sex-linked aggamaglobulinemia. And of the infections contacted by these subjects, respiratory infections were by far the most common, amounting to 56.6%. Generally speaking, the most frequently seen causative agent of respiratory infection occuring in germfree rooms is the cytomegalorivus, and a remedy to this situation is urgently wanted.

With regard to the experience from the protective isolation rooms at Tokai University where most of the cases were leukemia, it is interesting that not one case of respiratory infection was recorded. It is believed that the reasons for this are the selection of periods of remission for treatment, the use of gamma-globulin, and the fact that transfusions were made only from cytomegalovirus-negative donors. And it appears that the patients prefer the laminar air-flow room to the vinyl isolator.

Such are some of the problems surrounding the clinical applications of germfree technology which have drawn my attention and interest. I am fully aware that there are issues of greater importance which should have been covered, for instance in genetics, biochemistry and nutritional science. But I hope you will excuse this lop-sided presentation, stemming from the fact that I myself am a microbiologist, and an ancient microbiologist at that.

Thank you for your attention.

Germfree Research: Microflora Control and Its Application to the Biomedical Sciences, pages 17–23

THE FUTURE OF GERMFREE RESEARCH

Bengt E. Gustafsson, PH.D., M.D.
Department of Germfree Research
Karolinska Institutet
Box 60 400, S-104 01 Stockholm
SWEDEN

With the present shifts of responsibilities from all those scientists who started in this field 30 years ago now being in their 60ies, many of whom by the way are present in this audience, to younger investigators the question might well be asked: What are the future conditions for germfree research as we have defined it up till now.

MICROFLORA - GENERAL CONDITIONS

Germfree animal research has in the past been considered by many investigators to be difficult, costly and cumbersome. For those of us, who have been in this field for a long time, such a view-point is not easy to understand, as germfree animals nowadays are kept for generations without difficulties. As the practical problems have now been overcome, the germfree animal technique can be compared with other kinds of methods used in sciences to study problems in a wide range of disciplines, as the application of radioactive isotopes in metabolism studies, spectrophotometry, masspectrometry etc.

Will therefore germfree research be a service or technical branch only to all kinds of research and animal production facilities, or will it still be able to maintain its status an independent branch of sciences? The development in other fields depending on special techniques presents some examples which are not encouraging in this respect.

It is obvious, however, that the future of germfree research very much depends on to what extent we may pay attention to the simple fact that what is really studied in the germfree animal is the function on the normal microflora present in all living higher organisms from birth on.

The conventional animal is characterized by the harbouring of an ubiquitous microflora, the weight of which of around 1,000 gms in man must be considered to be equivalent to that of the larger organs of the body as the kidneys and the brain. Very little is known both concerning its composition and function. The mass is such that the total number of living entities easily outnumbers that of the total number of the host's own cells. It contains a great variety of strains, and the question arises which if any of these are essential to the host, which are pathogens and which are innocuous commensales or saprophytes (Figure 1). Another factor of importance is that 95% of the total number of intestinal bacteria are obligately anaerobic. These problems have to be taken into account when attempts are made to isolate such organisms. Some of the physiologically important strains are present in small numbers which again makes them hard to isolate.

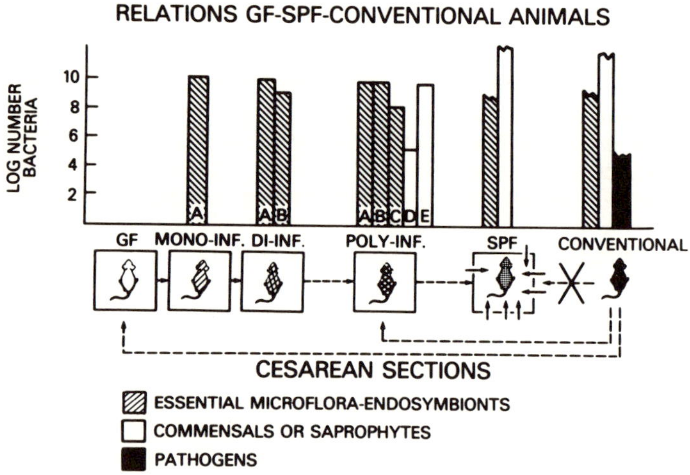

Figure 1. Relevant intestinal microflora conditions in experimental animals.

The germfree animal on the other hand represents the animal itself as a result of its own body cells and tissues and their functional capacities. Any differences from the conventional should be compensated by the introduction into the germfree animal of the conventional animal microflora. The next step is to isolate the pertinent microorganisms responsible for the recorded difference between the two types of animals. As is often the case not one organism is necessary but several often acting in a specific order or system. The four microorganisms necessary for the metabolism of wood by the thermites and our own dependence on different strains for the fermentation of lithocholic acid from the primary bile acids are examples of this.

The SPF animals are supposedly and hopefully free from pathogens but pose a problem for several reasons. First and obviously they are not too well defined as far as the inherent microflora is concerned. Although, they are better in this respect they are sharing this with the conventional. Furthermore, most of this research is performed to illustrate and study conditions in man, who is not pathogenfree. Germfree animal research can hardly be without these two controls - SPF and conventional animals.

Hopefully, the future will see polyinfected exgermfree animals kept in isolators harbouring known, well defined strains of microorganisms necessary to compensate for the physiological and other differences between germfree and conventional animals. This goal cannot be reached, however, until we are able to define and establish such a normal microflora in the germfree animal.

Table 1 Future areas of special interest in germfree research.

A. The normal microflora, i. e. composition, organization and structure
B. Intrinsic problems, i. e. effects of the microflora on the host
C. Extrinsic problems, i. e. effects of factors in the environment mediated by the microflora

Considering future areas of special interest in germfree research these could be divided in three parts as in Table 1. Most investigators in germfree research consider the conventional animal as the ultimate control to germfree animal research. But the investigator studying the normal microflora and its effect might as well consider the germfree animal as the control of the effect of body cells and tissues per se in the complex interactions in the conventional animal. An example of this is whether the effect of a carcinogen in a conventional animal is mediated by the normal intestinal microflora or is the result of the compound itself on the host tissues.

THE NORMAL MICROFLORA

In the past almost all our attention has been concentrated on the intestinal microflora, but as is evident from Table 2 even a modest appraisal places the microflora in several sites and being present in fair amounts. The 20 gms recorded in the mouth might be surprising to many of my listeners. We have to consider, however, that the surface of the tongue alone with all its papillae and irregularities might be that of a medium-sized dinner-table. One important feature with the mouth flora is the fact that itself and its products, making up to several gms per day are constantly swallowed and, therefore, easily absorbed within the intestinal tract. With this as background, many avitaminoses in humans are often correlated with changes in the mouth mucous membranes. The location of the normal microflora is therefore a problem we have to consider much more in the future. A fraction of a µg of microbial metabolites produced in the mouth could be much more important than several gms of the same compound in the colon.

Table 2 Appraised weights of the human ubiquitous microflora in different sites - gms wet weight

Eyes	1
Nose	10
Mouth	20
Skin	200
Lungs	20
Intestines	1,000
Vagina	20
Total	1,270

A future problem is also how many of the bacteria we

find in the upper parts of the intestinal tract are digested during the passage. Are the relatively low numbers encountered here due to this digestion or to rapid passage in connection with less optimal growth conditions. On the other hand, the important question might not be how many bacteria are present at a given moment. The turnover rate and subsequent digestion of microorganisms might be much more relevant from a functional point of view.

The intestinal bacteria are probably highly organized in relation to each other and in their contact with the body surfaces. The crypts of Lieberkühn harbour bacteria in what is obviously pure culture. The stability of such ecosystems seems to be considerable. The contents of the crypts of Lieberkühn from animals of the 46th inbred generation of a rat strain in our laboratory were compared in 1982 with contents from the same region in rats of the 26th generation in 1969. A striking similarity was observed in the microorganisms in these 2 specimens from the same strain, although investigated with a time difference of more than 10 years and a gap of 20 generations. The contents of the intestines and the bacteria seem to be transported inside a tube of densely packed microorganisms, intimately related to the host's tissues. This condition seems to be very stable as far as composition is concerned.

One might speculate whether the Lieberkühn's crypts are equivalent to the mycetoms in some insects. These are special organs derived from the intestinal tract to harbour single strains of microorganisms supporting the motabolism of the host. Strict behavioural patterns have been developed to propagate the symbiotic microorganism from the parent to the offspring. Are corresponding mechanisms or behaviour maintained in higher animals?

The applicability of results in germfree research to conditions in humans is again very much influenced by how much differences exist between different species. A lot more attention must in the future be directed to such organism as spirochetes, anaerobes killed by ambious oxygen and the growth requirements of these.

INTRINSIC PROBLEMS

In nutrition there are several problems which require attention in the future. Are microorganisms digested to such an extent that they supply the host with factors such as amino acids, which might be of importance to the host metabolism. It has been demonstrated recently that microorganisms are normally digested in the intestinal tract. Concerning the action of dietary fibers on the function on the intestinal tract and bile acid turnover it was easily demonstrated that fibers stimulated the turnover rate of bile acids also in germfree rats. The formation of the microbial plaque in the intestine is on the other hand highly dependent on the diet but also on the presence of the appropriate strains.

Defense systems. This is a field were techniques similar to those used for germfree animals has been applied also to human medicine. This has had an impact on monitoring treatment of cancer patients, and in the management of many other clinial problems including transplantation. One of the main intrinsic problems for the future will be the following. The germfree animal has an underdeveloped lymphatic and lymphoid system but with a full set of cells ready to act. From the present data it might be speculated whether or not certain members of the microbial flora not only have a better stimulating effect due to a great number of antigens but also due to a direct lymphocytopoetic effect without causing general symptoms in the animals.

Metabolism. The studies on bile pigments, bile acids and cholesterol demonstrate the importance of the germfree animal. It is from these studies, however, that the opinion has developed that it is sometimes hard to tell which animal - the germfree or the conventional - is the test animal and which is the control. In many instances they are both controls as the test object is the intestinal microflora or its components.

In relation to the interactions of the steroid hormones much more work remains to be performed in a field which has just opened. On the other hand, this might be one of the most important topics concerning interactions between the host and the microflora. The situation is extremely complex as both the hormones and their metabolites are excreted to the intestines, reabsorbed, reexcreted and at each intestinal passage acted upon by different microorganisms, where the end product after a second metabolism in the liver or other organs is reexcreted, acted upon by a new organism-reabsorbed etc. Many of the differences in structure and function now studied in this field could well have microbial action on steroid hormones as their deepest source. I assure the young scientists in the audience that the number of unsolved problems relevant to the intestinal microbial metabolism of steroids, according to our experience, is almost endless, but that they are rewarding due to their difficulty and importance.

EXTRINSIC PROBLEMS

The future will see the germfree animals used for the study of interactions of pathogens with certain members of the intestinal microflora in early childhood, in ulcerative colitis and Crohn's disease. This will also be the case in periodontal disease to investigate the anaerobic microorganisms, which might be of decisive influence. The possibility that the microflora metabolises primarily innocuous compounds to carcinogens has been recognized. Secondly, bile acids produced by the intestinal microflora might act as tumor-promoting factors in the colon. The xenobiotic effects of glutathione conjugates and their microbial metabolites have to be further studied. Germfree animal characteristics can easily be produced in animals treated with antibiotics and maintained also after the administration was terminated. Endogenous trypsin and elastase showed sharply increased fecal levels in normal students treated with antibiotics. The importance of these effects have to be further investigated.

With a slight travesty of a famous statement in the past this time does not mark the end of germfree research, as we define it as a separate branch of biological and medical science, it is not even the beginning of the end but it might be the end of the beginning, marking the wide use of the germfree animal in some relevant fields, but also a deeper knowledge of one of the largest organs of the body - the intestinal microflora.

Coates, ME, Gustafsson, BE (1984). "The Germfree Animal in Biomedical Research." London: Laboratory Animals Ltd.

Gustafsson, BE (1982). The physiological importance of the colonic microflora. Scand J Gastroenterol Suppl No 77:117.

II METHODOLOGY AND ANIMAL PRODUCTION

Germfree Research: Microflora Control and Its Application
to the Biomedical Sciences, pages 27–29

A NEW WORKBENCH ISOLATOR

Philip C. Trexler

Robens Institute
University of Surrey
Guildford, Surrey, England, United Kingdom

Ordinarily, two persons working within the same area, e.g. performing surgery, are placed across from each other when using an isolator though they usually work side-by-side on a bench or in a laminar airflow cabinet. A new type of sleeve has been developed which enables work to be carried on within an isolator in the same manner as on an open bench. The shoulder opening of the sleeve is a narrow slit rather than the usual circle or oval. To make it possible to turn the shoulders so that persons working side-by-side can face each other, the shoulder openings of the sleeves overlap, usually halfway. A circular sleeve must be either welded to the wall of the isolator or attached to a rigid support ring which is then fixed into an opening in the flexible isolator wall. The new sleeve may either be welded to the edges of a slit in the isolator wall or the sleeve may be formed from the isolator wall itself and a continuous weld made joining the side seams of the sleeves with a seam the length of the isolator wall. The top half of one sleeve may be formed from a sheet of film continuous with the upper portion of the isolator wall and the bottom half of the other sleeve is formed from film continuous with the bottom portion of the wall. A separate sheet of film is welded in place to complete the two sleeves and is simply folded where the two sleeves overlap. It is necessary to reinforce the corners where the sleeve joins the wall to prevent tearing as this is the point of greatest tension when arm movements exceed design limits.

Several different types of isolators have been made using the new sleeve design. Typically, four sleeves having shoulder openings 2 ft long will occupy 5 ft of wall space. The length of these sleeves is at least 4 ft to enable a worker to reach around within the isolator. The sleeves terminate in glove rings to which wrist-length gloves are attached. These gloves are usually held in place by rubber 'O' rings to enable them to be changed without breaking isolation. These sleeves may be placed on the front wall of an 8 ft isolator along with an entry port. Though these ports can be used with inside caps, they are not required; with containment isolation materials can be entered and removed through plastic bags attached externally and with exclusion isolation a peracetic acid isolator can be used with an outside cap only (Trexler 1980).

Isolators used for exclusion are operated with a positive internal pressure. This pressure presents no difficulty with the new sleeves provided it is not excessive. The positive pressure causes the sleeves to collapse around the operators arms which does not impede movements. A negative internal pressure, usually sued with containment isolation, may cause the sleeves to billow out which does interfere with movement. This interference can be readily corrected by placing rubber bands around the sleeves at about a worker's elbows. A negative pressure will also cause the front of the isolator above the sleeves to collapse thereby restricting the workspace within. This difficulty can be removed by keeping the top attachment of the sleeves taught by means of elastics attached to the end frame. This lateral tension can be easily adjusted by attaching the elastics to a series of holes welded in a fold of plastic which forms a ridge along the top of the sleeves at the line of attachment to the isolator.

The new type sleeves can be used as a single pair to increase the area which is available for manipulations. A pair of these sleeves has been placed on both sides of an isolator used for post-mortem examinations thereby enabling a pathologist to perform necropsies from a single position without the necessity to change positions and gloves (Trexler and Gilmour 1983).

While these sleeves enable an increase of the lateral movement within an isolator, they may limit vertical movement as compared with that obtained using large oval or circular sleeve attachments. For this reason, it is necessary to locate the vertical position of the sleeves carefully with regard to the manipulations that must be performed within the isolator. The use of these sleeves for some applications will reduce the manipulative restrictions associated with the use of isolators.

Trexler PC (1980). Rapid removal of peracetic acid fumes from isolators. Laboratory Animals 14:47.

Trexler PC, Gilmour AM (1983). Use of flexible film isolators in performing potentially hazardous necropsies. J Clin Pathol 36:527.

Germfree Research: Microflora Control and Its Application to the Biomedical Sciences, pages 31–34

RECENT ADVANCES IN ISOLATOR TECHNOLOGY: THE AUTO-BUBBLE

Bernard SAINT MARTIN

LA CALHENE Isotechnie Division
5rue E. Zola 95870 Bezons FRANCE

Usually, sterilization apparatus, excepting dry heat tunnels, are air-tight enclosures in which the product is put to be sterilized. When the sterilization is over, the product goes back into a contaminating area. So, the users package the products in stainless steel boxes, in welded plastic bags or in autoclavable paper bags. These boxes or bags allow the product to remain sterile when it is removed from the sterilisator. But the outside surfaces of those bags can be contaminated.

Today, many of the sterilisators are fitted with two doors, the first opening on the unprotected area, the second onto the aseptic area. They allow to introduce into the aseptic area the needed products, but only, into one aseptic area. In case of failure, the protected area can't be used any more.

The AUTO BUBBLE

This unit is developed to answer these previous issues and to increase the cost efficiency ratio of conventional sterilisators.

The autoclave is a conventional one, with two doors. One of the doors is directly connected at the inside of an air-tight, sterile, isolator. (2, 3, 4). So, when the inside of the autoclave is sterile, it's possible to introduce the sterilized products into the isolator which can be used as a pass through or a storage unit. The products can be transferred at any time, using the DPTE transfer system, to other isolators (1, 6, 8).

THE "AUTO - BUBBLE" (P)

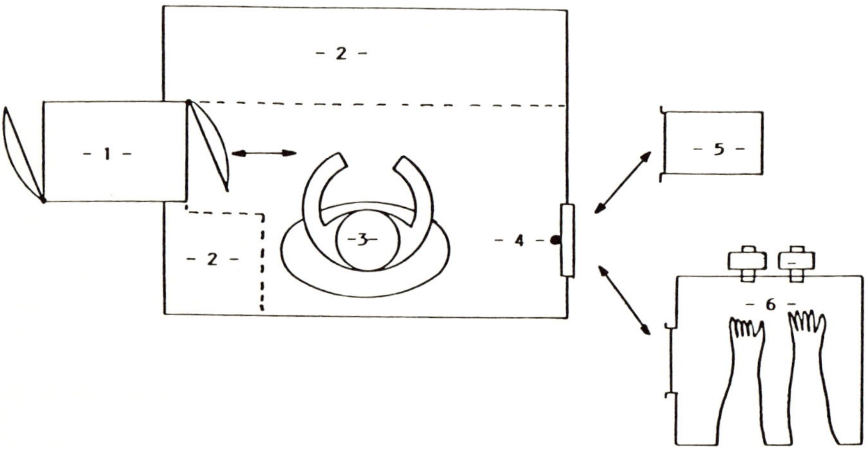

1 : sterilisator
2 : sterile storage area
3 : handling by half suit
4 : DPTE transfer door
5 : DPTE rigid container
6 : DPTE "shuttle" isolator on casters
(P) : Patented - La Calhène

The auto-bubble needs no infrastructure. It's able to serve different isolators or protected areas. This offers the opportunity to use the autoclave with maximum efficiency. In case of failure, every products which is stored in the Bubble can be transferred without any damage to their own sterility (7).

To set up one auto-bubble, we have to be careful with:

- the strict air-tight connection between the isolator and the cylinder of the autoclave.
- a lot of products: food, litter, linen, which can, if the vacuum is not duly drawn, produce a lot of steam. An exhaust HEPA filtered air system has to avoid any water condensation inside the isolator.
- the storage capacity of the shelves is increased because the products are without any envelope. So the conventional volume of storage needed for a batch is cut by factor 2.

TRANSFER OF THE PRODUCTS

We use the DPTE transfer system, described in former reports (1, 7, 8). But, as the quantities to be transferred increase more and more, the former container is now an air-tight isolator on casters, easily handled, and called the "shuttle". It's fitted with gloves and two HEPA air filters to sterilize it following the automatic process we use (1, 2, 7, 8).

CONCLUSION

The auto bubble in animal breeding units, can be used to sterilize and store: water, food, litter, cages, . . . (2, 3). The auto-bubble can be used also in industry and hospitals.

The auto-bubble presents different advantages:

- no prepacking: this means: no mistake or bad welding of bags, and economy: no disposable bags, saving time by no preparation.
- easy permanent control of the storage through the transparent film.
- controls of sterility which can be conducted in testing the sterility of enclosure, gloves, half suit, and products stored inside the unit. This method answers the Good Manufacturing Practices which prefer to check the total process and not only different samples of the products (5).

1. Ducluzeau R, Moreau C, et al. Improvement of techniques leading to time sparing in rearing gnotoxenic animals. Fliedner et al., Clin. and Experim. Gnotobiotics, Zbl, BAKT. Suppl 7, Gustav Fischer Verlag, Stuttgart, New York, 1979.
2. Gosse Ch, Dechambre RP, Ardouin P. Comparative efficiency and cost of different systems of protection against microbiological environment in mice. Recent Advances in Germ Free Research, Proceedings of the VIIth International Symposium on Gnotobiology, 1981, Tokyo, Japan, 75-78. S. Sasaki et al. eds., Tokai University Press, Japan.
3. Gosse Ch et al. Breeding and maintenance of nude mice in large sized isolators for preparing monoclonal antibodies with a view to human therapy. 13th International Cancer Congress, Part C. Biology of Cancer (2): 457, 1983, Alan R. Liss, Inc., NY 10011.
4. Guenet JL. Les isolateurs de grandes dimensions. Sci. Tech. Anim. Lab. Vol. 3, 1978, No. 2, 87.
5. Raibaud P, Ducluzeau R. Interet des techniques de la gnotoxenie pour la verification en continu de la sterilite des encientes utilisees dans l'industrie pharmaceutique. Labo. Pharma. Problemes et Techniques, No. 284, 1979.
6. Saint Martin B, Griscelli C, Lauvergeon. Amerlioration de la technologie des enceintes steriles permettant leur application a la medecine et a la chirurgie. rev. Franc. Gynec. 1975, 70, 10, 585-593.
7. Saint Martin B, Ducluzeau R, Ghnassia JC et al. Applications a la medecine et a la chirurgie des enceintes steriles pour animaux axeniques. Sci. Tech. Anim. Lab. 1976, Vol. 1, no. 3 , 152-157.
8. Saint Martin B. Technological improvement for sterilizing isolators by peracetic acid to avoid contacts between the acid and people inside or outside the isolator. 13th International Cancer Congress. Part C. Biology of Cancer (2). 459, 1983, Alan R. Liss, Inc., NY 10011.

Germfree Research: Microflora Control and Its Application to the Biomedical Sciences, pages 35–38

COST-EFFICIENCY RELATIONS OF ISOLATORSYSTEMS VERSUS BARRIER SYSTEMS IN VIEW OF LABORATORY ANIMAL SCIENCE

Willi P. O. Heine, Professor Dr.

Zentralinstitut für Versuchstierzucht
Hermann-Ehlers-Allee 57
D-3000 Hannover 91, F. R. Germany

Since Philip C. Trexler introduced in 1957 the "flexible apparatus for the rearing and use of germfree animals", gnotobiotechnic became widely used as a scientific method. The production costs as well as the maintenance costs of the Trexler type plastic isolator are considerably lower than of the customary Reyniers type or Gustafsson type steel isolators.

The vinyl film chamber can be built in different shapes and sizes. Their hygienic safety and reliability is at least as high as those of steel isolators. In consequence many scientific attempts have been made to adapt shape and size of the Trexler film isolator to a variety of different requirements of biomedical research.

It was also in 1957 when J.A. Reyniers recommended "the control of contamination in colonies of laboratory animals by use of germfree techniques". This idea became the most reliable and indispensable method to rear healthy laboratory animals.

During the past 25 years a considerable number of large isolators have been built - in addition to innumerable standard size isolators - to accomodate among other species large numbers of small laboratory rodents. Most of these isolators have been designed and built as jacket isolators with a plastic suit attached to the isolator chamber. Sometimes ingeneous techniques have been employed to arrange the cages in such isolators, e.g. suspension rails and paternoster shelves.

There is no doubt, proper operated isolators, steel isolators as well as plastic isolators are the most effective and reliable hygienic barriers. Compared with the closed or barriersystem and of course with the open or conventional system, the isolatorsystem (consisting of isolator and isolator running facilities) is the most reliable method to protect e.g. laboratory animals against microbial contamination from outside. It is therefore selfsuggesting to employ isolators, especially the less expensive film isolator not only for gnotobiotic animals but also for healthy animals, e.g. for specific pathogenfree animals.

But how about the cost-efficiency relations of isolatorsystems versus barriersystems in the view of laboratory animal science? It has been suggested that an isolatorsystem might be more economical than a barriersystem to rear and maintain larger colonies of healthy laboratory rodents. This discussion became revived during the 7th International Symposium on Gnotobiology in Tokyo. I doubt this supposition and I'll try to prove my doubts. Because a comparison of actual figures is difficult if not impossible due to the worldwide tremendous differences in the production costs for buildings, equipment, isolators etc. I shall compare only some of the main cost effective items. I presume the differences in the concepts and the technical details of the isolatorsystem versus the barriersystem as used in laboratory animal units are wellknown. I have discussed this matter already as long ago as at the International Symposium of Germfree Life Research in Inuyama in 1967. To recall it to mind the main differences between both systems is the direct contact of man and animal inside the barriersystem, while in the isolatorsystem there is always a germtight barrier between man and animal, either the gloves and suits mounted into the wall of the chamber or a complete, ventilated overall to walk in and to work in roomsize isolators.

The control of microbial effects on the laboratory animal by a proper hygienic barrier, the hygienic protection of healthy laboratory animals against undesired and uncontrolled microbial contamination from the environment however is just one of the problems laboratory animal science has to cope with.

Among the other wellknown effects which might interfere with the animal, the experiment and the result - see fig. 1 - are some which require for their proper and effective control considerable investments and running costs. Nevertheless these environmental effects have to be considered in context with animal experiments. Otherwise the results achieved might not be acknowledged by the international scientific community. And this can be most expensive in many ways.

```
  GENOTYPE OF THE EMPLOYED ANIMAL
+ CHEMICAL EFFECTS - food, gases, drugs etc.
+ PHYSICAL EFFECTS - climate, light, sound etc.
+ MACROBIOL.Eff. - man, other spec., husbandry etc.
+ PARASITES)
+ BACTERIA ) microbiological effects
+ VIRUS    )
___________________________________________________
  LABORATORY ANIMAL
+ EXPERIMENT - substances, methods etc.
___________________________________________________
= RESULT OF THE ANIMAL EXPERIMENT
=================================
```

Fig. 1

In fig. 2 some main cost-effective items are compiled for animal units, built and equipped according to the three discussed maintenance systems.

COST EFFECTIVE ITEMS (simplified)	maintenance systems		
	open	barrier	isolator
BUILDING			
animal rooms	+	+	+
side rooms	+	+	+
TECHN. EQUIPMENT			
aircond. system	+	+	+
HEPA-filter	o	+	o
boiler	+	+	+
autoclave	o	+	+
washing machine	+	+	+
cage racks	+	+	+
cages	+	+	+
watering system	+	+	+
waste disposal	+	+	+
shower	?	+	?
ISOLATOR	o	o	+

Fig. 2

Appropriate temperature and humidity control in an animal house is rather expensive but nevertheless indispensable, no matter whether it is an open system or a barrier system unit or an isolatorsystem. A boiler is required for heating the unit and for operating an autoclave. Larger units need washing machines for cages, bottles etc. Showers for the staff should be available in all three systems, but they are compulsory only in the barrier system.

Assuming that an open system unit needs neither showers for the staff nor sterilisation facilities like an autoclave - which is unlikely - this system differs from a barrier type animal unit by lack of HEPA-filters, of the autoclave and of the showers. In other words, compared with the barrier system the open system needs no investment and no running costs for these three items, i.e. for the built-in hygienic barrier.

The isolatorsystem requires all the listed items except HEPA-filters in the supply airducts of the airconditioning system of the animal unit. Showers might not be necessary for staff serving standard size isolators. However, showers are necessary for staff working in jacket isolators or in germtight overalls for roomsize isolators.

A comparison of the cost-effective items of the isolatorsystem versus the barriersystem shows that only the HEPA-filter can be saved in the isolatorsystem. But additional expenditures for investment and running costs for isolators, combined with the sophisticated operation procedures have to be taken into account either for roomsize isolators or for jacket isolators or for an adequate number of standard size isolators.

This leads to the conclusion that an isolatorsystem with an appropriate monitored environment as required for laboratory animals is more expensive than a barriersystem - taking as a basis an identical cage capacity. It might be the system of choice if hygienic protection and control is the one and only objective, like in gnotobiology and if cost-efficiency is of secondary importance.

Germfree Research: Microflora Control and Its Application to the Biomedical Sciences, pages 39–42

THE USE OF FLEXIBLE FILM ISOLATORS IN THE PRODUCTION OF GENETICALLY AND MICROBIOLOGICALLY DEFINED RATS AND MICE

Alan G. Peters, AIAT

Production Manager
OLAC 1976 Ltd
Shaw's Farm,Blackthorn,Bicester,Oxon, England

When the idea of producing animals in isolators at Olac was first discussed, it was with the view of having small reservoirs of microbiologically clean animals to be used as a back-up to the main barrier produced colonies in the event of a health breakdown.
When we began to formulate a policy for this, it became obvious to us that in the case of inbred strains of rats and mice it would make sense to protect the pedigree or foundation stock of each strain. This would give us a back-up of clean animals to start off a new colony when required, but more importantly the foundation stock of brother x sister pairs once established and proven to be genetically authentic would ensure the genetic integrity of the strain. The risk of accidental genetic contamination being removed by the presence of the isolator barrier.

We then embarked upon a programme the aim of which was to contain a pedigree breeding nucleus of each of our inbred strains of mice and rats in flexible film isolators. We decided to upgrade the health status of our pedigree stock by caesarean derivation to that of what we term an "associated" animal, initially starting with germ free foster mothers which were given a known gut flora in the form of a "cocktail" mix of organisms. We made the decision not to maintain our isolators in a germ-free state because almost all the animals that would be produced in the isolators would be transferred to our barrier breeding units where a germ free status could not be maintained.
Full term pregnant females from chosen lines were removed from the barrier units and taken to the surgery unit.

The hysterectomys were performed in a surgical isolator which is connected to a flexible resuscitation isolator via a dunk tank. Viable litters were fostered onto mothers of a different coat colour which were held in specially designed isolators. These small isolators hold 10 M2 mouse cages or 3 RB3 rat cages. The litters stay in the foster or holding isolators for a period of 9 weeks during which time two complete microbiological and virological screenings are performed, if the animals are given clearance by our Animal Health Department, they are then moved into larger isolators for breeding. Once the first few pairs of a strain are established, their progeny are mated until the required pedigree colony size is reached.

As found by many workers the isolator is also a useful tool for mutant strains which are especially susceptible to infection. For example, athymic nude mice are successfully produced and maintained in isolators and the athymic rnu rat although more robust than the nude mouse, nevertheless benefits from the stable microenvironment of the isolator. In our own situation we are aiming to maintain long term stability to enable workers to return to our colonies at intervals without any significant change in the animals.

Linked with the continuing increase in the use of inbred and congenic rat and mouse strains has been the incidence of genetic contamination (Festing 1982). With the number of strains we have in production it is impossible to house breeding colonies using coat colour as a distinguishing feature e.g.of the 28 congenic mouse strains we produce, 19 are black. The isolator gives one the opportunity to physically separate each strain whilst housing them in the same breeding area. Once a strain is established in an isolator and proven to be genetically authentic, brother x sister matings continue in total security as the only animal movement which takes place is when ex-breeders, or the progeny of the pedigree animals are transferred from the isolator. When a new strain is taken in the animals are authenticated wherever possible, using biochemical and immunological markers; when a colony is of sufficient size mandibles are sampled. Once a colony is established in an isolator the frequency of genetic sampling is reduced to once a year and this is a routine procedure to check for the incidence of genetic drift or spontaneous mutation.
During regular screenings of over 100 rat and mouse isolators during the past 5 years, using positive screening tech-

niques, microorganisms which commonly occur in barrier units of category 3 & 4 status (Townsend 1969) have never been isolated. During the same period, the isolator animals have shown negative titres for all of the common rodent viral pathogens including,

* Mouse hepatitis virus(MHV)	* Pneumonia virus (PVM)
* Sendai virus	
Lymphocytic Choriomeningitis virus (LCM)	
Ectromelia	Polyoma virus
Reo-virus type 3 (Reo 3)	Mouse Adenovirus
Minute virus of mice (MVM)	
Newborn Mouse Pneumonitis (K virus)	
Theiler's Mouse Encephalomyelitis (GDVII)	
Kilham rat virus (KRV)	
Sialodacryoadenitis virus (SDA)	
Rat Corona virus (RCV)	Toolan's H-1

The viruses marked with the asterisk are those which have caused us problems in our barrier units.

This pool of microbiologically clean animals, being produced on both a regular and long term basis, permits a high degree of flexibility when designing or modifying breeding programmes. Colonies can be re-directed to new areas, and new colonies into areas already in use, with the knowledge that there is minimal risk of infection. Once our initial "goal" of protecting every strain was achieved we began to examine the breeding records of the animals produced in the isolators. We produce over 80 strains of rats and mice in isolators and for almost every strain breeding performance is comparable with the animals bred in the barrier units. This led us to look at ways to increase the proportion of our colonies in the isolators. The largest single factor to overcome being the low number of breeding cages maintained by each isolator technician. We endeavoured to find ways of bringing it closer to the cage/technician ratio achieved in the barrier units. As we gained experience in using isolators we recognised areas where modifications in our management could increase the efficiency of our iso lator work.

Reduction in Health Screening - continuous negative results led us to reduce the isolator screening both in frequency and intensity.

Staff Selection - more care was taken in selecting staff for the isolator areas - note being taken of their compatibility and temperament.

Working Environment - re-organisation of the position of the isolators and their reduction in each area, crowding had caused stress and friction between staff members.

Technician Improvements - problem points in the isolator design were examined and altered e.g.sleeves changed from round to oval; an increased filter surface area resulted in a lower pressure differential making it easier for the technicians to work.

Other efforts were directed at the basic concept of isolator size and method of operation. Each of our technical and managerial improvements helped in their own way to increase the"cage maintained per technician"ratio, which has increased from 200 per technician in 1979 to our current 350 per technician. Despite our efforts we are still operating with some disadvantages in comparison with the barriered produced animals. The main disadvantages we have found being: higher animal cost, initial equipment cost, weakening in the technician/animal relationship. On a positive note some of the advantages of the isolator system are, long term stability both health and genetic, the ability to utilise low cost non-barrier buildings, no restriction on staff movement.

The flexibility that the isolator system gives to planning, development and maintenance of breeding programmes is of great value. Our experience confirms it is possible to maintain a high standard of animal health, and genetic authenticity can be guaranteed. The training and instruction of personnel is extremely important, with emphasis always being placed on the needs of the animals which are constant, regardless of the sophistication of their environment.

References:

Festing,M.F.W. (1982). Genetic Contamination of Laboratory Animal Colonies: An Increasing Serious Problem. ILAR News Vol. 15 No. 4.

Townsend,G. (1969). The Accreditation and Recognition Schemes for Suppliers of Laboratory Animals. Carshalton, M.R.C.

Germfree Research: Microflora Control and Its Application to the Biomedical Sciences, pages 43–45

COMBINED USE OF ISOLATORS WITH LAMINAR AIR FLOW CABINET FOR XENOGRAFTS IN NUDE MICE

Charles Gosse and Patrice Ardouin
Service d'Expérimentation Animale - Institut Gustave-Roussy 94805 Villejuif Cedex France

INTRODUCTION

Since seven years, S.P.F. Swiss-nude mice (from Carl Hansen, N.I.H. Bethesda) were bred in large-sized plastic isolators (Gosse et al., 1983 ; Raynal et al., 1978) and used for xenografts of human tumors (Guichard et al., 1977) and for preparing hybridomas and monoclonal antibodies (Gosse et al., 1984). Isolators were chosen for two reasons : i) they are the most reliable system of protection (absolute barrier) against microbiological contamination of animals ; ii) previous studies (Gosse et al., 1981) had concluded, for the maintenance of a few number of mice (up to 5,000), that isolators were no more expensive than little or medium-sized protected areas and much less expensive than two other tested systems (partial barrier) : laminar air flow (LAF) walls and cages with filter cap (CFC). Consequently it should have been logical that all the nude for experimentations were both treated and maintained in isolators. But, in fact, most of the time, CFC were used for experiments. Effectively if, in one hand, gnotobiologists did not jib at doing experiments in isolators, in the other hand, other experimenters, as medical people, were often scared by isolators, finding the absolute physical barrier as a too much awkward tool for their works in comparison with the easy handling image of CFC and LAF systems.

CFC WITH LAF SYSTEM

Nude mice were maintained in CFC but introduction of tumour cells (hybridoma or xenograft) and operations on mice (graft, sampling, taking of tumors) were made in LAF cabinet as well as daily controls and weekly changes and supplies.

In such a system : i) cost for the maintenance of mice in CFC was very expensive because two people were necessary for each manipulation of CFC in LAF (Gosse et al., 1981) ; ii) more assays were long more the reliability decreased opening after opening of the cage even in LAF (partial barrier). Therefore it was absolutely necessary to design a cheaper and more reliable system.

ISOLATOR WITH LAF SYSTEM

This system combined LAF cabinet with plastic standard isolators equipped with quick sterile transfer systems (french DPTE system) (Ducluzeau et al., 1979 ; Saint-Martin et al., 1975 ; Saint-Martin, 1976) allowing a safe, immediate and easy bidirectional connection between LAF cabinet and isolators. Only xenografts and sampling were made in the LAF cabinet and all other interventions (controls, cleaning, feeding and watering) in isolators.

CONCLUSIONS

As estimated, the combined use of isolators with LAF cabinet was more reliable and less expensive than CFC use for biological experiments. This system can offer to scientific and medical people who are not trained in gnotobiology a first step and a good apprenticeship for experimentation in isolators. Maybe some day all operations (including xenografts and sampling) will be made in isolators and here and now gnotobiologists know that it is possible.

Ducluzeau R, Moreau C, Corpet D, Tancrede C (1979). Improvement of techniques leading to time sparing in rearing gnotoxenic animals. Fliedner and all., Clin. and Experim. Gnotobiotics, Zbl. Bakt. Suppl. 7, Gustav Fisher Verlag, Stuttgart, N.Y.

Gosse C, Dechambre RP, Ardouin P (1981). Comparative efficiency and cost of different systems of protection against microbiological contamination of mice. In Recent Advances in Germfree Research, Proceedings of the VIIth International Symposium on Gnotobiology, 1981, Tokyo Japan. pages 75-78, S. Sasaki (Eds) Tokaî University Press.

Gosse C, Faure E, Ardouin P, Boumsell L, Bernard A (1983). Breeding and maintenance of nude mice in large-sized isolators for preparing monoclonal antibodies with a view to human therapy. 13th International Cancer Congress. Part C, Biology of Cancer (2), page 457. Alan R. Liss Inc. New-York, NY 10011

Gosse C, Faure E, Bretou G, Ardouin P (1984). Utilisation de souris nude produites en isolateurs de grandes dimensions pour la préparation d'hybridomes et d'anticorps monoclonaux. Biosciences, 3 : 3-8.

Guichard M, Gosse C, Malaise EP (1977). Survival curve of a human melanoma in nude mice. J. Natl. Cancer Inst., 58 : 1665-1669.

Raynal B, Chaigneau H, Eveno J (1978). Utilisation d'isolateurs, de cages à couvercle filtrant et de systèmes à flux laminaire horizontal pour la production et le maintien de souris nude. Sci.Tech. Anim. Lab., 3 : 119-126.

Saint-Martin B (1976). Applications à la médecine et à la chirurgie des enceintes stériles pour animaux axéniques. Sci. Tech. Anim. Lab., 1 : 152-157.

Saint-Martin B, Ducluzeau R, Ghnassia JC (1975). Amélioration de la technologie des enceintes stériles permettant leur application à la médecine et à la chirurgie. Rev. Franç. Gynéc., 70 : 585-593.

Germfree Research: Microflora Control and Its Application to the Biomedical Sciences, pages 47–50

PROGRESS OF GNOTOBIOLOGY IN CHINESE ACADEMY OF MEDICAL SCIENCES

Gu Fangzhou

Chinese Academy of Medical Sciences (CAMS)

9 Dong Dan San Tiao, Beijing, China

The information of gnotobiology has been spread in China for many years. In 1977 and 1981 Dr. Pollard visited China, subsequently Prof. Trexler, Prof. Miyakawa and Dr. Sumi visited Beijing giving reports, introducing techniques on germfree (GF) and specific pathogen free (SPF) animals.

The Medical Laboratory Animal Center (MLAC), CAMS, began to remodel the plastic isolator in 1979. A first germfree isolator was made of plastic membrane, and passed through the examination and identification on bacteriological monitoring, physical examination, ventilation rate, pressure and expansion tests. All of these examination parameters proved to be suitable for rearing GF and SPF animals. At the same time caesarean section on pregnant rabbits (Wang, Sun, Wang, He 1984) and guinea pigs (Zhang, Wang, Chu, Su, Jing, Zhao, Wu, He, Sun 1984), artificial feeding of neonatal animals and a series of bacteriological examination methods were accomplished. In addition, to hatch SPF chickens, the new double isolator has been developed successfully in our animal center. Its temperature can be regulated by a heating apparatus which was linked with a transformer and kept at a higher level (38-42°C).

Since the VII International Gnotobiology Symposium we have devoted ourselves to developing gnotobiotic science in our research work.

Department of Gnotobiology, MLAC, reared GF rabbits and guinea pigs. 28 GF rabbits were obtained from 6 pregnant rabbits through caesarean section. These young rabbits were fed by means of artificially combined diet in plastic isolators. All the baby rabbits grew in good condition and maintained with solid diet L-478 without weaning. The growth curve of seven of 28 axenic animals showed as follows (Fig.1).

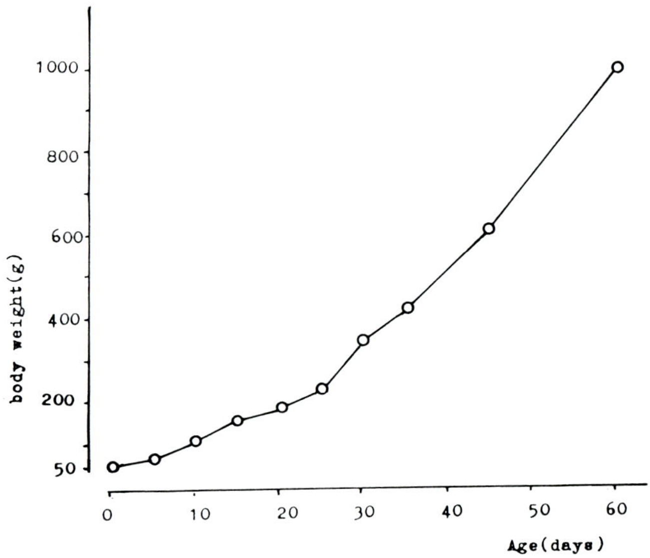

Fig. 1 Growth curve of germfree rabbits

35 GF baby guinea pigs were obtained from 10 pregnant guinea pigs by hysterectomy and fed with milk diet during the first few days after operation. Five days later they were fed with a combination of milk and solid diet L-477. The weaning survival rate was about 94%. The growth curve of 11 axenic guinea pigs is shown in Figure 2.

Dr. Jing of our Animal Center and his cooperator Dr. Zhang from the Institute of Epidemiology and Microbiology National Center of Preventive Medicine studied the establishment of gnotobiotic state in guinea pigs and the relationship between *Shigella flexneri* IIa and normal intestinal flora in gnotobiotic (GN) guinea pigs. The results of observation revealed that a high resistance to *S. flexneri* IIa infection had developed in GN guinea pigs in comparison with those of control animals (mono-associated with *Peptococcus indolicus*).

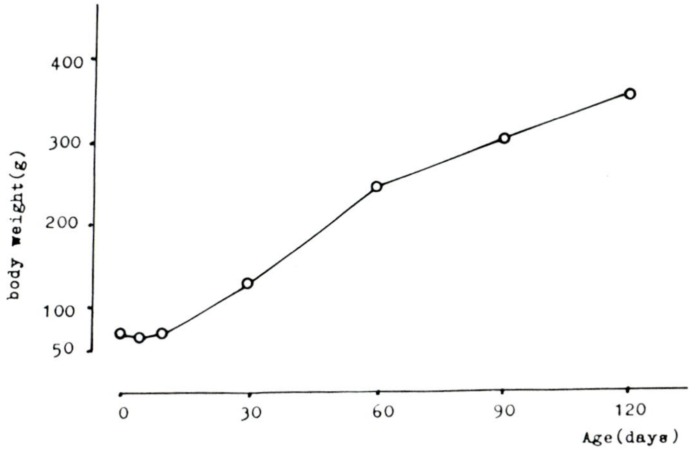

Fig. 2 Growth curve of germfree guinea pigs

Some progress was made in the field of applying nude mice to oncological research.

Dr. Gao in Department of Pathology, Institute of Basic Medical Sciences CAMS and Dr. Wei in MLAC observed the mechanism of invasion of cancer cells in nude mice using human esophageal carcinoma cells (Eca 109) (Gao, Xue, Li, Wei, Zhang, Xu - prepared for publication). Tumor cells (5 10^6 and 1 10^7) were injected either subcutaneously or intraperitoneally in different regions. The research results indicated that the invasions of cancer cells were evidently different in the subcutaneous and peritoneal cavity. For study of cancer invasion, the intraperitoneally transplanted tumor cells were of more utility than those transplanted subcutaneously.

The nude mice in our animal center were also used in studies on nasopharyngeal carcinoma, osteosarcoma, cancer of the liver and ovarian tumor. Some encouraging progress has been made in these studies.

Gnotobiological study is going on well and prosperously in CAMS. I am sure that the prospect of gnotobiological research would be promising. May I extend our great thanks to Drs. Pollard and Sumi and Professors Trexler and Miyakawa for their help and support.

REFERENCES

Gao J, Xue K, Li B, Wei L, Zhang Z, Tian S, Xu Z. The characteristics of growth and invasiveness of cell line from human esophageal carcinoma. (Prepared for publication)

Wang Y, Sun S, Wang A-J, He T (1984). Rearing germfree caesarean-born rabbits. Shanghai Laboratory Animal Science 4(2):92.

Zhang Y, Wang A, Chu C, Su S, Jing S, Zhao M, Wu J, He T, Sun H (1984). Initial study of rearing method on germfree caesarean-born guinea pigs. Shanghai Laboratory Animal Science 4(1):17.

Germfree Research: Microflora Control and Its Application to the Biomedical Sciences, pages 51–54

ABSENCE OF PNEUMOCYSTIS CARINII IN LOBUND GERMFREE AND CONVENTIONAL RAT COLONIES

Morris Wagner, Ph.D.

Lobund Laboratory
University of Notre Dame
Notre Dame, IN 46556 U.S.A.

Dr. Linda Pifer *et al.* attempted to use germfree (GF) rats as an experimental model for the study of *Pneumocystis carinii* (PC) pneumonia in human cancer patients undergoing immunosuppressive drug therapy (Pifer, Lattuada, Woods, Edwards 1981). Untreated Sprague Dawley GF rats purchased from two different commercial animal breeders were found to carry unprovoked PC infection as demonstrated by histological and serological examination. Matched conventionally-reared (CV) rats from one of the breeders were also positive for PC. They suggested that neither GF nor CV rats may be suitable for experimental PC studies because of their carrier state.

The above report raised concern regarding the microbiological pedigree of the Lobund GF rat colonies since specific tests for PC had not been performed. Since a number of foreign and domestic GF rat colonies, including those of commerical breeders, have descended either directly or indirectly via foster-nursing on Lobund GF dams, it became important to determine whether the PC found in commerically available GF rats might have come from their Lobund progenitors or foster mothers. Therefore, a study was undertaken to search for evidence of PC in Lobund GF and CV rats.

Sprague Dawley and Lobund Wistar rats from the Lobund GF and CV colonies were sacrificed by exsanguination via cardiac puncture. The centrifuged serum was stored at -25C for subsequent serological testing. Sagittal cuts were made in the lung lobes and touch smears were prepared from the cut surfaces. Squash preparations were also prepared by crushing a piece of lung between two glass slides. The slides were fixed for 5 minutes in absolute methanol prior to staining. The remaining lung tissue was fixed in 10% neutral formaldehyde prior to embedding, sectioning and staining.

Duplicate touch and squash smears were stained respectively with Giemsa and with toluidine blue O stain (Chalvardjian, Grawe 1963). Lung sections were stained for PC cysts with methenamine silver (Mahan, Sale 1978). Human lung tissue from cancer patients with diagnosed PC pneumonia was used as a positive control for the staining procedures.

GF and CV rat sera was tested for PC antigen by counterimmunoelectrophoresis (Pifer, Hughes, Stagno, Woods 1978). The sera were shipped frozen in dry ice to Dr. Pifer's laboratory for the serological analysis.

Rats and mice reportedly carry PC subclinically but can be provoked to clinical infection with immunosuppressive drugs (Barton, Campbell 1969; Walzer, Powell, Yoneda 1979). Therefore, Lobund's Sprague Dawley CV rats were immunosuppressed with cyclophosphamide (CPA) implanted subcutaneously in sealed silastic tubes (48-50mg/tube) at ages 2,4, and 6 months. The rats were sacrificed at age 7 3/4 months.

Clusters of dark staining PC cysts in control human lung are seen in Figure 1.

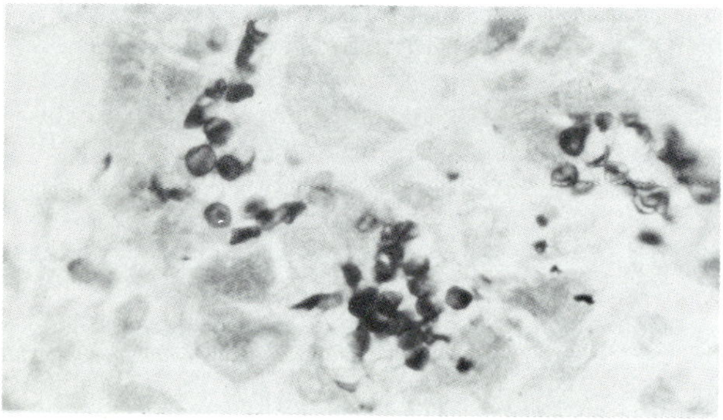

Fig. 1. Clusters of PC cysts stained dark with methenamine silver. Specimen is lung tissue from human cancer patient with PC pneumonia. Cysts were absent in all rat lungs.

The results of examining GF and CV rats for PC are shown in Table 1. All the animals were histologically and serologically negative regardless of rat strain, microbiological status, age or treatment with CPA.

Table 1

Tests for Pneumocystis carinii in Germfree and Conventional Rats

Strain	Status	Age	Treatment	Histology	Serology
SD (10)[a]	GF	21d	none	neg	neg
SD (5)	CV	21d	none	neg	neg
SD (10)	GF	47d	none	neg	nd[b]
SD (3)	CV	47d	none	neg	neg
LW (7)	GF	22d	none	neg	neg
LW (2)	GF	adult	none	neg	neg
LW (3)	CV	22d	none	neg	neg
SD (9)	CV	230d	CPA	neg	nd
SD (1)	CV	230d	none	neg	nd

a = number of rats
SD = Sprague Dawley
LW = Lobund Wistar
CPA = cyclophosphamide
b = not done
GF = germfree
CV = conventional

Pifer et al. (Pifer, Lattuada, Woods, Edwards 1981) reported that cysts and trophozoites of PC are more readily evident in rats weighing less than 50 grams. For this reason, we examined rats in the 21-22 day age group (ave. wgt. = 46 g). The Pifer group (Pifer, Woods, Edwards 1982) also reported that cysts are only abundant in rats less than 50 g and become rare with increasing age and rise in antibody titer. Furthermore, fulminant infection usually occurs only when the host's immunological defenses are impaired so that the cysts can proliferate. However, the results on the young as well as older rats reported here, including the rats immunosuppressed with CPA, were negative. It is concluded from the above data that the GF and CV rat colonies maintained at Lobund do not harbor PC. This information extends the germfree pedigree of the Lobund stock.

Barton EG, Campbell WG (1969). Pneumocystis carinii in lungs of rats treated with cortisone acetate. Amer J Pathol 54:209.

Chalvardjian AM, Grawe LA (1963). A new procedure for the identification of Pneumocystis carinii cysts in tissue sections and smears. J Clin Pathol 16:383.

Mahan CT, Sale GF (1978). Rapid methenamine silver stain for Pneumocystis and fungi. Arch Pathol Lab Med 102:351.

Pifer LL, Hughes WT, Stagno S, Woods D (1978). Pneumocystis carinii infection: Evidence for high prevalence in normal and immunosuppressed children. Pediatrics 61:35.

Pifer LL, Lattuada CP, Woods DR, Edwards CC (1981). Germ-free rat model for Pneumocystis carinii in cancer patients: Spontaneous infection in untreated animals. Abstracts of the 19th Annual Meeting of the Association for Gnotobiotics, Roswell Park Memorial Institute, Buffalo, NY.

Pifer LL, Woods DR, Edwards CC (1982). Life cycle of Pneumocystis carinii in germ-free rats: An immunologic compromise. Abstracts of the 82nd Annual Meeting, Amer Soc Microbiol, Atlanta, GA.

Walzer PD, Powell RD Jr, Yoneda K (1979). Experimental Pneumocystis carinii pneumonia in different strains of cortisonized mice. Infect Immun 24:939.

The author gratefully acknowledges the aid of Dr. Linda Pifer for doing the serum tests for PC antigen, Dr. Morris Pollard and Phyllis Luckert for supplying the lungs from CPA immunosuppressed animals, and Douglas Ranschaert for the preparation of the histological sections.

Germfree Research: Microflora Control and Its Application to the Biomedical Sciences, pages 55–58

A SEVEN YEARS BREEDING OF S.P.F. NUDE MICE IN ISOLATORS EXPERIENCE : EFFECTS OF VITAMIN K DEFICIENCY IN DIET

Patrice Ardouin, Charles Gosse and Michel Huard
Service d'Expérimentation Animale - Institut
Gustave-Roussy 94805 Villejuif Cedex France

BREEDING DATA

Since seven years, S.P.F. swiss nude mice (from Carl Hansen, N.I.H. Bethesda) were bred in large-sized plastic isolators (Gosse *et al.*, 1981 ; Gosse *et al.*, 1983 ; Gosse *et al.*, 1984 ; Raynal *et al.*, 1978). The maintenance permanently of 200 monogamous pairs (nu/+ female x nu/nu male) allowed an annual production of about 6,700 nude mice. The productivity index (nude weaned/dam/week) was 0.65 and remained rather regularly at this level during more than seven years. But for six months (during 1980) it was observed an important and rapid decrease of productivity index : from 0.65 to 0.40. This decrease happened just before a mortality of 41.5 % of nude male breeders which presented most of time, haemorrhagic symptoms of thoracic cavity. Owing to haemorrhagic observed symptoms and taking into account the available literature data (Festing and Blackmore, 1971 ; Gustaffson, 1959 ; Ley *et al.*, 1969), vitamin K deficiency in the irradiated food was suspected. The incriminated batch of diet was removed and reserved for analytical and experimental controls. After changing of diet batch, the nude male breeders mortality stopped and the productivity index reached rapidly its initial level (0.65).

ANALYTICAL DATA

It was not possible to find vitamin K in the suspected diet batch. But a lot of analytical controls allowed to know the evolution, for over three months, of vitamin K rate in standard diet, after pelletisation only and after pelletisation and irradiation (40 kiloGray).

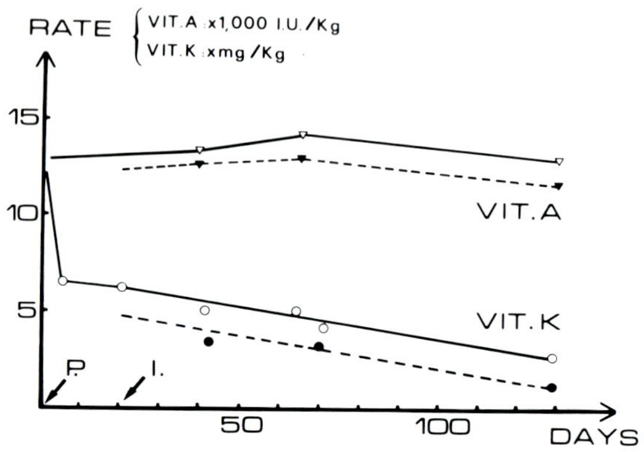

Fig. 1. Evolution of vitamin K rate in standard diet after:
pelletisation only (P)
Pelletisation and irradiation (I.) at 40 kiloGray. (Vitamin A was used as control).

It was surprising that vitamin K decrease was more important after pelletisation only than after irradiation (Fig. 1). Because it is very difficult to determine vitamin K chemically in diet (Ley *et al.* 1969) the low levels of vitamin K found after 2 months, in this study, were not significant, but they were probably near to the critical rate (2 mg/kg) for germfree rats (National Academy of Sciences New-York 1978).

EXPERIMENTAL DATA

It was possible to reproduce these symptoms and mortality both in germfree C3H male mice and in S.P.F. male swiss nude mice by using two different diets sterilized by gamma-radiation at 40 kiloGray : i) a free-vitamin K semi-synthetic diet, ii) the deficient vitamin K standard diet (remnants of criminated batch).
The response-time difference between axenic mice and S.P.F. nude mice fed with free-vitamin K diet was due to the intestinal bacterial flora in nude mice (Fig. 2).

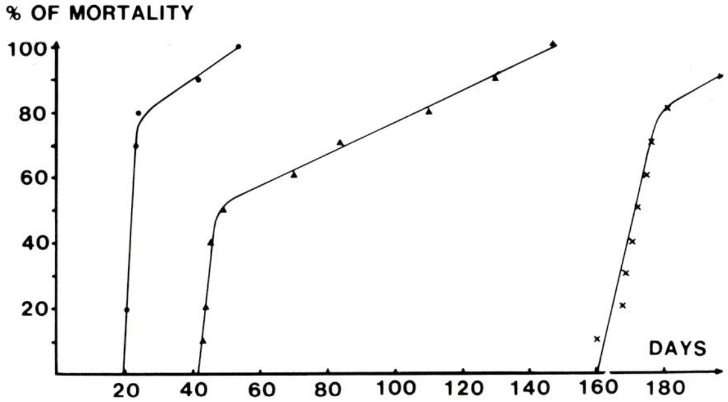

Fig. 2. Evolution of the mortality in :
—●— 10 germ-free male C3H mice fed with free-vitamin K semi-synthetic diet
—▲— 10 SPF male swiss nude mice fed with free-vitamin K semi-synthetic diet
—x— 10 SPF male swiss nude mice fed with deficient vitamin K standard diet (criminated batch).
It was not possible to feed germfree C3H male mice with the deficient vitamin K standard diet because remnants of incriminated batch were not sufficient.

It must be emphasized that the presence or absence of *Enterobacteriaceae* in the nude mice intestinal microflora did not modify the response time (preliminary and unpublished data). In addition the difference of mortality delay between nude males fed with a free vitamin K or a vitamin K deficient diet (Fig. 2) was due to presence, at low level, of vitamin K in the incriminated batch (deficient vitamin K diet) although it was not possible to find vitamin K in this batch by chemical determination.

CONCLUSION

This breeding mischance as well as these experimental results confirmed that male mice can be depleted in vitamin K to death. As it is very difficult, if not impossible, to determine vitamin K chemically in diets (Ley *et al.*, 1969), it would be interesting to develop biological controls for determining the fate of vitamin K in diet, and so to be able to prevent such breeding failure.

Festing MFW, Blackmore DK (1971). Life span of specified pathogen-free (MRC category 4) mice and rats. Lab. Anim. 5 : 179-192

Gosse C, Dechambre RP, Ardouin P (1981). Comparative efficiency and cost of different systems of protection against microbiological contamination of mice. In Recent Advances in Germ-free Research, Proceedings of the VIIth International Symposium on Gnotobiology, 1981 Tokyo Japan 75-78, S. Sasaki and all (Eds) Tokai University press.

Gosse C, Faure E, Ardouin P, Boumsell L, Bernard A (1983) Breeding and maintenance of nude mice in large-sized isolators for preparing monoclonal antibodies with a view to human therapy. 13th International Cancer Congress Part C Biology of Cancer (2) page 457. Alan R. Liss Inc. New-york NY 10011.

Gosse C, Faure E, Bretou G, Ardouin P (1984). Utilisation de souris nude produites en isolateurs de grandes dimensions pour la préparation d'hybridomes et d'anticorps monoclonaux Biosciences 3 : 3-8.

Gustafsson BE (1959). Vitamin K deficiency in germfree rats. Ann. New-York Acad. Sci. 78 : 166 - 174

Ley FJ, Bleby J, Coates ME, Paterson JS (1969). Sterilization of laboratory animal diets using gamma radiation. Lab. Anim. 3 : 221 - 254

National Academy of Sciences. New-York (1978). Nutriments requirements of laboratory animals. 3rd revised edition page 22.

Raynal B, Chaigneau H, Eveno J (1978). Utilisation d'isolateurs, de cages à couvercle filtrant et de systèmes à flux laminaire horizontal pour la production et le maintien de souris nude. Sci. Tech. Anim. Lab. 3 : 119 - 126.

Germfree Research: Microflora Control and Its Application to the Biomedical Sciences, pages 59–60

WEIGHING PROBLEMS IN GERMFREE ENVIRONMENT

Bernard A. Teah

Lobund Laboratory
University of Notre Dame
Notre Dame, Indiana 46556 U.S.A.

For years, people involved in germfree work have been working to develop easier methods to weigh various objects inside of isolators. Some techniques have involved (a) electrically-charged magnets, (b) hooks fastened to the top of the plastic isolator, and (c) other mechanical connections.

At Lobund, we have used various types of triple beam balances which were steam sterilized. This ruined the finish on the balances, corroded the springs and the fine adjustment, and ruined the balance rod. The balances were difficult to adjust, and after being adjusted they were difficult to read. The manufacturer then replaced the agate edge with plastic, which made them impossible to sterilize.

We then turned to electronic digital balances, which raised the question: How would we sterilize them? We had three choices: steam, peracetic acid, or ethylene oxide. We knew that steam would rust the integral parts; peracetic acid would corrode any minute surface contacts; but, we did not know what ethylene oxide would do, but it probably would have some deleterious effects.

We obtained an inexpensive electronic hand calculator, which was treated with ethylene oxide for approximately sixteen hours. To the surprise of all involved, it still functioned accurately.

The next problem was to enter this equipment into a germfree environment. We procured some polyethylene envelopes 10"x12"x.0035". The calculator was placed in this plastic container and was then strip heat-sealed at both ends. This was done in a Vertrod Thermal Impulse Heat Sealer Model 14C, with the heat dial at 4½, and dwell time of eight seconds.

The calculator, which had been sterilized previously, was again subjected to the same ethylene oxide exposure and again it worked fine.

A Sartorius Electronic Digital Balance was then purchased and again subjected to treatment with ethylene oxide. It worked accurately. We then placed the balance in a polyethylene envelope, with spore strips which were placed in strategic places. It was sealed and sterilized with ethylene oxide and the spore strips were negative. By the same procedure, a Polaroid Electronic Digital Balance did not prove to be accurate. An Ohaus Electronic Digital Balance was tested using the same procedures as used for the previous balances. It also weighed accurately after sterilization.

We then learned of a little known Hanson Electronic Digital Balance which we tested. The same procedures were used for the other balances. It performed accurately within 1%.

The three balances which we examined for sterilization and functioning status were well adapted to isolator use for weighing of animals and feed. They all required approximately the same space, with the Hanson balance requiring slightly less space. They are all approximately equal in accuracy. The sterilization procedure was the same for all. The cost comparison of the three balances is: Sartorius - $500. Ohaus - $150. Hanson - $70-80.

In conclusion, the Hanson balance was judged superior on the basis of cost, accuracy, convenience, availability and finally dependability.

Germfree Research: Microflora Control and Its Application to the Biomedical Sciences, pages 61–63

STERILIZATION ISSUES AND UPDATE FOR GNOTOBIOLOGY RESEARCH

Frank E. Halleck, Ph.D.
Corporate Director of Scientific Affairs
American Sterilizer Company
2222 West Grandview Boulevard
Erie, Pennsylvania 16514

Although sterility is a basic attribute of Gnotobiological research, continuous problems arise in achieving and maintaining asepsis because sterilization processes have been relegated to trial and error methods by researchers rather than development of standardized procedures that are derived from definitive research investigations. This presentation covered some sterility and sterilization issues that are basic and mandatory to Gnotobiological research and offered suggestions and recommendations for monitoring the attainment and maintenance of sterility of processed items.

STERILIZATION OPTIONS

The criticality of sterility in germfree research obviates the employment of any arbitrary and untested sterilization process in any particular application. However, wherever possible, the standard and usually readily available sterilization processes are preferred. Two of these sterilization systems are dry heat and steam cycles. Thermal processes are, by far, the most advantageous procedures because of their normal availability in research areas and laboratories.

Materials and objects that cannot withstand thermal processing conditions may require other sterilization processes such as gases, liquids or vapors and irradiation. Plastic devices, complex electronic equipment and instrumentation, natural materials and mixtures and chemicals are amenable to one or more "cold" sterilizing processes provided that physical, chemical, biological and toxicological changes do not occur that will affect their use. Gaseous sterilants are by far the more reliable than other forms because of the ease of control, operation and permeation characteristics. The two gaseous systems that have worldwide acceptance are those employing ethylene oxide or formaldehyde.

Other gases such as propylene oxide have been employed as sterilants but various deficiencies in their processing parameters have resulted in abandonment of them to the two gases previously discussed.

The use of liquid chemosterilants for various research applications have been cited as effective sterilizing agents and some of them are in routine use. The primemost agent is glutaraldehyde of which there are various commercial formulations containing different pH levels. All have been registered by EPA and meet the claims for disinfection and sterilization as prescribed by the EPA efficacy tests using AOAC test methodologies.

Other chemical solutions such as peracetic acid, formalin, etc. are commonly utilized for surface sterilization of objects, walls, etc. but these chemical formulations are fraught with the same problems as glutaraldehyde solutions and many are also highly corrosive and damaging to materials which mitigate their use.

Irradiation processes utilizing either Cobalt 60 or electron beam offer little potential for the Gnotobiologist because of the high facility cost and extensive operating controls.

STERILITY MONITORING

All sterilization processes, whether programmed by the equipment manufacturers or by cycle development research, should be validated to assure sterility attainment from cycle-to-cycle under regimented and documented procedures that take into account all the factors that could reduce the probability of sterility attainment and maintenance.

Wherever possible, biological indicators best serve to monitor sterility. There are many types of BIs which can be tailored to be applicable to specific areas and systems of sterilization.

STERILIZATION PROBLEMS

Feed and Other Nutrients

One of the biggest problems in Gnotobiotic animal research is the assurance that the feed or nutriment can be provided consistently in the sterile state. Any sterility breach during an experimental phase automatically results in time and energy set backs. Animal feeds can be sterilized by more than one sterilization process but there are always compromises as to which method is preferred or advantageous.

The use of thermal sterilization processes is generally more than others because of the major advantage of each of control and minimal sterilant residues.

Bedding

Animal bedding of wood chips, ground or chopped corn cobs, etc. require the same advance study of sterilization processes as feed. The problems of sterility failures are more easily solvable because the functional stability of the bedding is all that is required. Pressurized steam is far better for this than other forms of sterilization because of the limited physical changes that occur during steam processes.

Fluids

Fluids consisting of water and other liquid preparations such as nutrients should be researched very carefully before acceptance of a sterilization process for any experimental program. Specialized procedures have been developed for fluids of all types by sterilization technologists involved in drug and food research that take into account conditions required for attainment and maintenance of sterility and other factors such as chemical quality and physical retention and operator safety.

Germfree Research: Microflora Control and Its Application to the Biomedical Sciences, pages 65–68

STEAM QUALITY AND EFFECTIVE STERILIZATION

Robert S. Sedlacek and Edward F. Rose

Department of Radiation Medicine
Massachusetts General Hospital
Boston, Massachusetts 02114

ABSTRACT

Faced with using steam from a commercial utility having boilers > 5 miles distant and being the last user on the system resulted in ineffective sterilization.

A three phase testing program was established utilizing:

1. Direct physical measurements - an Ellison model 915A portable steam calorimeter.
2. Direct microbiology - Autoclaved feed pellets were aseptically placed in fluid thioglycolate medium and incubated at 37°C.
3. Indirect microbiology - Feces from "defined flora" mice fed the autoclaved pelleted feed were tested.

Colorimetric measurements verified that the steam sometimes contained > 5% entrained water. During periods of wet steam it was impossible to maintain consistent sterility of the mouse pellets even using a cycle of 126°C for 60 minutes. One spore-forming Gram positive rod, *Clostridium perfringens* type D was the predomenant bacterium isolated.

Lactating mice, or mice stressed experimentally came down with diarrhea within days of eating pellets treated with wet steam (calorimetric measurements) and a subsequent positive culture. These mice voided stools predominantly showing *Clostridium perfringens* type D.

INTRODUCTION

Even though it is textbook knowledge that superheated and/or supersaturated steam are poor media for sterilization, most infection control officers are satisfied by being able to kill bacterial endospores on a string carrier.

Sterilizing a solid biomass e.g., rodent pellets, presents a much harder challenge. When a droplet of condensate hits a pellet it: a) tries to evaporate, thus cooling the product, or b) forms a glaze or armour coating to interfere with further heat penetration.

METHODS AND MATERIALS

An 8" sampling tube was placed in a 12" downward verticle flow steam line having a pressure ranging between 130 and 150 psig. This sampling tube was fitted with a 1/8" throttling orfice to allow steam flow at a rate of $\cong$ 100 lbs. per hour. The Ellison 915A portable steam calorimeter (Cal Research, Ransomville, N.Y.) was attached to this sampling tube. Temperature and pressure were recorded and physical steam characteristics computed from a nomogram by George V. Vosseller Power Engineering/September, 1980.

Cages were assembled with lid, feed and bedding in place. These cages were sterilized in a AMSCO DT-2 industrial sterilizer using 3 pre-vacuums interrupted by two chamber steams at 126.6°C for 60' followed by a 10 minute dry time.

Three diet pellets from one cage on each load were aseptically placed in a 250 ml straight neck flask fitted with a 38 mm Morton closure and containing 125 ml sterile fluid thioglycolate medium. These samples were incubated at 37°C for 14 days. Anaerobic spot checks were also made by placing a flask in an anaerobic jar, with chopped meat broth and/or Schaedler's broth at 37°, 42° and 55°C.

Fresh mouse fecal pellets were aseptically placed in 2 ml sterile physiological saline, taken to the bacteriology area and quickly streaked onto TSA with 5% sheep blood agar plates (Gibco) for aerobic and anaerobic incubation. Fecal smears were gram-stained for direct microscopic examination.

A comparison of line pressure position on the steam system and distance to the boiler as a function of effective sterilization between two mouse colonies established by the senior author were made.

RESULTS

On moving to our present quarters (B) it quickly became apparent that the sterilizing cycle using 132 C for 4½ minutes as in our former quarters (A) was not achieving sterility of the mouse pellets. Furthermore, cages in the top row in a load became very wet, with up to 5 cm of water collecting in the cages on the worst loads.

Times of sterilization were increased by 1/2' increments at 132°C until 12 minutes were reached without achieving consistent sterility. Due to cage deterioration, the temperature was dropped to 126°C and time increased to 40-50 and finally 60' without improved sterility.

Calorimetric readings showed that slugs of water passed through the calorimeter in excess of the quantity measurable by throttling techniques at the pressure we received the steam.

The amout of water collected from the calorimeter over a predetermined time period was based on the following average steam flow equation: pounds water collected/pounds of steam flow/hr x 100 = % excess water by separating techniques. The excess water in the steam determined by throttling techniques (% reabsorbable for a given pressure) plus the water collected by separating techniques gave on occasion 10% super-saturation or total excess water in our sterilization procedure.

When feed came through the sterilizer with a wet look ("armour coated"), the % of feed samples producing growth in thioglycolate increased. These growths were clostridia, mostly C. perfringens type D; and Bacillus sp. Within days of feeding these samples to stressed mice or to lactating females, diarrhea developed. Bacterial Gram stains showed that these feces consisted almost entirely of large blunt spore-forming Gram(+) rods identified as C. perfringens, type D.

A comparison was made between the two colonies; A, where consistent sterilization was easily obtained and B, the present colony. Both colonies were established and run by the senior author. Calorimetric measurements were not made in Colony A. However, we made a comparison where data existed (Table 1). Successful sterilization in Colony A was accompanied by higher line pressure, dynamic (always moving steam) piping, and close location to the boiler. Also, the cycles were much less severe in Colony A.

Table 1

Comparison of Steam Conditions and Sterilizing Cycles Between Colony A Having Consistent Effective Sterilization and Colony B Where Consistent Effective Sterilization was Unattainable

	Colony A Successful	Colony B Unsuccessful
Line Pressure PSIG	80	55
Piping System	Dynamic	End of Line
Distance to Boiler	$\cong$300'	>5 miles
Isolator Cycle	20' 126°C	40' 132°C
Cage Cycle	4.5' 132°C	60' 126°C

From this we have concluded that:

a. Steam quality should be 99%.
b. Place sterilizers on a dynamic steam line - this maintains line temperature.
c. Have boiler as close to sterilizer as possible to reduce condensation.
d. Bring steam to the sterilizer at 80 psig thereby having extra superheat available to bring the load up to temperature.

Germfree Research: Microflora Control and Its Application
to the Biomedical Sciences, pages 69–72

AMBIENT PRESSURE ETHYLENE-OXIDE STERILIZATION OF FLEXIBLE FILM GERM FREE ISOLATORS

E. D. Dennis, P. J. Matthews and M. T. Jensen

Departments of Animal Supply and
Environmental Health and Safety
National Animal Disease Center
P. O. Box 70, Ames, Iowa 50010

INTRODUCTION

Robie (1969) reported on the successful ETO sterilization of plastic film isolators using a conventional ETO sterilizer. Installing a large ethylene-oxide (ETO) sterilizer to sterilize isolators in our facilities at the National Animal Disease Center (NADC) was not feasible because of initial cost and installation difficulties. Therefore, we sought an alternate sterilizing technique. Schley (1960) and Smith (1968) reported on simpler and less expensive methods of ETO sterilization, neither of which required a vacuum or pressure chamber. These techniques were adapted and modified to develop ambient pressure ETO sterilization of flexible film germ free isolators.

MATERIALS AND METHODS

Facilities and Equipment

1. Negative pressure room ventilated to the atmosphere.
2. Tank of 12% ETO and 88% dichlorodifluoromethane
3. ETO entry device
4. Platform scales.
5. Venting manifold assembly.
6. Psychrometer.
7. Bottle of sterile water.
8. Full face gas mask approved for ETO usage.
9. Heat lamps.
10. Peracetic acid, sprayer and pump.

11. *Bacillus subtilis* test spores ([#1244 Attest Biological Indicators for ETO sterilization], 3M, Eagan, MN.)
12. Chemical ETO sterilization indicators ("SteriGage EO", Info-Chem, Fairfield, NJ).

Procedure

A flexible film isolator is located in a negative pressure room, equipment and supplies are placed in the isolator and arranged to give greatest exposure to the ETO. ETO test indicators and spores are placed at various locations within the isolator. The ETO entry device and all other stoppers or attachments attached to isolator orifices are sprayed with peracetic acid and installed. The venting manifold is attached and the connections are taped. Inner port caps are not put on until after degassing is completed. Heat lamps are used to achieve an isolator temperature of 90-100°F. Sufficient sterile distilled water is added to the floor of tub isolators or to a pan in vinyl chamber isolators to give a relative humidity of 55-65%. Latex tubing is attached to the ETO pipe nipple inside of the isolator and placed so that the ETO will flow directly onto the isolator tub floor or into the pan in a vinyl chamber isolator because liquid ETO allowed to come into contact with the chamber or other plastic materials may damage them.

During the following procedures, a full face respiratory mask approved for ETO usage is worn by personnel. The ETO tank is connected to the ETO valve installed in an isolator flare. Isolator gloves are placed in the interior position, venting manifold valves are closed and ETO is discharged into the deflated isolator to a minimum final ETO concentration of 235 mg/L volume of an inflated isolator. Exhaust and supply air filters are purged with ETO for 5-10 seconds. After twenty hours, ETO indicators are checked, test spores are removed and the isolator is vented directly to outdoor atmosphere. A temperature of 90° F to 100° F is maintained within the isolator through the sterilizing and degassing periods. Entry to the room is restricted to personnel wearing proper protective equipment until the isolator has been degassed for 24 hours.

RESULTS

Over 300 isolators have been successfully sterilized in the past 3-1/2 years at significant cost and labor savings.

Table 1. Cost Comparison of Isolator Sterilizing Methods.

	COSTS	
Sterilizing Method	*Operational (per unit)	Initial Equipment
Peracetic Acid	$ 32.79	$ $359.80
ETO-Sterilizer	$ 45.00 (est)	$112,668.35 (Plus installation)
ETO-Ambient Pressure	$ 24.41	$ 1,250.76

*Includes labor, power and chemicals.

Positive test spores occurred twice outside of the filter barrier within the air filter cases, once within the finger of an extended isolator glove and once under a plastic mat. All failures were believed to be the result of inadequate temperature and have been corrected by appropriate adjustments.

Monitoring of exhaust gasses from ventilating isolators with a Moran 1A infra red gas spectrophotometer set to moniotor at 3.29 μ indicated 11 to 100 hours of ventilation are required to reduce ETO levels to 1 ppm in the isolator exhaust. Room concentration of ETO was monitored during five routine sterilizing cycles. The range of maximum concentration was 2.7 to 5.2 ppm. The time elapsed from beginning degassing until room concentration of ETO was down to 1 ppm was 85-196 minutes for these 5 sterilizing cycles.

DISCUSSION AND CONCLUSION

Robie (1969) has sited many advantages of ETO sterilization over peracetic acid. Advantages of this method over sterilization in a conventional ETO Sterilizer include: lower initial and operating costs, no risk of damage from vacuum or chamber pressure and capacity is not restricted to the volume of a sterilizer chamber.

The Federal Register June 22, 1984, standard 1910.1047 established a permissible exposure level of 1 ppm as an 8-hour time-weighted average. To assure compliance with safety procedures, refer to this standard.

The time to achieve 1 ppm ETO concentration in an isolator exhaust varies depending on the isolator temperature during venting and the type of materials included in the isolator (Roberts 1972, White 1977). It is necessary to establish, by representative monitoring, the ETO levels during and after sterilization to assure safe conditions are being achieved in any individual situation.

We have found the described technique for ambient pressure sterilization of flexible film isolators to be highly effective, cost saving and labor efficient. ETO is a chemical that can be hazardous to the health of personnel and animals, but if properly managed, can be utilized safely.

ACKNOWLEDGEMENTS

The technical assistance of Roger J. Spaete and the guidance and advice of Joseph R. Songer is gratefully acknowledged.

REFERENCES

Roberts RB, Rendell-Baker L (1972). Aeration after ethylene oxide sterilisation. Anaesthesia 27:278.

Robie DM, Walburg, Jr. HE (1969). Ethylene oxide sterilization of plastic film isolators: A six-year study of operational efficiency. In Mirand EA, Back N (ed): Germ Free Biology Experimental and Clinical Aspects. New York: Plenum Press, p 357.

Schley DG, Hoffman RK, Phillips CR (1959). Simple improvised chambers for gas sterilization with ethylene oxide. Applied Micro 8:15.

Smith RM, Young JA (1968). Simplified gas sterilization: A new answer for an old problem. British J. of Anaesthesia 40:909.

White JD (1977). Standard aeration for gas-sterilized plastics. J. Hyg., Camb. 79:225.

III PHYSIOLOGY, METABOLISM, AND NUTRITION

Germfree Research: Microflora Control and Its Application to the Biomedical Sciences, pages 75–76

INCREASED LIFE SPAN AMONG GERMFREE RATS

Morris Pollard and Bernard S. Wostmann

Lobund Laboratory
University of Notre Dame
Notre Dame, Indiana 46556 U.S.A.

Compared to conventional counterpart rats, the average life span of male Lobund-Wistar (L-W) rats was extended by 1/3 among those rats that were maintained on the same feed (L-485) but under germfree (GF) conditions. The latter rats (average life span 36 months) manifested over 90% incidence of benign tumors in the medullary areas of the adrenal glands; and significant but lower incidences among other endocrine glands (pituitary, thymus, thyroid) and hormone-regulated organs. The latter changes were additional to findings of tumors in the breasts and uteri among aged female GF rats. Approximately 90% of GF male rats over age 30 months manifested a spectrum of liver lesions from hepatomas to carcinomas (Pollard, Luckert 1979). About 10% of rats over age 30 months had developed adenocarcinomas of the prostate gland (Pollard 1973). Deaths and illnesses among aged GF L-W rats were associated with hemorrhagic pituitary tumors, thymomas, hepatic tumors, and adrenal medullary tumors.

A group of 10 GF male L-W rats were maintained from weaning age to age 38 months on 12 grams of diet L-485, fed individually at daily intervals. The dietary restriction was estimated at 25-30%. It was necessary to kill them for examinations because of an accidental bacterial contamination in the isolator system. The rats were in excellent physical condition, appearing alert and aggressive. The body weights were 25-30% below weights of comparable male GF L-W rats on ad libitum dietary regime (Pollard, Wostmann 1984; Wostmann, Pollard 1983).

For examinations, the rats were subjected to deep ether

anesthesia and exsanguinated from the exposed heart. Gross examinations revealed no evidence of disease: the visceral organs appeared "normal". Tissues were fixed in Bouin's solution, embedded in paraffin, and thin sections thereof were stained with hematoxylin and eosin. On the basis of microscopic examinations, there was no evidence of lesions among them; the key tissues examined included pituitary, adrenal, thymus, thyroid, parathyroidm and pancreas glands; plus kidneys, liver, lungs, and prostates. This was in sharp contrast to the changes that had been observed in over 90% of the ad libitum-fed aged GF rats, especially in the adrenal glands and livers (Pollard, Luckert 1979).

Conventional rats have been considered aged at 24-30 months. The recorded causes of deaths were usually obscure, but they included infectious processes in the lungs and intestines, degenerative changes in the kidneys (nephrosis), and neoplasms. At average 36 months the GF rats on the ad-libitum dietary regime were free of infections and the kidney lesions were minimal, but they developed benign lesions in the breasts and uteri and malignant neoplasms in the livers and prostates. This complex of diseases was not observed among the GF rats that were maintained for 38 months on the restricted dietary regime.

There was no evidence that the GF rats on the restricted diet schedule were diseased; so, their ultimate life span is open to speculation. It is likely that the criteria for classifying rats as "aged" will require revision. It is likely that the aging process will be associated with endocrine mechanisms.

References:

Pollard M (1973). Spontaneous prostate adenocarcinomas in aged germfree Wistar rats. Jr Nat'l Cancer Inst 51:1235.

Pollard M, Luckert PH (1979). Spontaneous liver tumors in aged germfree Wistar rats. Laboratory Animal Science 29:74.

Pollard M, Wostmann BS (1984). Aging in germfree rats: the relationship to the environment, diseases of endogenous origin, and to dietary modification. Proc Int'l Symp on Laboratory Animal Science (VIIIth Symposium ICLAS), in press.

Wostmann BS, Pollard M (1983). Prevention of spontaneous hepatic neoplasms among aged germfree (GF) Lobund Wistar (L-W) rats by dietary restriction. Federation Proceedings 42:801 (Abstr).

Germfree Research: Microflora Control and Its Application to the Biomedical Sciences, pages 77–79

ESTABLISHMENT OF A COLONY OF GERMFREE SPONTANEOUSLY HYPERTENSIVE RATS

T.R. Kingsley, PhD. and Morris Pollard, PhD

South Bend Center for Medical Education,
Indiana University and Lobund Laboratory,
University of Notre Dame,
Notre Dame, Indiana 46556

The Wistar-Kyoto spontaneously hypertensive rat (SHR) is a valuable model for human hypertension (Okamoto 1963; Okamoto et al 1964). A substantial body of evidence indicates that the immune system in these animals is abnormal and may be a factor which contributes to the hypertensive syndrome. Various aspects of T cell function (blastogenic response, delayed type hypersensitivity, allograft reactions) are depressed (Takeichi et al 1980a; Takeichi et al 1981), and the presence of a natural thymotoxic antibody has been reported (Takeichi et al 1980b). Manipulation of the immune system by administration of anti-rat thymocyte serum (Bendich et al 1981) or by thymus grafts (Ba et al 1982) modulates the hypertensive condition.

Further elucidation of the relationship between immune system abnormalities and hypertension in SHR will require the use of a variety of techniques, some of which are feasible only in a germfree (GF) environment. For this reason we have established a colony of GF-SHR, and have evaluated several parameters to determine whether the hypertensive syndrome develops in the same manner under GF conditions as it does under conventional (C) conditions. Also included in this study is data from GF animals which were sacrificed when contamination with a single microorganism was discovered. These animals, designated gnotobiotic SHR (GN-SHR) may be considered to be essentially the same as GF-SHR. Our comparison of these groups included: observation of general health and ability to breed in the GF environment; measurement of body weight and systolic blood pressure (via tail cuff sensor, IITC, Inc.); and histological examination of tissues, primarily the kidney, as an estimate of the development of vascular pathology.

Findings indicate that SHR, derived and raised according to standard GF technology, exhibit growth and reproduction patterns similar to C-SHR. Litter size, through 7 inbred generations, has averaged 9 pups per litter. There are no significant differences in body weights among male C, GF, and GN-SHR at ages up to 1 year (Table 1). The time course of development of hypertension is similar in all three groups, i.e. elevation of systolic blood pressure is detectable at 6 weeks of age, and approaches 200 mmHg. in male SHR by 12 weeks (Table 2).

Table 1. Comparison of body weights of male C, GF, and GN spontaneously hypertensive rats. (gm)

Age	C-SHR	GF-SHR	GN-SHR
6 wks	104 ± 7.4 (14)*	101 ± 1.4 (2)	114 ± 1.1 (3)
8 wks	129 ± 3.7 (18)	160 ± 6.4 (2)	125 ± 2.5 (5)
12 wks	224 ± 4.8 (18)	229 ± 9.9 (2)	238 ± 2.1 (2)
8 mos	370 ± 10.6 (3)	336 ± 18.1 (2)	---
1 yr	392 ± 9.1 (6)	---	353 ± 9.8 (6)

Table 2. Comparison of systolic blood pressures of male C, GF, and GN spontaneously hypertensive rats (mmHg)

Age	C-SHR	GF-SHR	GN-SHR
6 wks	145 ± 2.7 (14)*	139 ± 6.2 (2)	142 ± 7.8 (3)
8 wks	171 ± 2.3 (18)	175 ± 1.2 (2)	150 ± 4.1 (5)
12 wks	190 ± 1.8 (18)	185 ± 5.4 (2)	195 ± 4.0 (2)
8 mos	207 ± 4.9 (3)	193 ± 0.5 (2)	---
1 yr	213 ± 3.0 (6)	---	216 ± 2.6 (6)

*Mean, ± S.E. (N). Student T-test revealed no significant differences ($p = \leq .05$) in either body weight or blood pressure when GF or GN groups were compared to age-matched conventional SHR.

Comparison of the histological features of kidneys of GF-SHR vs. C-SHR reveals that neither group exhibits marked pathological changes prior to 8 months of age. In animals younger than 8 months, moderate thickening of arteriolar walls and narrowing of arteriolar lumina is fairly common. Larger vessels exhibit no major pathological changes. Renal corpuscles are generally normal, although occasional glomerular swelling and reduction of Bowman's space may be observed. After 8 months of age, it is common to observe pronounced thickening and hyalinization of the media and complete occlusion of the lumina of aterioles. Larger vessels characteristically exhibit extensive leukocytic infiltration of the periarterial connective tissue. Glomeruli show extensive hyalinization, and may degenerate completely, leaving large areas of fibrous connective tissue. Tubules are frequently grossly dilated and filled with casts. Again, there are no obvious differences between the kidneys of age-matched C-SHR and GF-SHR.

These studies indicate that it is feasible to maintain SHR in a germfree environment without modification of their hypertensive syndrome. The GF-SHR is therefore a valid experimental model, useful for examining physiological parameters in SHR. It may be further concluded that, since leukocytic infiltration (periarteritis nodosa-like lesions) are observed in GF animals, no bacterial process is necessary to precipitate this type of lesion.

Ba D, Takeichi N, Kodama T, Kobayashi H (1982). Restoration of T cell depression and suppression of blood pressure in spontaneously hypertensive rats (SHR) by thymus grafts or thymus extracts. J Immunol 128:1211.

Bendich A, Belisle EH, Strausser HR (1981). Immune system modulation and its effects on the blood pressure of the spontaneously hypertensive male and female rat. Biochem Biophys Res Comm 99:600.

Okamoto K, Aoki K (1963). Development of a strain of spontaneously hypertensive rats. Jpn Circ J 27:282.

Okamoto K, Aoki K, Nosaka S, Fukushima M (1964). Cardiovascular disease in spontaneously hypertensive rats. Jpn Circ J 28:943.

Takeichi N, Ba D, Kobayashi H (1980). Natural cytotoxic autoantibody against thymocytes in spontaneously hypertensive rats. Cell Immunol 60:181.

Takeichi N, Suzuki K, Kobayashi H (1981). Characterization of immunological depression in spontaneously hypertensive rats. Eur J Immunol 11:483.

Takeichi N, Suzuki K, Okayasu T, Kobayashi H (1980). Immunological depression in spontaneously hypertensive rats. Clin Exp Immunol 40:120.

Germfree Research: Microflora Control and Its Application to the Biomedical Sciences, pages 81–85

ON GUT MEMBRANE TRANSPORT OF TRYPTOPHAN IN THE GERMFREE MOUSE

Arthur W. Phillips, Sc. D.

Biological Research Laboratories
Department of Biology
Syracuse University
Syracuse, New York U.S.A. 13210

Microorganisms in the alimentary tract of animals are major stimuli necessary for maturation of a variety of host cells and tissues (Hentges,1983). One important tissue is the absorptive epithelial membrane of the small intestine where profound differences are observed between germfree (gf) and conventional (cv) animals (Abrams, 1983). Although gf animals possess less total absorptive gut epithelium than cv animals (Gordon, Bruckner-Kardoss 1961), the absorption efficiency of the gf epithelium is usually greater than that of cv animals for xylose (Heneghan 1963), glucose (Ford, Coates 1971), vitamins (Ford, Coates 1971), minerals (Reddy et al 1969), and amino acids (Phillips, Newcomb 1965). This report describes the increased absorption rate from the gut into the blood circulation of ^{14}C and ^{3}H from radioactively labelled tryptophan given perorally to gf mice vis-à-vis cv mice.

Methods and Materials

The gf and cv mice were taken from our colonies of Swiss Webster MF-1 strain originally obtained from Manor Farms Inc., Staatsburg, New York, and ND4 strain originally obtained from Lobund Laboratory, University of Notre Dame. L-Tryptophan-^{3}H (Spec act 8.95 mc/mM) was from Volk Radiochemical Co., New York; and DL-tryptophan-3-^{14}C side chain label (Spec act 0.01 mc/mM) obtained from New England Nuclear Corp., Boston, Mass. Each animal was given per os the following solution of tryptophan (trp) in Krebs Ringer buffer, pH 7.4. A volume of 0.20 ml containing 9.6 μc of DL-tryptophan-^{14}C and 96 μc of L-tryptophan-^{3}H was administered by syringe with a 23-gauge, 2-inch, cannula into the stomach (zero time). At 1-min intervals after peroral injection of tryptophan, 0.01 ml blood specimens were removed from a tail vein and placed in plastic planchets containing 2 N NH_4OH. Planchets were dried 24 hrs at 50 C. Weighed tissue specimens of small intestine were similarly treated after rinsing with Krebs Ringer buffer, homogenized in 0.1N HCl and proteins precipitated by 10% $ZnSO_4$ in 1 N NaOH. Washed protein was treated as above. Samples were counted in a 2-channel automatic liquid scintillation counter.

Results

The rates of absorption of trp radioactivity into the blood circulation of gf and cv mice are shown in Fig. 1. While the initial rates of absorption were quite similar in gf and cv mice, the absorption of ^{3}H and ^{14}C radioactivities during the initial 10-min period were higher in the gf mice than in cv animals. The peak absorption of radioactivity in both gf and cv mice was reached in 10 min after peroral injection of trp. Similar results were obtained with MF-1 and ND_4 mice. Groups of gf and cv mice were given perorally radioactive trp as described above. Fifteen minutes later the mice were killed by ether anesthesia, the small intestine excised, washed, homogenized, and total protein isolated as described. The dried samples of protein were assayed for ^{14}C and ^{3}H as described for the blood specimens. The radioactivity levels were higher in the cv mouse tissue than in gf gut tissue, indicating a greater rate of trp incorporation into protein of cv mice than into the gf mice (TABLE 1).

Discussion and Summary

The increase we observed in the absorption rate of ingested trp radioactivity in gf mice may be due to the increased number of mature absorbing cells in villi of gf mice since the rate of replacement of epithelial mucosa in cv animals is more rapid than in gf animals (Abrams 1983). Also, the increased 5-HT levels in small intestinal tissue of gf animals (Phillips, Newcomb, Smith, Lachapelle 1961;

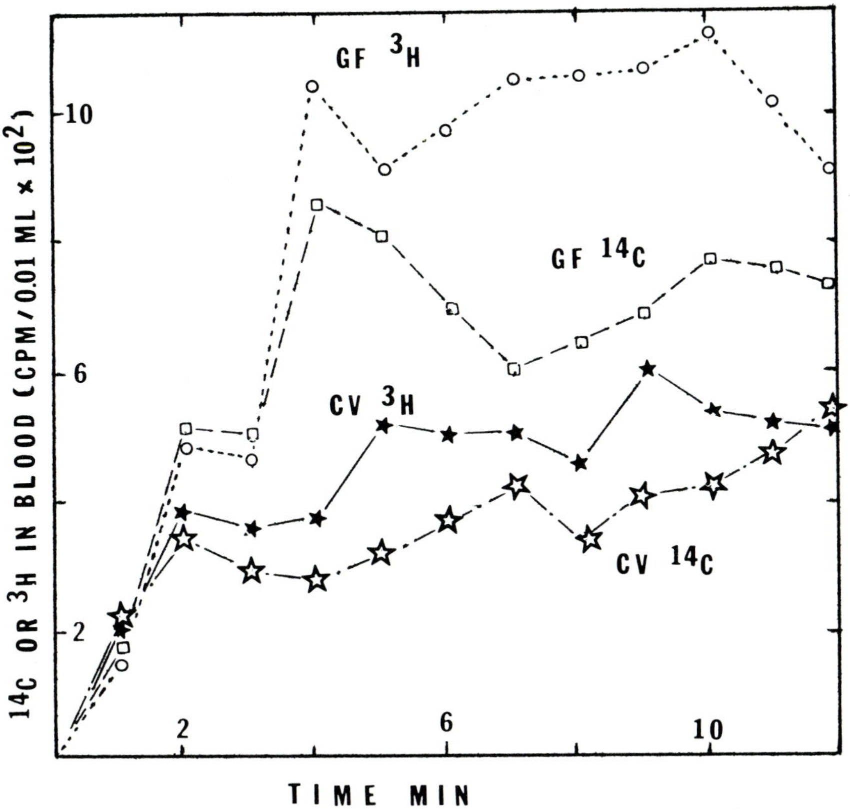

Fig. 1. Absorption rates into the blood circulation of ^{14}C and ^{3}H from DL-tryptophan-3-^{14}C and L-tryptophan-^{3}H from the gastrointestinal tract of adult germfree (gf) and conventional (cv) Swiss Webster mice after tryptophan administration through stomach tube. Each animal was given 0.2 ml of Krebs Ringer buffer pH 7.4 containing 9.6 μc of trp-^{14}C and 96 μc of trp-^{3}H. Blood specimens were taken from a tail vein and treated as described in the text. Radioactivity was measured in a liquid scintillation counter. Data are shown for MF-1 Swiss mice; similar results were observed with the ND4 strain of Swiss mice.

Beaver, Wostmann 1962; Phillips, Newcomb 1965) may be responsible for increased absorption rates in these animals. The increased incorporation of trp labels into tissue protein (15-min) of small intestine in cv vs gf mice (Table 1) corresponds with a greater renewal rate of mucosal cells in cv vs gf mice (Gordon, Bruckner-Kardoss 1961; Abrams 1983; Khoury et al 1969). In summary, the gf mouse manifests a higher rate of absorption into the bloodstream of peroral radioactivity from labelled trp than the cv mouse. In contrast, the cv mouse had enhanced incorporation of trp labels into tissue protein of the small intestine. Possible relationships of the absorption data to serotonin functions in membrane transport were discussed.

TABLE 1. Incorporation of ^{14}C and ^{3}H from radioactively labelled tryptophan into tissue protein of small intestine in germfree (gf) and conventional (cv) mice.

Mice	Radioactivity (cpm/g protein)	
	^{14}C	^{3}H
cv mice	9,987 ± 150*	4,382 ± 64
gf mice	7,273 ± 90	3,478 ± 73

*SD (12 mice per assay)

Acknowledgments

This work was supported by grants from the N.I.H. and Bristol Laboratories, Syracuse, N.Y. I thank Dr. Harvey R. Newcomb for help and Dr. Morris Pollard and Bernard Teah for germfree mice stock (Lobund Laboratory, University of Notre Dame). I thank Georgia Ventura and Patricia Kiernan for typing the manuscript.

References

Abrams GD (1983). Impact of the intestinal microflora on intestinal structure and function. In Hentges DJ (ed): Human Intestinal Microflora in Health and Disease", New York: Academic Press, p 292.

Beaver MH, Wostmann (1962). Histamine and 5-hydroxytryptamine in the intestinal tract of germfree animals harboring one microbial species and conventional animals. Brit J Pharmacol 19:385.

Ford, DJ, Coates ME (1971). Absorption of glucose and vitamins of the B-complex by germfree and conventional chicks. Proc Nutr Soc 30:10A.

Gordon HA, Bruckner-Kardoss E (1961). Effect of normal microbial flora on intestinal surface area. Am J Physiol 201:175.

Heneghan JB (1963). Influence of microbial flora on xylose absorption in rats and mice. Am J Physiol 205:417.

Hentges DJ (ed) (1983). "Human Intestinal Microflora in Health and Disease", New York: Academic Press.

Khoury KA, Floch MH, Hersh T (1969). Small intestinal mucosal cell proliferation and bacterial flora in the conventionalization of the germfree mouse. J Exp Med 130:659.

Phillips AW, Newcomb HR, Smith JE, Lachapelle R (1961). Serotonin in the small intestine of conventional and germfree chicks. Nature 192:380.

Phillips AW, Newcomb HR (1965). Absorption of amino acids in germfree and conventional mice. Bacteriol Proc. p 64.

Reddy BS, Pleasants JR, Wostmann BS (1969). Effect of intestinal microflora on calcium, phosphorus and magnesium metabolism in rats. J Nutr 99:353.

Germfree Research: Microflora Control and Its Application to the Biomedical Sciences, pages 87–90

SUPERIOR REPRODUCTION BY BALB/C VS C3H/HeCr GERMFREE MICE FED CHEMICALLY DEFINED, ULTRAFILTERED, ANTIGEN-FREE DIET

Julian Pleasants and Bernard S. Wostmann

Lobund Laboratory
University of Notre Dame
Notre Dame, IN 46556 USA

The combination of the germfree (GF) animal with a chemically defined (CD), low molecular weight, water-soluble, ultrafiltered diet provides the first complete control of nutrient and antigen intake. Such a model provides baseline data for nutrition, immunology, physiology, pharmacology and toxicology. Immunological applications of this model are reported in 4 articles of this symposium.

The nutritional adequacy of a chemically defined diet for GF animals shows that all essential nutrients are now known, and validates the normalcy of the animals for other studies. The diet had proved its qualitative adequacy (provision of every essential nutrient) by supporting reproduction through 5 generations of CFW mice (Pleasants et al 1970), 3 generations of ICR mice (Hashimoto et al 1978), and 6 generations of C3H/HeCr mice (Pleasants et al 1981). However, quantitative limitations of some nutrient(s) were indicated by declining reproductive efficiency of C3H/HeCr females producing successive litters within one generation.

A similar reproductive pattern had been reported in rats fed a low selenium diet (Halverson, 1974). To determine if Se might be the limiting nutrient, a new diet containing 0.44 ppm Se in contrast to the previous concentration of 0.1 ppm, was fed to GF C3H/HeCr mice through successive litters and generations. To determine if another strain of mouse would respond to CD diet in the same way as the C3H strain, GF Balb/C mice supplied by Dr. Edward Balish of the Univ. of Wisconsin, were fed the higher selenium diet. All mice were maintained in plastic and stainless steel cages on strips of ash-free filter paper bedding, which the mice regularly consumed. Balb/C litters were reduced to 4 and were weaned at 24 days of age. Complete details of their husbandry have been provided in a review by Pleasants (1984). The water-soluble part of the diet, whose composition is given in Table 1, was dissolved in 192 ml of water and ultrafiltered to remove impurities above 10,000 daltons molecular weight. Table 2 gives the composition of the separately filtered lipid portion (this was incorrectly given by Pleasants et al (1981)). Each mouse received daily in a stainless steel dish 0.25 ml (0.22 g) of lipid. However, lactating females received twice that amount. Weanling C3H mice received no lipid from day 21 to 27, then 0.125 ml daily from day 28 to 34. Balb/C mice received 0.125 ml lipid daily from day 25 to day 31.

Table 1. Chemically Defined Diet L489E14 and L489E14Se*

g/100 g water-soluble solids

Leucine	1.9	Asparagine	1.03
Phenylalanine	0.74	Arginine HCl	0.81
Isoleucine	1.08	Threonine	0.74
Methionine	1.06	Lysine HCl	1.77
Tryptophan	0.37	Histidine HCl·H_2O	0.74
Valine	1.23		
Glycine	0.30	Alanine	0.59
Proline	1.48	Na Glutamate	3.40
Serine	1.33	Tyrosine-Ethyl·HCl	0.62
Ferrous Gluconate	0.05	NaCl	0.086
Ca glycerophosphate	5.22	Choline Cl	0.31
Mg glycerophosphate	1.43	K acetate	1.85
$CaCl_2 \cdot 2H_2O$	0.185	α-D-Dextrose	71.40

mg/100 g water-soluble solids

Mn(acetate)$_2 \cdot 4H_2O$	55.4	$NiCl_2 \cdot 3H_2O$	0.37
Co(acetate)$_2 \cdot 4H_2O$	0.11	$SnSO_4 \cdot 2H_2O$	0.37
$ZnSO_4 \cdot H_2O$	40.6	Na_2SeO_3	0.022*
NaF	2.1	Na_3VO_4	0.22
Cu(acetate)$_2 \cdot H_2O$	3.7	KI	0.68
Cr(acetate)$_2 \cdot H_2O$	2.5	$(NH_4)_6Mo_7O_{24} \cdot 4H_2O$	0.37
Thiamine HCl	1.23	Riboflavin	1.85
Pyridoxine HCl	1.54	Niacinamide	9.2
Biotin	0.25	i-Inositol	61.6
Folic Acid	0.37	Ca pantothenate	12.3
Vitamin B-12 (pure)	1.44		

* Diet L489E14Se was prepared from diet L489E14 by dissolving in the final solution an additional 0.074 mg Na_2SeO_3.

Table 2. Daily Lipid Supplement LADEK 69E6

Purified soy triglycerides[a]	0.22 g
Vitamin A (retinyl) palmitate[b]	4.3 μg (7.8 I.U.)
Vitamin D3 (cholecalciferol)	0.0192 μg (0.77 I.U.)
DL-α-tocopherol	2.2 mg
DL-α-tocopheryl acetate	4.4 mg
Vitamin K-1 (phylloquinone)	48.0 μg

[a] 11.8% tripalmitin, 1.5% tristearin, 23.8% triolein, 54.8% trilinolein, 8.0% trilinolenin. Prepared from soybean oil by saponification, vacuum distillation of methyl esters, and transesterification with glycerol (Nu-Chek Prep, Elysian, MN).

[b] Supplied in pure form by Hoffmann-LaRoche, Inc., Nutley,NJ.

Results of the reproduction study are given in Table 3.

Table 3. Comparative reproduction of Balb/C and C3H/HeCr mice through successive litters in one generation.

		1st	2nd	3rd	4th	5th	6th
No. born	Balb/C	2.9 (10)*	5.0 (10)	4.7 (9)	5.4 (8)	3.9 (7)	6.0 (4)
per litter	C3H	4.0 (5)	5.2 (5)	4.0 (5)	3.2 (5)	1.8 (4)	--
% born dead	Balb/C	7	2	0	2	15	16
	C3H	5	19	15	0	57	--
% of live-born lost by 21 d.	Balb/C	11	4	0	7	9	0
	C3H	5	5	41	31	100	--

* Number of females bearing litters

The C3H mice fed the 0.44 ppm selenium diet showed the same pattern of reproductive decline through successive litters shown by C3H mice fed the 0.1 ppm selenium diet. The Balb/C mice, however, showed no decline in reproduction through successive litters or generations. Unlike C3H mice, they reproduced as well as their conventional controls fed natural ingredient diet. Weanling Balb/C mice did not show the seborrhea sporadically shown by C3H mice as they adjusted to the intake of lipid supplement. Therefore, the Balb/C weanlings could be given twice as much lipid by day 35 as was given to C3H mice. However, the Balb/C mice proved susceptible to intestinal blockage by trichobezoars, a hazard which had been eliminated in C3H mice by providing ingestible filter paper bedding at all times. Although the Balb/C mice had bedding available continuously, most Balb/C males died of intestinal blockage by fur balls between 2 and 6 months of age. The females were able to pass the fur balls and survive.

The results demonstrate both qualitative and quantitative adequacy of this CD diet for the demands of reproduction in Balb/C mice. The reason for the inferior performance of C3H mice is unknown. The development of large trichobezoars in Balb/C mice draws attention to one unphysiological aspect of the present diet: the mice consume their entire daily supply of lipid in a few minutes. Polysorbate-80 cannot be used to disperse the lipid in the water-soluble portion of the diet because of its hepatotoxicity (Reyniers et al 1985). A more physiological diet which would combine the water-soluble part with one of the commercially available intravenous feeding emulsions is being investigated. This might prevent formation of trichobezoars and would assure that all mice housed in a group receive their share of lipid. It could also improve diet intake by pregnant, lactating, and newly weaned mice, as well as add to the convenience of husbandry.

REFERENCES

Halverson AW (1974). Growth and reproduction with rats fed selenite-Se. In Hoekstra WG, Suttie JW, Ganther HE, Mertz W (eds): "Trace Element Metabolism in Animals", Baltimore: University Park Press, p 587.

Hashimoto K, Handa H, Umehara K, Sasaki S (1978). Germfree mice reared on "antigen-free" diet. Lab Animal Sci 28:38.

Pleasants JR (1984). The germ-free animal fed chemically defined diet. In Coates ME, Gustafsson BE (eds): "The Germ-Free Animal in Biomedical Research", London: Laboratory Animals Ltd., p 91.

Pleasants JR, Reddy BS, Wostmann BS (1970). Qualitative adequacy of a chemically defined liquid diet for reproducing germfree mice. J Nutrition 100:498.

Pleasants JR, Bruckner-Kardoss E, Bartizal KF, Beaver MH, Wostmann BS (1981). Reproductive and physiological parameters of germfree C3H mice fed chemically defined diet. In Sasaki S (ed): "Recent Advances in Germfree Research", Tokyo: Tokai University Press, p 333.

Reyniers JP, Machado EA, Farkas WR (1984). Hepatotoxicity of chemically defined diets containing polysorbate-80 in germfree and conventional mice. In this symposium.

Germfree Research: Microflora Control and Its Application to the Biomedical Sciences, pages 91–94

HEPATOTOXICITY OF CHEMICALLY-DEFINED DIETS CONTAINING POLYSORBATE-80 IN GERMFREE AND CONVENTIONAL MICE

Jon P. Reyniers, Ph.D., Emilio A. Machado, M.D., and Walter R. Farkas, Ph.D.

Texas A&I University, Kingsville, TX 78363, and The University of Tennessee, Knoxville, TN 37920

Chemically-defined (CD) liquid diets are used experimentally (Reyniers *et al* 1981; Pleasants *et al* 1981) and clinically for parenteral and enteral support of human patients (Kark 1974) for strict control of nutritional parameters. Incorporation of non-ionic solubilizing agents, generally considered to be non-toxic, facilitates miscibility of the aqueous and lipid soluble components of CD-diets (Greenstein *et al* 1960), enabling the diets to be used as a single-component formulation. Polysorbate-80 is widely used for this purpose in CD-diets and other medicinal products (Gleason *et al* 1969; Merck Index 1960). In earlier studies which established a dietary requirement for the purine derivative, queuine (Farkas 1980; Reyniers *et al* 1981), we used Polysorbate-80 in a CD-diet. Since the intestinal microflora of animals can synthesize queuine or its derivatives (Farkas 1980; Reyniers *et al* 1981), we used both germfree (GF) and conventional (CV) mice. On removal of the livers from the GF and CV mice, for extraction of transfer RNA, we noted their pallid appearance. In the present study, the livers of GF and CV mice fed Polysorbate-80 in a CD-diet were histologically examined. We found that hepatic changes in GF and CV mice were due to Polysorbate-80, not the GF status of the mice or the presence or absence of queuine in the CD-diet.

MATERIALS AND METHODS

Male CD-1 outbred Swiss GF and CV mice were obtained (Charles River Laboratories, Inc., Wilmington MA). GF mice were maintained in flexible-film isolators and CV mice were maintained in a typical animal facility. All mice were housed in trios in polycarbonate cages (28x13x18 cm), fitted with wire tops. Cages with CV mice had filter caps. The mice were fed a CD-diet *ad libitum* in separate aqueous and lipid components. When used, queuine was dissolved in the aqueous component and Polysorbate-80 in the lipid component of the CD-diet. Water was supplied *ad libitum* to all mice. The feces of GF mice were monitored for contamination (Wagner 1959). The CD-diet was prepared as previously described (Reyniers *et al* 1981). The aqueous and lipid components of the CD-diet were separately filter-sterilized, or, when combined, the components were made miscible by the addition of 4.5g of Polysorbate-80 to 1295 ml of aqueous diet prior to sterilization. Some groups of mice were fed a natural (N) pelleted autoclavable diet (Purina). Polysorbate-80 was adsorbed to the surface of the pellets at a level previously determined to be equivalent to that consumed by mice fed the CD-diet. At weekly intervals, mice were anesthesized by etherization, bled by cardiac puncture, and specimens of liver, kidney, intestine, spleen, and pancreas were obtained. Tissues were fixed in 10% buffered formalin (pH 7.4), embedded in paraffin, sectioned, and stained by hematoxylin and eosin (H&E), periodic acid-Schiff (PAS), or methyl green-pyronin prior to examination by light microscopy. Cryostat sections were prepared and stained for lipids by oil red-O. Selected specimens were fixed in 3% gluteraldehyde in 50mM Na-cacodylate (pH 7.2), post-fixed in 2% osmium tetroxide in 0.1M Na cacodylate, embedded in Epon, and examined by transmission electron microscopy.

RESULTS

In GF or CV mice fed the N-diet plus Polysorbate-80, fine cytosplasmic vacuoles were seen, predominately in those hepatocytes near the periportal spaces and extending into the medial area of the hepatic spaces (Fig. 1A). Similar changes were seen in GF or CV mice fed Polysorbate-80 in the CD-diet. The vacuoles gradually increased, becoming confluent by the 10th week of treatment. Both the hepatocyte cytoplasm and nucleus were affected. The sinusoids were collapsed (Fig. 1B). In mice fed the N-diet, the sinusoids of the lumen remained patent, and the endothelial and Kupffer cells were only slightly hyperplastic (data not shown). In mice fed Polysorbate-80, the cytoplasm of hepatocytes were consistently positive for multiple lipid droplets by oil red-O staining. An increase in lipid content of the cells was accompanied by a diffuse or zonal type lipid degeneration. Some hapatocytes adjacent to affected cells were swollen, but were essentially free of droplets (Fig. 1B). The intrahepatocyte distribution of lipid droplets was irregular. Tissue necrosis was not evident in mice fed the CD-diet plus Polysorbate-80 for 10 weeks. Differences in GF or CV hepatocyte vacuolation was minimal. The severity of the lesion was time-dependent on Polysorbate-80 exposure. Control mice fed the CD-diet devoid of Polysorbate-80 rarely displayed hepatic abnormalities, all of which were within normal limits (data not shown). The livers of mice fed queuine-supplemented CD-diet were normal. The ultrastructure of hepatocyte cytoplasm of mice fed the CD-diet plus Polysorbate-80 showed a marked overload of lipids and early signs of cell injury, as revealed by autophagosomes and myelin figures (data not shown). Many Kupffer cells had large phagocytic vacuoles with cell membrane debris and lipid droplets. Mitochondria were unaffected. Even at 10 weeks. the hepatic changes were considered to be reversible.

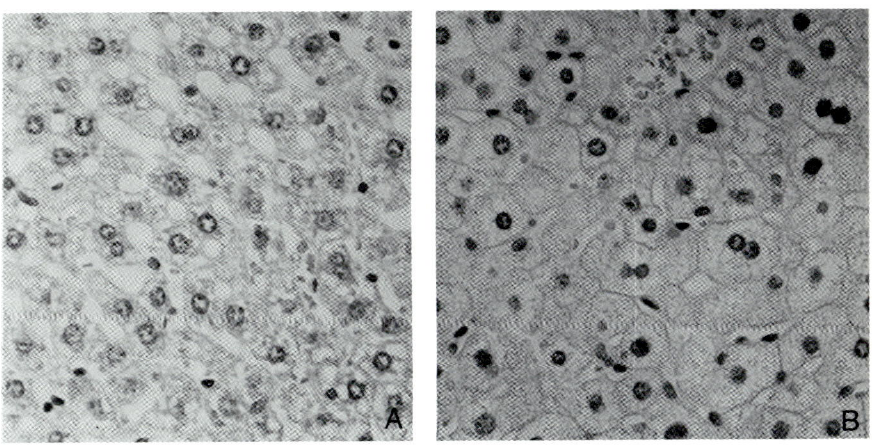

Fig. 1. Light microscopy of hepatocytes of GF and CV mice fed Polysorbate-80: A. GF mouse fed N-diet (H&E, 180x); B. GF mouse fed CD-diet for 10 weeks (H&E, 180x).

DISCUSSION

We have shown that hepatotoxic changes were induced in GF and CV mice fed Polysorbate-80 in their diets up to 10 weeks. The hepatic changes were not attributed to GF status or the presence or absence of queuine in the diet.

Polysorbate-80 is generally considered to be non-toxic and little literature exists on the topic. In this study, the dosage of Polysorbate-80 given to the mice was, on a weight basis, equivalent to that deemed acceptable for medicinal use in humans (Merck Index 1960). Polysorbate-80 is also used as a food additive (Kilgore and Li 1980). The hepatic changes we observed were considered to be reversible. However, prolonged consumption or increased dosage would likely result in further degenerative changes and possible hepatonecrosis.

Farkas WR (1980). Effect of diet on the queuosine family of tRNAs of germfree mice. J Biol Chem 255:6832.
Gleason MN, Gosselin RE, Hodge HC, Smith RP (1969). Clinical Toxicology of Commercial Products. 3rd Edition, The Williams and Wilkins Co., Baltimore MD p 116.
Greenstein JP, Oteg MC, Birnbaum SM, Winitz M (1960). Quantitative nutritional studies with water-soluble, chemically defined diets. X. Formulation of a nutritionally complete liquid diet. J Nat Can Inst 24:211.
Kark RM (1974). Liquid formula and chemically defined diets. J Amer Diet Assoc 64:476.
Kilgore WW, Li MY (1980). Food additives and contaminants. In Toxicology 2nd edition, J Doull, CD Klaasen, MO Andus eds., Chpt 23, MacMillan Publishing Co., N.Y.C.
Merck Index (1960). PG Stecher, 7th edition, Merck & Co. Rahway, NJ p 883.
Pleasants JR, Schmitz HE, Wostmann BW (1981). Immune response of germfree C3H mice fed a chemically defined ultrafiltered antigen-free diet. In Advances in Germfree Research, Sasaki ed., University Press, p 535.
Reyniers JP, Pleasants JR, Wostmann BW, Katze JR, Farkas WR (1981). Administration of exogenous queuine is essential for the biosynthesis of the queuosine-containing transfer RNAs in the mouse. J Biol Chem 256:11591.
Wagner M (1959). Determination of germfree status. Ann NY Acad Sci 78:89.

Supported by NIH grants ES01323, HD14062, and RR08107.

Germfree Research: Microflora Control and Its Application to the Biomedical Sciences, pages 95–98

COMPARISON OF BLOOD CHEMISTRY AND HORMONE LEVELS FROM CONVENTIONAL AND GERMFREE RATS

David L. Snyder and Bernard S. Wostmann

Lobund Laboratory
University of Notre Dame
Notre Dame, Indiana 46556 U.S.A.

We are currently conducting a long-term study of aging in the germfree (GF) rat and of the life-extending ability of restricted diet intake. The GF condition provides an aging model free of infectious complications, offering a true picture of the degenerative processes which occur during aging. However, the GF rat has many metabolic differences from the conventional (CV) rat (Wostmann 1975; Wostmann 1981). Therefore, this study compares blood chemistry and hormone levels in CV and GF rats in order to define normal values.

METHODS

Orbital blood samples were taken from 14 CV, 27 GF, and 12 germfree but restricted (GR) Lobund-Wistar rats. *Actinomyces sp.* had previously contaminated the isolators of these GF rats. However, fecal smears did not indicate growth in the intestinal tract. Restricted intake was 12 grams per day (equal to 70% of *ad lib.* intake). Restriction began at six weeks of age. All rats were male, 8 to 12 months old, and fed natural ingredient diet L 485 (Kellogg, Wostmann 1969). Blood samples were collected between 10 A.M. and 12 P.M., under light halothane anesthesia, and after an overnight fast. Serum samples were assayed on a Technicon SMAC (Computerized Sequential Multiple Analyzer). Hormone levels were determined in serum samples obtained under the same conditions except that blood was collected by cardiac puncture. Total thyroxine (T4), total triidiothyronine (T3), and testosterone (Test.) were measured using Diagnostic Products Corporation Coat-A-Count assay kits (5700 West 96th St., Los Angèles, CA 90045). Means were compared using student's T-test.

RESULTS AND DISCUSSION

Average body weights at the time of sacrifice were CV 430g, GF 385g, and GR 269g. The following serum values showed no significant difference between CV, GF and GR rats: uric acid (1.8 mg/dl), creatinine (0.4 mg/dl), calcium (10.9, 10.4 and 10.4 mg/dl), triglycerides (115, 81 and 72 mg/dl), cholesterol (84, 83 and 77 mg/dl), total bilirubin (0.1 mg/dl), albumin (3.8 g/dl) and the enzymes alkaline phosphatase (134, 93 and not determined IU/l), GGTP (5.1, 2.3, and 3.3 IU/l), SGOT (131, 144 and 152 IU/l), and SGPT (50, 39 and 58 IU/l). These values are similar to those listed by Mitruka and Rawnsley (1981) for CV male Wistar rats. Comparable values for GF rats were reported for calcium (Reddy et al. 1969), triglycerides and cholesterol (Kawai et al. 1981), and SGOT and SGPT (Sumi, Miyakawa 1983). Significant differences are given in Table 1.

Table 1. Differences in Blood Chemistry of Adult Male Conventional and Germfree Rats. (Mean ± SE) (N>12)

	Glucose (mg/dl)	BUN (mg/dl)	Chloride (meg/l)	CO_2 Content (meg/l)	In. Pi (mg/dl)	Total Protein (g/dl)	Globulin (g/dl)
CV	140	17.6	104	24.9	7.1	6.3	2.5
	± 2	± 0.5	±0.3	± 0.5	±0.1	±0.1	±0.05
GF	114*	14.5*	108*	19.8*	6.8	5.8*	2.0*
	± 2	± 0.4	±0.5	±0.5	±0.1	±0.1	±0.03
GR	98**	16.4**	111**	17.4	7.8**	5.6**	1.7**
	± 2	± 0.4	±0.4	±0.6	±0.2	±0.1	±0.03

* GF significantly different from CV at $P < .01$
** GR significantly different from GR at $P < .01$

The blood glucose levels suggest that GF rats have slightly lower blood glucose when compared to CV rats and that the condition is intensified by diet restriction. The lower BUN levels in GF and GR rats may be due to the lack of microbially produced ammonia entering the liver from the gut. At least 20% more urea is produced in CV animals than in comparable GF animals (Saheki et al. 1980). The increase in serum Cl in GF and GR rats agrees with the hypothesis that in GF animals more Cl is absorbed from the gut and excreted in the urine (Lev et al. 1970). CO_2 content is a measure of the buffering capacity of the blood; the low values in GF and GR rats suggest a possible difference in acid-base regulation. The lower globulin content in GF and GR rats is due to reduced antibody production in the GF condition.

Table 2. Hormone Levels of Adult Male Germfree and Conventional Rat. (Mean ± SE) (N > 12)

	Thyroxine (ug/dl)	Triiodothyronine (ng/dl)	Testosterone (ng/ml)
CV	6.2 ± 0.2	114.6±4.6	2.1 ± 0.3
GF	5.5 ± 0.2*	122.0±6.5	3.4 ± 0.4*
GR	5.6 ± 0.3	130.0±8.0	7.8 ± 1.3**

* GF significantly different from CV at $P < .05$
** GR significantly different from GF at $P < .01$

We find lower serum T4 in GF rats regardless of intake but T3 levels may be slightly higher. Ukai and Mitsuma (1978) also reported lower plasma T4 and higher plasma T3 levels in 75-day-old GF rats when compared to CV rats, and Wostmann and Bruckner-Kardoss (1981) using 9 months old 30% restricted GF rats found T3 levels equal to those reported here. The reduction in thyroid hormones seen in studies of dietary restriction and life extension by Ooka et al. (1978) may not have occurred because our restriction was only 30%. Yet its life extending property has been established (Pollard, Wostmann 1984).

The dramatic difference in Test. levels between GR and GF rats suggests a new area of investigation. Though body weights of GR rats were only 75% of GF rats, GR rats maintained testes sizes similar to GF rats, which may account for part of the Test. increase. Similar Test. levels for CV rats age 6 to 9 months have been reported (Chambers, Phoenix 1984; Ghanadian et al. 1975). Merry and Holeman (1981) did not show a rise in Test. levels in CV rats on a 50% dietary restriction, but rats on restricted intake did maintain their peak levels of Test. longer than fullfed rats. Dietary restriction of even 30% appears sufficient to modify certain hormonal patterns which in turn may lead to the extended lifespan observed in these animals.

REFERENCES

Chambers KC, Phoenix CH (1984). Testosterone and the decline of sexual behavior in aging male rats. Behav Neural Biol 40:87.

Ghanadian JR, Lewis JG, Chisholm GD (1975). Serum testosterone and dihydrotestosterone changes with age in rat.

Steroid 25:753.
Kawai Y, Suegara N, Yazawa K (1981). Intestinal microflora and aging: Age-related change of lipid metabolism in germfree and conventional rats. Mech Ageing Dev 16:149.
Kellogg TF, Wostmann BS (1969). Stock diet for colony production of germfree rats and mice. Lab Anim Care 19:812.
Lev M, Alexander RH, Levenson S (1970). Impaired water metabolism in germfree rats. Proc Soc Exp Biol Med 135:700.
Merry BJ, Holehan AM (1981). Serum profiles of LH, FSH, testosterone and 5α-DHT from 21 to 1000 days of age in ad libitum fed and dietary restricted rats. Exp Gerontol 16:431.
Mitruka BM, Rawnsley HM (1981). "Clinical biochemical and hematological reference values in normal experimental animals and normal humans", New York: Masson Publishing USA, p 157.
Ooka H, Segall PE, Timiras PS (1978). Neural and endocrine development after chronic tryptophan deficiency in rats: II. Pituitary-thyroid axis. Mech Ageing Dev 7:19.
Pollard M, Wostmann BS (1985). Increased life span among germfree rats. In Wostmann BS (ed): "The Proceedings of the VIII International Symposium on Germfree Research", New York: Alan R. Liss, p .
(From the publication to which this manuscript is submitted)
Reddy BS, Pleasants JR, Wostmann BS (1969). Effect of intestinal microflora on calcium, phosphorus and magnesium metabolism in rats. J Nutr 99:353.
Saheki T, Veda A, Hosaya M, Katsunuma T, Onishi N, Ozawa A (1980). Comparison of the urea cycle in conventional and germfree mice. J Biochem 88:1563.
Sumi T, Miyakawa M (1983). Susceptibility of germ-free rats to the hepatotoxic effects of dimethylnitrosamine or dimethylamine plus sodium nitrite administered orally. Cancer Res 43:2942.
Ukai M, Mitsuma T (1978). Plasma triiodothyronine, thyroxine and thyrotrophin levels in germfree rats. Experientia 34:1095.
Wostmann BS (1975). Nutrition and metabolism of the germfree mammal. World Rev Nutr Diet 22:40.
Wostmann BS (1981). The germfree animal in nutritional studies. Ann Rev Nutr 1:257.
Wostmann BS, Bruckner-Kardoss E (1981). Functional characteristics of gnotobiotic rodents. In Sasaki S, Ozawa A, Hashimoto K (eds): "Recent Advances in Germfree Research, Proceedings of the VIIth International Symposium on Gnotobiology," Tokyo: Tokai University Press, p 321.

Germfree Research: Microflora Control and Its Application to the Biomedical Sciences, pages 99–101

BILE PIGMENTS IN GERMFREE AND CONVENTIONAL RATS

Henri Saxerholt, B.S.
Tore Midtvedt, PH.D., M.D.
Bengt E. Gustafsson, PH.D., M.D.
Department of Germfree Research
Karolinska Institutet, Stockholm, Sweden

Bilirubin is a yellow tetrapyrrolic bile pigment derived from the heme catabolism. Conjugation of bilirubin takes place in the liver and conjugates containing sugars such as glucuronic acid, glucose and xylose are secreted with the bile (Heirwegh et al 1974). In the intestine, the bilirubin conjugates undergo deconjugation and reductive conversions to urobilinogens. The deconjugation of the conjugates is performed by microbial as well as mammalian enzymes whereas the formation to urobilinogens entirely seems to be caused by microbial enzymes (Gustafsson et al 1960). The urobilinogens are eliminated with the feces or reabsorbed and reexcreted by the liver or the kidney. Previously, we were able to confirm the presence of conjugated bilirubin in feces from germfree (GF) rats by qualitative methods, whereas no conjugated bilirubin was detected in conventional (CONV) rat feces (Saxerholt et al 1983). The methods included chromatographic purification of the fecal bilirubin conjugates followed by formation of dipyrrolic ethyl anthranilate azopigment derivatives of the conjugates. The azopigments were separated by thin-layer chromatography (tlc) and the results indicated the presence of bilirubin glucuronides and other conjugates of unknown structure in the feces from GF rats. The use of coupling with diazotized ethyl anthranilate has been widely accepted for the analysis and quantitation of conjugated bilirubin in bile and other body fluids (Heirwegh et al 1974). So far, however, this method has not been adopted for quantitation of conjugated bilirubin in feces. Recently, we developed quantitative methods for the determination of conjugated bilirubin in rat feces (Saxerholt et al 1984). One of the methods can be described as follows: feces → purification (extraction, tlc) of bilirubin conjugates → formation of azopigments (by coupling with diazotized ethyl anthranilate) → spectrophotometrical quantitation of azopigments (calculation of the amount-of-substance content of conjugated bilirubin in feces) → separation of the azopigments (tlc) → identification of the separated azopigments (comparison of their Rf values with the Rf values of reference azopigments prepared from rat bile or rat small intestinal contents).

In one experiment, the purpose was to study deconjugation of bilirubin conjugates and urobilinogen formation in GF, CONV and exgermfree (EXG) rats conventionalized with feces suspension from CONV rats (Saxerholt et al 1984). Conjugated bilirubin and urobilinogens were determined in feces before and after the conventionalization. In another experiment, the effect of antibiotics on the bile pigment metabolism in man was studied (Midtvedt et al. This volume).

CONCLUSIONS

Rat. Feces from GF rats contained significant amounts of conjugated bilirubin (glucuronides and other conjugates of unknown structure). At least 8 different azopigments could be detected. The amount of conjugated bilirubin in feces decreased rapidly after conventionalization and the number of detectable azopigments was successively reduced. No urobilinogens were present in feces during the GF condition, but these metabolites appeared in feces one day after conventionalization and increased during a few days to a CONV level. Three to five weeks after conventionalization, only 2 azopigments were detected, namely the unconjugated and glucuronic acid conjugated azopigments of bilirubin, respectively, thus indicating the presence of only one bilirubin conjugate, the monoglucuronide. Ten weeks after conventionalization, no azopigments could be detected thus indicating a CONV pattern of the bilirubin conjugates.
): The establishment of the intestinal microflora in the EXG rats seems to take place as a step by step procedure yielding a successive deconjugation of the various bilirubin conjugates. Man. Deconjugation of bilirubin conjugates and urobilinogen formation was markedly reduced after administration of clindamycin whereas ampicillin and metro-

nidazole had much less effect on the bile pigment metabolism.

Treatment	Before T. CB	Before T. U	Day 6 of T. CB	Day 6 of T. U	6 weeks after T. CB	6 weeks after T. U
Clindamycin	1.4	1.1	15.4	0.1	2.6	0.9
Ampicillin	1.2	1.2	3.7	0.5	3.1	1.0
Metronidazole	1.9	1.5	2.6	0.7	2.0	0.9

Table. Conjugated bilirubin (CB, µmol/kg feces) and urobilinogen (U, mmol/kg feces) in feces from healthy subjects receiving antibiotics. Median values are given. T = therapy.

The findings were normalized 6 weeks after therapy.

Gustafsson BE, Swenander-Lanke L (1960). Bilirubin and urobilins in germfree, ex-germfree and conventional rats. J Exp Med 112:975.
Heirwegh KPM, Fevery J, Meuwissen JATP, De Groote J, Compernolle F, Desmet V, Van Roy FP (1974). Recent advances in the separation and analysis of diazo-positive bile pigments. Methods Biochem Anal 22:205.
Saxerholt H, Midtvedt T, Gustafsson BE (1983). Detection of bilirubin conjugates in faeces of germfree rats. Scand J clin Lab Invest 43:477.
Saxerholt H, Midtvedt T, Gustafsson BE (1984). Methods for determination of conjugated bilirubin in rat faeces. Scand J clin Lab Invest 44:in press.
Saxerholt H, Midtvedt T, Gustafsson BE (1984). Deconjugation of bilirubin conjugates and urobilin formation by conventionalized germfree rats. Scand J clin Lab Invest 44:in press.

Germfree Research: Microflora Control and Its Application to the Biomedical Sciences, pages 103–106

FORMATION OF HYODEOXYCHOLATE FROM β-MURICHOLATE IN GNOTOBIOTIC RATS ASSOCIATED WITH CLOSTRIDIUM HDCA-1

H. Eyssen, G. De Pauw and J. Van Eldere

The Rega Institute, University of Leuven
B-3000 Leuven, Belgium

The major primary bile salts in germfree (GF) rat bile are conjugates of cholate (CA), chenodeoxycholate (CDA), α-muricholate (α-MCA), and β-muricholate (β-MCA). In addition, conventional (CV) rat bile contains bacterially modified bile salts, e.g. deoxycholate (DCA) and hyodeoxycholate (HDCA) formed from CA and β-MCA through 7-dehydroxylation. DCA is found in the intestines of most mammalian species and DCA producing microorganisms have been isolated (Midtvedt, 1967). HDCA is formed from the MCA's in CV rats, but not in CV mice (Eyssen *et al.*, 1976; Sacquet *et al.*, 1979; Wostmann *et al.*, 1977).

In this study we now report on the first isolation of a HDCA forming microorganism. This strictly anaerobic *Clostridium* HDCA-1 was isolated from CV rat feces and converts the MCA's and HCA into HDCA.

METHODS

Culturing was done in an anaerobic chamber at 37°C in an atmosphere of 10 % H_2/90 % N_2. Meat Medium contained per tube: 125 mg of freeze-dried powdered beef meat, 3 mg HCA + 0.3 mg of CA in 0.5 ml H_2O, and 5 ml of broth containing : Brain Heart Infusion = 0.4 %, Proteose Peptone = 0.5 %, Bacto-Tryptone Yeast = 1 %, $K_2HPO_4.3H_2O$ = 1 %, $Na_2CO_3.10H_2O$ = 0.1 %, Trizma pH 7.6 = 0.15 %, $MgSO_4.7H_2O$ = 0.1 %, cysteine = 0.04 %, hemin = 2.5 mg %, vit K_1 = 0.1 mg %. Cultures were also grown in PC Medium, i.e. Meat Medium without meat powder, Na_2CO_3 or Trizma buffer. To support growth of pure cultures of the HDCA producing strain, both media had to be supplemented with 6-8 drops of filter-sterilized or boiled (5 min) medium on which *Peptostreptococcus productus* had grown for 24 to 48 h.

RESULTS

1. Isolation and characteristics of Clostridium HDCA-1

CV rat feces (250 mg in 5 ml of prereduced Meat Medium) were heated for 20 min at 80°C. Anaerobic subcultures converted β-MCA and HCA into HDCA; in contrast, no HDCA was produced in aerobic subcultures. These results suggested that the HDCA producing strain was an anaerobic spore-former. It will be further referred to as *Clostridium* HDCA-1.

The mixed cultures could be maintained in Meat Medium with β-MCA or HCA, but the activity was invariably lost after one or two subcultures on the same medium without bile salts. Hence, *Clostridium* HDCA-1 was bile salt dependent. The MCA's and HCA were better substrates than CA, CDCA or ursodeoxycholate (UDCA). Therefore, the culture media were routinely supplemented with HCA or β-MCA.

Several types of basal medium supported the HDCA producing activity in the mixed cultures, provided that the pH of the grown culture did not decrease below pH 6.8. The activity was invariably lost upon addition of more carbohydrate as the pH then dropped below pH 6.8. The sensitivity to weakly acid pH was confirmed in later studies with pure cultures.

When active mixed cultures were plated on Meat Medium plus 2 % agar, none of the pure cultures isolated produced HDCA. Some impure isolates, however, demonstrated 7-dehydroxylating activity. Among these was a culture containing no more than two species. One species was identified as *Peptostr. productus*; grown as a pure culture, however, it did not produce HDCA. The second species, termed *Clostridium* HDCA-1, formed micro-colonies (0.05 - 0.10 mm) located in the vicinity of the larger colonies (1 mm) of *Peptostr. productus*. Under the microscope the micro-colonies presented as very flat, apparently consisting of a monolayer of small bacilli. Initially, all attempts to grow this *Clostridium* HDCA-1 in pure culture failed. By contrast, the strain developed well and produced HDCA in mixed culture with *Peptostr. productus*, indicating that the *Peptostreptococcus* produced an essential growth factor. Finally, *Clostridium* HDCA-1 could be grown as a pure culture when Meat Medium or PC Medium were supplemented with 5 to 8 drops of a filter-sterilized culture of *Peptostr. productus*. The growth factor could not be autoclaved, but resisted boiling for 5 min under anaerobic conditions and it could be kept for 3 to 4 months at -20°C. Among a limited number of other anaerobic bacteria tested, a few produced the growth factor; a number of coenzymes and vitamins had no effect.

Our studies also showed that Clostridium HDCA-1 was lost when glucose was omitted from the medium. Hence, the strain required a carbohydrate for growth, e.g. 0.1 % of glucose, although addition of glucose had no measurable pH decreasing effect. Even under these "optimal" conditions, growth was poor and less than 10^6 cells/ml were obtained.

2. Transformation of bile salts by Clostridium HDCA-1

Growth of Clostridium HDCA-1 was supported only by those bile salts which could be 7-dehydroxylated, suggesting that strain HDCA-1 required the hydroxyl groups as hydrogen acceptors. Strain HDCA-1 did not deconjugate bile salts, and conjugated bile salts were not dehydroxylated and did not support growth. HCA (= 3α,6α,7α-tri-OH), β-MCA (= 3α,6β,7β-tri-OH), α-MCA (= 3α,6β,7α-tri-OH), Ω-MCA (= 3α,6α,7β-tri-OH) supported growth equally well and were more than 90 % converted into HDCA after 3 days of incubation. CA (= 3α,7α,12α-tri-OH), supported growth less well and was for 50 to 70 % converted into DCA (= 3α,12α-di-OH). Ursodeoxycholate (3α,7β-di-OH) and CDCA (= 3α,7α-di-OH) were for 25 to 50 % converted into lithocholate (= 3α-OH) (LA).

3. Association of gnotobiotic rats with Clostridium HDCA-1

Strain HDCA-1 could not be established as a monoassociate in GF rats, as well because of its anaerobic character as because of its requirements for an unconjugated 3,6,7-tri-OH bile salt and the heat-labile growth factor. The latter was found in filtered extracts from CV rat feces, but not, or in too low amounts, in filtered extracts from GF feces. When the rats were first associated with Peptostr. productus, which only produced growth factor, and with Clostridium perfringens ATCC 19574, which partially deconjugated the bile salts (± 30 %), Clostridium HDCA-1 became easily established and converted 40 % of the β-MCA, 22 % of CA, and 61 % of CDCA, into HDCA, DCA and LA, respectively. When the bile salt desulfating and deconjugating Clostridium S_1 (Huijghebaert et al., 1982) was added to this microflora, the fecal bile salts were more than 90 % deconjugated and more than 75 to 90 % of the β-MCA, CA and CDCA were 7-dehydroxylated. These studies demonstrated that Clostridium HDCA-1 efficiently produced HDCA in gnotobiotic rats.

DISCUSSION

The present investigations show that HDCA can be formed from the MCA's by the fastidious Clostridium HDCA-1 and that Ω-MCA does not seem to be an obligate intermediate in this reaction. The observation that growth of strain HDCA-1 is dependent on the presence of 3,6,7-tri-OH bile salts in the culture medium, can explain why previous attempts to isolate the microorganism have failed. The fact that strain HDCA-1 does not develop at pH values below pH 6.8 also can explain why the conversion of β-MCA into HDCA is suppressed in rats fed lactose-containing diets. In these animals more β-MCA is converted into Ω-MCA (Wostmann et al., 1977), a process which appears to be less sensitive to slightly acid pH. The influence of the microbial conversion of the MCA's and CA into the respective 7-dehydroxylated derivatives on bile salt pools and turn-over is currently being investigated in gnotobiotic rats.

REFERENCES

Eyssen HJ, Parmentier GG, Mertens JA (1976). Sulfated bile acids in germ-free and conventional mice. Eur J Biochem 66: 507.

Huijghebaert SM, Mertens JA, Eyssen HJ (1982). Isolation of a bile salt sulfatase-producing Clostridium strain from rat intestinal microflora. Appl Environm Microbiol 43:185.

Midtvedt T (1967). Properties of anaerobic gram-positive rods capable of 7α-dehydroxylating bile acids. Acta Path et Microbiol Scandinav 71:147.

Sacquet EC, Raibaud PM, Méjean C, Riottot MJ, Leprince C, Léglise PC (1979). Bacterial formation of ω-muricholic acid in rats. Appl Environm Microbiol 37:1127.

Wostmann BS, Beaver M, Chang L, Madsen D (1977). Effect of autoclaving of a lactose-containing diet on cholesterol and bile acid metabolism of conventional and germ-free rats. Am J Clin Nutr 30:1999.

Germfree Research: Microflora Control and Its Application to the Biomedical Sciences, pages 107–110

LACK OF 7 α-DEHYDROXYLATION IN GNOTOBIOTIC GERBILS ASSOCIATED WITH AN OCTAFLORA INCLUDING CLOSTRIDIUM SORDELLII

B.S. Wostmann, M. Beaver-Johnson and M. Wagner

Lobund Laboratory
University of Notre Dame
Notre Dame, Indiana 46556 U.S.A.

The intestinal microflora greatly influences cholesterol and bile acid metabolism. Research with germfree (GF) rats indicates that cholesterol absorption is enhanced and both cholesterol and bile acid excretion via the feces are reduced in the absence of a "normal" microflora (Kellogg, Wostmann 1969; Wostmann 1973). In the absence of microbial deconjugation and secondary modification of the biliary bile acids (BAs), ilial reabsorption is enhanced and BAs accumulate in the enterohepatic circulation (Wostmann 1973). The microflora thus promotes removal of cholesterol from the body, largely via increased excretion of microbially modified BAs. However, the microbial species involved in producing these quite well defined fecal BA patterns are largely unknown, especially those involved in 7 α-dehydroxylation.

More recently, we have used the gnotobiotic (GN) gerbil to study these microbial effects. Gerbils, like man, produce only cholic acid and chenodeoxycholic acid as major primary BAs. In the rat, the liver converts chenodeoxycholic acid to the α- and β-muricholic acids which give rise to hyodeoxycholic acid and ω-muricholic acid via microbial modification. The latter BAs are much more easily excreted (Wostmann, Beaver, Chang, Madsen 1977) and the rat will switch to muricholic acid production to alleviate the potential effects of high cholesterol intake (Gustafsson, Einarson, Gustafsson 1975). Man and gerbil do not have this ability.

Our most recent studies concentrated on obtaining information on the microbial species involved in the modification of BA in the conventionally-reared CV gerbil, and to try to compose a defined microflora that would mimic these secondary changes, especially the formation of the large amounts of deoxycholic acid found in the feces of gerbils and man.

MATERIALS AND METHODS

Since caesarian-derived GF gerbils did not reproduce, they were associated with a hexaflora consisting of *Lactobacillus brevis*, *Streptococcus faecalis*, *Staphylococcus epidermidis*, *Bacteroides vulgatus*, *Enterobacter aerogenes* and a *Fusibacterium sp*. GN gerbils thus associated reproduced well (Bartizal, Beaver, Wostmann 1982). To this, a fusiform-shaped *Clostridium sp*. was added by accidental contamination (heptaflora). Recently, to promote 7 α-dehydroxylation, a strain of *Clostridium sordellii* (ATCC 9714) was added (octaflora). The two latter strains established well and were observed in large numbers in fecal smears.

Fecal BAs were determined as described earlier (Madsen, Beaver, Chang, Bruckner-Kardoss, Wostmann 1976).

RESULTS

Fecal BA analyses of hexa-, hepta-, and octaflora associated gerbils are given in Table 1. The original hexaflora was chosen because, associated with rats, it produced substantial deconjugation, but little further modification of BA. In the gerbil, around 90% of BAs were deconjugated, but only 10% were further modified. Very little 7α-dehydroxylation occurred. Results obtained with the heptaflora changed somewhat with time in that cholic acid production was increased at the expense of chenodeoxycholic acid. The addition of the fusiform-shaped clostridium resulted in extensive keto-acid formation, but only slightly higher 7α-dehydroxylation. In the CV gerbil on the other hand, 7α-dehydroxylation obviously was the main microflora-associated modification. The addition of the *C. sordellii* strain brought no increase in 7 α-dehydroxylation capacity. Rather, with a decrease in keto-acid formation, the fecal BA pattern resembled that of the hexaflora-associated gerbils.

GF gerbils obtained via caesarian section produced approximately 35% chenodeoxycholic acid and 65% cholic acid. Upon microbial association, production shifted further towards cholic acid. Also, the GN gerbils did not excrete any cholestanol with the feces.

DISCUSSION

7 α-dehydroxylation is the major microbial BA modification in gerbils and man. However, it has proven difficult to link this reductive function to specific microbial species. Recent studies by Hirano et al. (Hirano, Nakama, Tamaki, Masuda, Oda 1981) suggest that strains of C. sordellii demonstrate this function in vitro.

The accidental addition of a fusiform-shaped clostridium to the hexaflora imparted to the ensuing heptaflora an increased potential for keto-acid formation, essentially an oxidative function. Little increase in 7 α-dehydroxylation occurred. Neither was any cholesterol reduced to cholestanol. The introduction of C. sordellii did not provide the anticipated reductive changes, although the keto-acid forming potential of the resulting octaflora appeared to be diminished. Hexa-, hepta- and octaflora all deconjugate the majority of BAs.

C. sordellii showed little 7 α-dehydroxylating capacity in our in vitro tests, but in vitro testing may not indicate in vivo potential, especially in context with other microbial species. Eyssen (REGA Institute, Louvain, Beligum, personal communication) found no 7 α-dehydroxylation in rats monoassociated with C. sordellii ATCC 9.714. However, as a monoassociate this strain caused little or no deconjugation. This presumably would have prevented further microbial modification of primary BAs. In our case, the same strain of C. sordellii demonstrated no 7 α-dehydroxylation capacity in a microbial environment which guaranteed extensive deconjugation and, as demonstrated by copious numbers of fusiforms in the ceca of these gerbils, a redox potential comparable to that of a CV animal.

REFERENCES

Bartizal KF, Beaver M, Wostmann BS (1982). Cholesterol metabolism in gnotobiotic gerbils. Lipids 17:791.
Gustafsson BE, Einarson K, Gustafsson JA (1975). Influence of cholesterol feeding on liver microsomal metabolism of

steroids and bile acids in conventional and germfree rats. J Biol Chem 250:8496.

Hirano S, Nakama R, Tamaki M, Masuda N, Oda H (1981). Isolation and characterization of thirteen intestinal microorganisms capable of 7α-dehydroxylating bile acids. Appl Environ Microbiol 41:737.

Kellogg TF, Wostmann BS (1969). Fecal neutral steroids and bile acids from germfree rats. J Lipid Res 10:495.

Madsen D, Beaver M, Chang L, Bruckner-Kardoss E, Wostmann BS (1976). Analysis of bile acids in conventional and germgree rats. J Lipid Res 17:107.

Wostmann BS (1973). Intestinal bile acids and cholesterol adsorption in the germfree rat. J Nutr 103:982.

Wostmann BS, Beaver M, Chang L, Madsen D (1977). Effect of autoclaving of a lactose-containing diet on cholesterol and bile acid metabolism of conventional and germfree rats. Am J Clin Nutr 30:1999.

Germfree Research: Microflora Control and Its Application to the Biomedical Sciences, pages 111–114

URINARY 3-METHYLHISTIDINE EXCRETION IN GERMFREE AND CONVENTIONAL RATS MAINTAINED ON 70% OF AD LIB DIET INTAKE

Shi Shun-di, Margaret Beaver-Johnson, David L. Snyder, and Bernard S. Wostmann
Lobund Laboratory, University of Notre Dame, Notre Dame, Indiana 46556 U.S.A.

3-methylhistidine (3-MH) is formed by post-translational methylation of histidine and is a constituent of actin in muscles and myosin in white muscle fibers. In rats, about 90% of whole body 3-MH occurs in skeletal muscle (Haverberg, Omstedt, Munro, Young 1975). When muscle protein is degraded, 3-MH is released but is not reutilized nor enters other metabolic events (Young, Haverberg, Bilmazes, Munro 1973). Both man and rat excrete 3-MH quantitatively, man as the free amino acid (Bilmazes, Uauy, Haverberg, Munro, Young 1978), rats in part as its N-acetyl derivative (Omstedt, Kihlberg, Tingvall, Shenkin 1978; Ward, Buttery 1980). Therefore, urinary 3-MH excretion has been widely used to measure the *in vivo* rate of muscle protein degradation.

Lobund Laboratory is conducting a long-term study of the effects of dietary restriction on general health and lifespan. Our data show that in the rat a 30% reduction in *ad* lib. intake causes only a 25% reduction in body weight. Since energy requirement is a function of body weight, a 25% reduction in body weight would suggest a reduction in intake of 18%, not 30%. Restricted rats apparently utilize diet more efficiently. Since protein turnover may be affected by both restricted diets and age (Fischer, Canolty 1983; Millward 1978), measurement of 3-MH in urine may be useful in detecting changes in protein turnover in our studies.

The germfree (GF) condition avoids infectious complications and allows aging to be measured only as a direct degenerative process. However, GF rats and mice have a resting metabolic rate lower than their conventional (CV) counterparts

(Wostmann, Bruckner-Kardoss, Knight 1968), and many facets of intermediary metabolism are quantitatively different in the GF rat (Reddy, Pleasants, Wostmann 1973; Sewel, Wostmann 1975). Therefore, 3-MH excretion must be compared between GF and CV rats before this measurement can be used in the study of protein turnover during aging in a GF system.

MATERIALS AND METHODS

Adult male CV and GF Lobund Wistar rats were fed natural ingredient diet L 485 ad lib., or 12 grams per day, which is equivalent to a 30% reduction in feed intake. Although the isolators of these GF rats were contaminated with Actinomyces sp., fecal smears did not indicate growth of these organisms in the intestinal tract. Urine was collected for 3 days in metabolism units after a 3 day adjustment period. The procedures for determination of 3-MH were modified from Radha and Bessman (1982). Urine samples were deproteinized by adding TCA to a final concentration of 10%. The supernatants were combined with equal volumes of concentrated HCl to hydrolyze the N-acetylated methylhistidine and then evaporated to dryness. The residue was resuspended in 0.1 M disodium phosphate buffer, and the supernatant absorbed on a Dowex-50 column. 3-MH was eluted using 0.1 N $NaHCO_3$ while most ammonia remained on the column. The eluate was evaporated to dryness and reconstituted with 0.1 M citrate to pH 5. After addition of ortho-phthaldialdehyde, the optical density was read one hour later to allow part of the unstable histidine-ninhydrin OPT complex to disappear. Readings at 410Å, 450Å, and 490Å were taken and used in calculations to eliminate interference by histidine. Means were compared using student's T-test.

RESULTS AND DISCUSSIONS

3-MH excretion in ad lib. fed rats is given in Table 1.

Table 1. Urinary 3-methylhistidine excretion in ad lib. fed rats. (mg/day/kg)

	CV	GF	CV	GF	CV
Age	10 Weeks	11 Weeks	3-4 Months	8 Months	> 2 Years
N	6	6	6	8	8
Mean ± S.E.	2.09 ± 0.12	1.93 ± .15	1.17 ± 0.03	1.19 ± 0.10	1.18 ± 0.07

In order to avoid differences in 3-MH due to body weight, 3-MH is expressed as mg/day/kg. 3-MH excretion was not different between CV and GF rats at 10 weeks or older, indicating that microbial status does not affect protein degradation. No age related differences were found for animals 3-4 months or older. Values in ten week old GF and CV rats were significantly elevated ($P < 0.01$) over older rats of the same condition. This agrees with the findings of Millward (1978) that turnover rates decrease after weaning to reach stable values by two to three months. These levels are then maintained for most of adult life.

The results for restricted intake both in 8 month old GF and 2 year old CV rates are given in Table 2. The difference between dietary restriction and ad lib. are 0.16 mg/day/kg and 0.13 mg/day/kg, in CV and GF respectively. Only the difference between old rats is significant at the 5% level. A reduction in protein turnover may be a reasonable assumption since restricted rats appear to utilize diet more efficiently.

Table 2. Comparison of urinary 3-methylhistidine excretion in ad lib. and restricted fed rats. (mg/day/kg)

	Germfree	Conventional
	8 Months	> 2 Years
Ad Lib.	1.19 ± 0.10 (N=8)	1.18 ± 0.07 (N=8)
Restricted	1.06 ± 0.04 (N=7)	1.01 ± 0.06 (N=6)

Problems in estimating muscle protein breakdown by 3-MH have been mentioned by several investigators (Lobley, Harris 1977). Consideration must be given to differences in muscle and fat content, and to the actual source of 3-MH in differing age groups before 3-MH can be used as an accurate measure of protein degradation. However, since fat depots in restricted animals are almost nonexistent, lean body mass and muscle mass will be relatively larger in these animals, which will accentuate the differences in Table 2.

In conclusion, microbial status does not affect muscle protein degradation; age does not seem to affect muscle protein degradation in mature rats when 3-MH is expressed as mg/day/kg/ and 30% dietary restriction apparently slows down muscle protein degradation by 10-15%.

REFERENCES

Bilmazes C, Uauy R, Haverberg L, Munro H, Young V (1978). Muscle protein breakdown rates in humans based on N-methylhistidine (3-methylhistidine) content of mixed proteins in skeletal muscle and urinary output of N-methylhistidine. Metabolism 27:525.

Fischer J, Canolty N (1983). Influence of age on protein and energy utilization in adult rats. J Nutr 113:1095.

Haverberg L, Omstedt P, Munro H, Young V (1975). N-methylhistidine content of mixed proteins in various rat tissues. Biochim Biophys Acta 405:67.

Lobley G, Harris C (1977). Problems in estimated turnover. In "Proceedings of the Second International Symposium on Protein Metabolism and Nutrition", Wageningen: Centre for Agricultural Publishing and Documentation, p 29.

Millward D (1978). The regulation of muscle-protein turnover in growth and development. Biochem Soc Tran 6:494.

Omstedt P, Kihlberg R, Tingvall P, Shenkin A (1978). Effect of dietary protein on urinary excretion of 3-methylhistidine in rat. J Nutr 108:1877.

Radha E, Bessman SP (1982). A rapid calorimetric method for 3-methylhistidine in urine. Anal Biochem 121:170.

Reddy B, Pleasants J, Wostmann B (1973). Metabolic enzymes in liver and kidney of the germfree rat. Biochim Biophys Acta 320:1.

Sewell D, Wostmann B (1975). Thyroid function and related hepatic enzymes in the germfree rat. Metabolism 24:695.

Ward L, Buttery P (1980). Dietary protein intake and 3-methylhistidine excretion in the rat. Br J Nutr 44:381.

Wostmann B, Bruckner-Kardoss E, Knight P (1968). Cecal enlargement, cardiac output and O_2 consumption in germfree rats. Proc Soc Exp Biol Med 128:137.

Young V, Haverberg L, Bilmazes C, Munro H (1973). Potential use of 3-methylhistidine as an index of progressive reduction in muscle protein catabolism during starvation. Metabolism 22:1429.

Germfree Research: Microflora Control and Its Application to the Biomedical Sciences, pages 115–118

STUDIES ON THE EFFECT OF THE GUT MICROFLORA ON DIETARY NITROGEN DIGESTIBILITY IN PIGS

B. Ratcliffe and A.G. Low

National Institute for Research in Dairying

Shinfield, Reading, RG2 9AT, U.K.

INTRODUCTION

The processes of protein digestion and amino acid absorption have been investigated for many years. However it has only recently become clear that a major part of the proteins and amino acids in faeces of a wide variety of species are in the form of bacteria or are endogenous secretory products. So only a small part of faecal proteins and amino acids is of undigested dietary origin. These findings challenge the validity of long-established methods of protein evaluation based on the difference between dietary input and faecal output, corrected for endogenous losses by feeding protein-free diets, giving values of the so-called true digestibility of proteins. The two fundamental problems with this approach are (a) that it does not take account of the bacterial component of faeces, which depends on the type of protein fed and on the available energy for bacterial growth, and (b) that the amount of endogenous protein secreted varies greatly according to the type of protein fed and the other components of the diet. From a practical viewpoint, reliable measurement of the digestibility of proteins, while of great importance, is impossible to achieve at present.

The digesta passing through the large intestine support the growth of a microflora including a wide range of bacteria. This microflora is able to digest and metabolize the dietary and endogenous residues extensively so altering their respective contributions to the final composition of the faeces. It is known that bacterial enzymes are potentially capable of catabolizing all amino acids (Michel, 1966). Furthermore, it has been shown that virtually all soluble proteins or amino acids introduced into the caecum of pigs through a cannula are rapidly digested and absorbed (Presumably as ammonia, amines etc.) and the nitrogen (N) derived from them is very quickly and almost completely excreted in the urine (Zebrowska, 1973).

The significance of the microflora in the digestion of protein residues in pigs has also been indicated in experiments which have shown discrepancies between the apparent digestibilities of amino acids measured at the terminal ileum or in the faeces (Holmes *et al*, 1974; Low, 1979).

It has been estimated that only 6% of faecal N is of undigested dietary origin in pigs while the remainder is of bacterial or endogenous origin (Low, Sambrook and Yoshimoto, 1978). It was therefore of interest to attempt to estimate which proportions of faecal N, derived from a barley-soya diet, are of either bacterial, endogenous or undigested dietary origin. The principle of the studies was to compare germ-free (GF) and conventional (CV) pigs, so that the bacterial component of faeces could be measured.

METHODS

Four GF and four CV Göttingen miniature pigs were used from approximately 8 to 16 weeks of age. Miniature swine were chosen because they are small enough to be kept in our isolators for periods up to 20 weeks.

The GF pigs were derived by post-natal decontamination using a modified version of that described by Ducluzeau *et al*, (1976) as reported by Ratcliffe and Fordham (1984). They were housed in metabolism cages in stainless steel isolators of the Gustafsson type. The CV pigs were housed in identical metabolism cages within controlled environment rooms.

All pigs were given a sterilized, barley-soya diet (27.2 g N/kg diet) at about 4% body weight per day, with water supplied *ad libitum*.

Weekly 5-day collections of faeces were made. The contents of N in the whole faeces and in the acid detergent fibre (ADF) fraction were analysed by the Kjeldahl method. Fractionating the faeces with acid detergent solution allows all bacterial N, if present, and all endogenous N to pass into the filtrate. The N associated with the ADF residue was considered to provide an estimate of undigested dietary N (Mason, 1969; Low *et al*, 1978).

RESULTS

There was no difference in apparent digestibility of N between the two groups of pigs (see Table 1) and the values for digestibility were very similar to those found by other workers using this same diet for conventional Large White pigs (Partridge *et al*, 1982).

From the results shown in Table 2, however, it can be seen that undigested dietary N formed a greater proportion of faecal N in the CV pigs and this difference was highly significant.

TABLE 1 Apparent digestibility of N (mean ± se)

Germ-free	0.86 ± 0.01	no significant difference
Conventional	0.85 ± 0.01	

TABLE 2 Acid detergent fibre N as a % of faecal N (mean ± se)

Germ-free	3.6 ± 0.1	significance of difference
Conventional	7.4 ± 0.9	$P<0.001$

DISCUSSION AND CONCLUSIONS

The apparent digestibility of N was the same in both CV and GF pigs and so the presence of the microflora appears to alter the distribution of N in faeces but not the total amount. In the GF pigs, 99.5% of the dietary N appears to have been digested, while 98.9% seems to have been digested in CV pigs. The value for dietary N digestibility in CV pigs corresponds well with the value of 98.9% for conventional Large White pigs given similar diets by Low *et al*, (1978). The lower level of ADF N in GF pigs may be the result of the much longer time available for hydrolysis of dietary proteins by endogenous enzymes, since these enzymes were not degraded by a microflora. The use of GF pigs did not allow satisfactory estimation of bacterial and endogenous components of faecal N.

REFERENCES

Ducluzeau R, Raibaud P, Lauvergeon B, Gouet Ph, Riou Y, Griscelli C, Ghnassia J-C (1976). Immediate postnatal decontamination as a means of obtaining axenic animals and human infants. Can J Microbiol 22:563-566.

Holmes JHG, Bayley HS, Leadbeater PA, Horney FD (1974). Digestion of protein in small and large intestine of the pig. Br J Nutr 32:479-489.

Low AG (1979). Studies on digestion and absorption in the intestines of growing pigs.6.Measurements of the flow of amino acids. Br J Nutr 41:147-156.

Low AG, Sambrook IE, Yoshimoto JT (1978). Studies on the true digestibility of nitrogen and amino acids in growing pigs. EAAP Annual Meeting, Stockholm, Paper PV9.

Mason VC (1969). Some observations on the distribution and origin of nitrogen in sheep faeces. J agric Sci, Camb 73:99-111.

Michel MC (1966). Metabolisme de la flore intestinale du porc degradation des formes L et D les acides amines. Ann Biol anim Biochim Biophys 6:33-46.

Partridge IG, Keal HD, Mitchell KG (1982). The utilization of dietary cellulose by growing pigs. Anim Prod 35:209-214.

Ratcliffe B, Fordham JP (1984). A technique for rearing germ-free piglets obtained without operative surgery. Lab Anim (in press).

Zebrowska T (1973). Digestion and absorption of nitrogenous compounds in the large intestine of pigs. Roczn Nauk roln 95B(3):85-90.

The authors acknowledge the technical assistance of Mr JP Fordham.

Germfree Research: Microflora Control and Its Application to the Biomedical Sciences, pages 119–122

PRODUCTION OF AMINES IN MONOXENIC CHICKEN INOCULATED WITH A LACTOBACILLUS STRAIN ISOLATED FROM HOLOXENIC (CONVENTIONAL) COCK CROP.

Nugon-Baudon Lionelle, Odette Szylit, *M. Chaigneau, **N. Dierick and P. Raibaud.
Lab. Ecologie Microbienne INRA - CNRZ
78350-JOUY-en-JOSAS (France)
*Université René Descartes 75270-PARIS (France)
**Rijksuniversiteit GENT (Belgique)

Lactobacilli are the most numerous bacteria in the intestinal tract of chickens but little is known about their physiological role. An amylolytic strain of lactobacilli sp. LEM 207, isolated from the crop of a holoxenic cock had a high level of proliferation and could colonize the epithelium crop of monoxenic chickens (Champ et al. 1981). This strain was a particular interest because it differed from other species in producing carbone dioxyde in a carbohydrate free medium. When bacteria ferments amino-acids the decarboxylation leads to amines which may have a large influence on physiology of the host. The nature of formed amines was first investigated in vitro and then in axenic chickens monoassociated with the LEM 207 strain.

Material and methods : Strain LEM 207 belonged to the subgenus thermobacterium, ressembled L.acidophilus and possessed an α amylase (Champ et al. 1983). Production of D and L lactic acids from carbohydrates and breakdown of starch (Champ et al. 1981) were observed in vitro and in vivo.

Culture methods were already described (Champ et al. 1983). In in vitro studies, lactobacilli were cultured in the LAPT carbohydrate free medium (Difco Yeast Extract 1 %, Tryptone 1 %, Peptone 1.5 %, Tween 80 0.1 %, pH 6.5 at 37°C. Total CO_2 (Chaigneau et al. 1965) pH and bacterial growth measured at 620 nm were determined. Acidity of the sample was then lowered to pH 3 with 6 N HCl. Samples were centrifuged (10.000 g, 10 min.) and the supernatant was analysed for amine concentration by ion-exchange chromatography (Dierick 1976).

Axenic (germ-free) chickens were obtained by previously described methods (Le Coz et al. 1977) and divided into two groups. One remainded axenic, the other (monoxenic chickens) was inoculated per os with 0.1 ml of culture. They were fed a semi-purified diet[230 g fish meal, 659 g maize starch, 40 g dextrins and maize oil, 20 g cellulose/kg,salt vitamine and methionine supplement]sterilized by gamma-irradiation. At the end of a 4 week period, chickens were killed by chloroform and the crop and cecal content were withdrawn for pH and bacterial examination. Digestive samples and livers were homogenized (Ultra-Turrax) in 0.1 N HCl (final pH 3.0) and centrifuged. Analysis of amines was made in the supernatant.

Results : In vitro amine production (Table 1)

TABLE 1 - PH , GROWTH RATE , CO_2 AND AMINE PRODUCTION IN CARBOHYDRATE-FREE BROTH CULTURE INOCULATED WITH LACTOBACILLUS STRAIN 207

CULTURE LENGTH	PH	DO (620 nm)	CO_2 (%)	AMINES (µMOLES/L CULTURE BROTH)		
				TYRAMINE	HISTAMINE	OTHER AMINES (TRACES)
T 0	6.40	0.053	0	198	3.74	N D
T 4 H	5.96	0.228	1.40	210	4.07	PUTRESCINE CADAVERINE
T 18 H	5.45	0.397	2.10	540	4.03	PUTRESCINE CADAVERINE

. STRAINS WAS CULTURED IN ONE LITER OF MEDIUM AT 37°C. EACH VALUE REPRESENTS THE AVERAGE OF TWO ANALYSIS.
. N D = NOT DETECTED.

A low concentration of tyramine and histamine was found in the culture broth. During a 18 hour aerobic incubation, the strain growth was low while pH medium decreased by one unit and a carbone dioxyde production was observed (2.1 %). Amines appeared between the fourth and the eighteenth hour. The tyramine concentration reached (540 µmoles/l, histamine concentration remained low and traces of putrescine and cadaverine were detected.

In vivo amine production (Table 2)

TABLE 2 - pH AND AMINE PRODUCTION (μMOLE OF CONTENT) IN CROP AND CECA OF AXENIC AND MONOXENIC CHICKENS

Chickens	CROP				CECA			
	pH	Tyr	Cad	Put	pH	Tyr	Cad	Put
Monoxenic chicken LEM 207 (n = 5)	5.79	0.254	0.579	0.259	7.02	0.159	ND	ND
	5.63	0.288	0.060	0.038	6.61	0.100	ND	ND
	5.53	0.447	0.142	0.096	6.15	0.183	ND	ND
	4.26	0.775	ND	0.060				
	3.71	0.850	ND	0.011				
Axenic chicken (n = 2)	6.21	0.099	ND	ND	6.78	0.210	ND	ND
	5.69	0.064	ND	ND	7.08	0.141	ND	ND

Tyr = tyramine ; Cad = cadaverine ; Put = putrescine
ND = not detected.
n = Animal numbers.

The only amine we found in the crop and cecal contents of axenic chicken was tyramine of endogenous origin.

In the crop of monoxenic chickens the strain developed at a high level (10^8/g of content) and there was a production of tyramine, putrescine and cadaverine. The concentration of cadaverine and putrescine at its highest at the beginning of digestion, decreased with the increasing acidity of the content whereas the level of tyramine progressively increased. Its concentration could be 7 fold higher than in the axenic crop. Histamine was not formed in any of the samples. In the ceca of monoxenic chickens, where pH was always high, tyramine concentration was similar to that found in axenic chickens. No other amines were detected. There was an accumulation of tyramine in the liver: 1.05 μmole/g of liver in the monoxenic chicken versus 0.15 μmole/g in the axenic one.

Conclusions : The amylolytic lactobacillus strain, LEM 207, produced amines. Increase in tyramine concentration observed in vivo studies may be explained by the fact that amino acid decarboxylases are more active at low pH values. Indeed due to carbohydrate fermentation, the pH value decreases rapidly in the crop of monoxenic chickens. To our knowledge there is no explanation on the influence of pH on qualitative and quantitative distribution of putrescine and cadaverine.

Production of tyramine by a lactobacillus strain of digestive origin is exceptional. Rodwell (1953) was the only one to mention a histamine producing strain of lactobacilli, isolated from sheep and horse. These strains were responsible for a high mortality rate. Likewise growth disorders, cannibalism, a marked excitability and mortality were observed in monoxenic 207 animals (Champ 1978).

Thus, bacteria belonging to commensal intestinal microflora may possess a pathogenic potential. The influence of nutrients upon the expression of this potential is being studied.

Chaigneau M, Bory J, Le Mohan G, Labarre G (1965). Composition des gaz dégagés par Clostridium septicum cultivé en milieu thioglycolate. Note C R Acad Sci Paris t261:2403-2406.

Champ M (1978). Rôle de deux lactobacilles amylolytiques dans la dégradation de l'amidon chez le poulet monoxénique. Thèse de 3ème cycle. Univ de Paris-Sud, Orsay n° d'ordre 253.

Champ M, Szylit O, Raibaud P & Aït-Abdelkader N (1983). Amylase production by three Lactobacillus strains isolated from chicken crop. J Appl Bact 55:487-493.

Dierick NB, Vervaeke IJ, Decuypere JA, Henderickx HK (1976). Determination of biogenic amines in intestinal contents by ion exchange chromatography. Note à Elsevier Scient. publishing company Chrom 9458.

Le Coz Y, Szylit O, Ducluzeau R (1977). Mise au point d'isolateurs souples destinés à des études nutritionnelles sur poulets axéniques et gnotoxéniques. Sci Tech Anim Lab 2 (1):27-33.

Rodwell AW (1953). The histidine decarboxylase of a species of Lactobacillus ; apparent dispensability of pyridoxal phosphate as co-enzyme. J Gen Microbiol 8:233-237.

Germfree Research: Microflora Control and Its Application to the Biomedical Sciences, pages 123–125

NITRATE REDUCTION IN GERMFREE RATS

F.W. Ward, M.E. Coates and R. Walker

Robens Institute,
University of Surrey
Guildford, GU2 5XH, Surrey, U.K.

INTRODUCTION

The presence of nitrate in the mammalian organism can be of toxicological significance through the formation of methaemoglobin (MetHb) and carcinogenic N-nitroso compounds. These reactions proceed via the reduction of nitrate to nitrite in the alimentary tract. Current thought only implicates the microbial flora in the reduction of nitrate although Fritsch et al. (1976) hinted at the presence of a mammalian nitrate-reducing system in the gastrointestinal tract. The ability of germfree (GF) rats to reduce orally administered nitrate to nitrite, with blood MetHb values being used as an index of nitrite formation, is reported here. In addition, evidence for direct microbial involvement in the nitrosation reaction *in vivo* is presented.

METHODS

2% w/v KNO_3 in distilled H_2O was given for 6d to 2 male and 2 female gnotobiotic (GN) BDIX rats, 5 months old, monocontaminated with an unknown yeast, and their conventional (CV) counterparts. 4 male GF Wistar rats, 3 months old, and their CV counterparts were treated in a similar manner for 9d. MetHb in samples of blood taken from the tail was determined by a modification of the method of Evelyn et al. (1938).

Following the procedure of Fritsch et al. (1976), mucosal scrapings were taken under GF conditions from the stomach and small intestine of GF rats and incubated anaerobically in filter-sterilized Krebs-Ringer phosphate buffer (pH 7.4), containing 2% w/v KNO_3 and 2% w/v glucose, for 24h at 37°C. A similar sample was immersed in boiling H_2O for 10 min before incubation. Nitrite was determined by the method of Shechter et al. (1972).

A further step in the in vivo metabolism of nitrate is the formation of N-nitroso compounds. In a preliminary study, 2 GF and 2 CV female Wistar rats were given 2% w/v $NaNO_3$ and 1% w/v proline in distilled H_2O to drink, fresh solutions being prepared daily. 24h urine samples were collected in 1M NaOH and the N-nitrosoproline excreted quantitated by a gas-liquid chromatograph coupled to a thermal energy analyser.

RESULTS

In both experiments, MetHb values after oral ingestion of nitrate were similar in GF and CV rats (Figs. 1 and 2). The two experiments were done on rats of different age and strain, which may account for the difference in rate of development of methaemoglobinaemia.

FIGURE 1

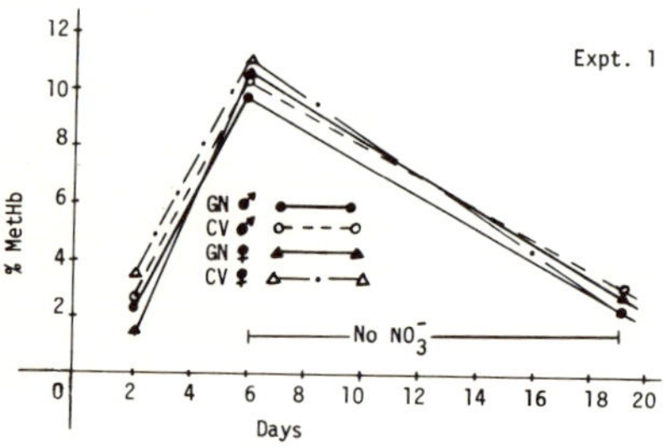

FIGURE 2

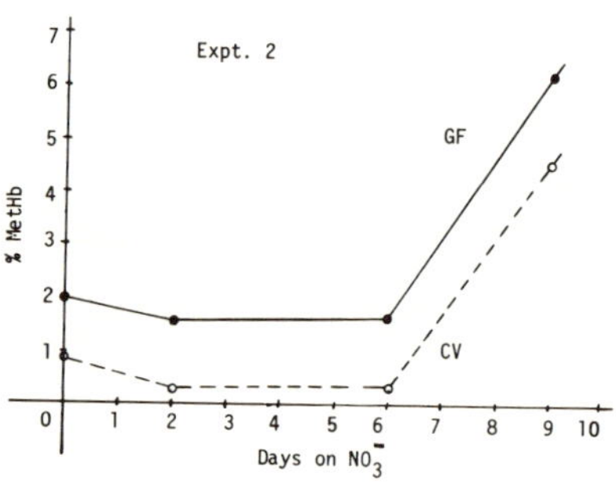

The mucosal scrapings exhibited high nitrate-reducing activity which was heat labile in the small intestine. The effect was less clear in the stomach. (Table 1).

The higher excretion of N-nitrosoproline by CV rats than their GF counterparts after ingestion of precursors is outlined in Table 2. No N-nitrosoproline was detected in the dosing solution.

TABLE 1

Rat No.	Treatment of Sample	Amount of NO_2^- Formed (μg)	
		Stomach	Small Intestine
GF1	Unheated	15.65±0.00	407.78±55.87
GF2	Unheated	49.63±0.11	293.49± 0.63
GF3	Heated	14.23±0.00	73.29± 0.11

TABLE 2

Status	N-Pro Excreted (ng)	
	Day 0	Day 6
GF1	10.7	ND
GF2	20.7	143.5
CV1	98.7	1190.8
CV2	209.5	2209.6

ND = Not detected

DISCUSSION

The extent of MetHb formation in GF rats indicates that nitrate reduction is not confined solely to the microbiota. From the results of the in vitro incubations it can be inferred that a nitrate-reducing system is present in the gastrointestinal mucosa of rats. It is heat labile and so probably enzymic in nature. Although the formation of N-nitrosoproline from proline and nitrate occurred in the GF rat, much more was formed in the CV rat. Since nitrate reduction occurred equally in GF and CV rats these preliminary findings suggest that endogenous nitrosation results mainly from microbial action, the mechanism of which is under investigation.

REFERENCES

Evelyn KA, Malloy HT (1938). Microdetermination of oxyhemoglobin, methemoglobin, and sulfhemoglobin in a single sample of blood. J Biol Chem 126:655.

Fritsch P, de Saint Blanquat G, Canal M-T (1976). Formation des nitrites a partir des nitrates dans le tube digestif. Ann Nutr Aliment 30:793.

Shechter H, Gruenar N, Shuval HI (1972). A micromethod for the determination of nitrite in blood. Anal Chim Acta 60:93.

Germfree Research: Microflora Control and Its Application to the Biomedical Sciences, pages 127–131

BIOLOGICAL SIGNIFICANCE OF THE ANAEROBIC METABOLISM OF METRONIDAZOLE

Peter Goldman, M.D., Tin-Chuen Yeung, Ph.D., Ronald L. Koch, Ph.D., Bernard B. Beaulieu, Jr., M.S. and Martha A. McLafferty, M.S.

School of Public Health, Harvard University, Boston, MA 02115

Both metronidazole's bacterial mutagenicity (Rosenkranz & Speck, 1975) and its microbiocidal activity apparently are mediated by a derivative of metronidazole formed during nitro group reduction. Although there is evidence for the formation of several intermediates during reduction of a nitro group, these intermediates are too labile to isolate from biological systems (Goldman, 1982). As a result it has not been possible to associate any of them directly with metronidazole's biological activity.

On the other hand, it has been shown that metronidazole is metabolized under anaerobic conditions to yield acetamide (Koch et al., 1979). It seemed logical, therefore, to postulate that the labile reactive metabolite responsible for metronidazole's biological activity might also be involved in the formation of this stable metabolite. Thus, the labile intermediate may have one of the following fates: It might interact with a macromolecule, such as DNA, to initiate a biological effect or it might interact, for example, with water to form a stable metabolite such as acetamide. This proposal is outlined in the scheme shown below.

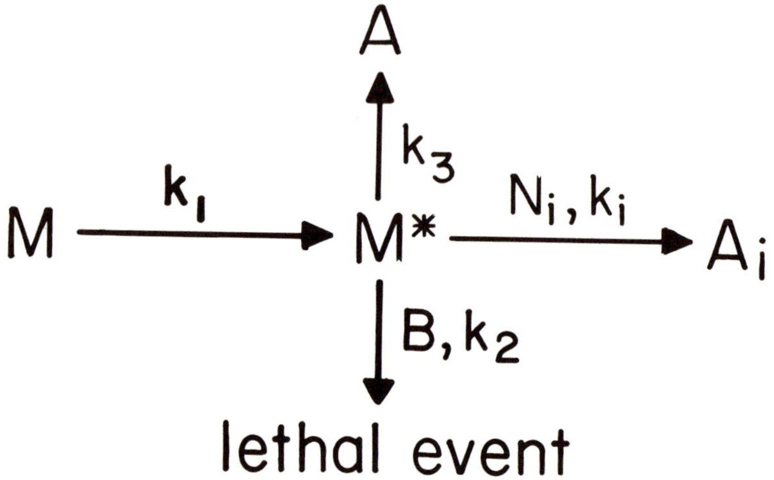

According to this scheme the bacterial reduction of metronidazole (M) initially yields the labile reactive intermediate M^*. This intermediate may then either interact with a bacterial target to cause a lethal event or it may interact, for example, with other molecules (N_i) to yield stable metabolites (A_i) (Chrystal et al., 1980). In our analysis we focussed on the stable metabolite acetamide (A) which arises, we assume, as the result of an interaction of M^* with water.

If it is assumed that M^* forms at a constant rate, which depends only on the redox potential of the bacterial culture, then the formation of acetamide from M^*, which depends only on the concentration of water, becomes a measure of the exposure of the bacterial culture to M^*. If M^* exerts its bactericidal effect in proportion to its concentration in the medium, then the accumulation of acetamide should be related linearly to the log of bacterial survival. This relationship is found for B. fragilis and Clostridium perfringens, which are sensitive to metronidazole (Chrystal et al., 1980), as well as for several strains of E. coli (Yeung et al., 1984) which, as discussed below, have variations in their sensitivity to metronidazole.

That metronidazole is a bacterial mutagen (Voogd et al., 1974) suggests that it interacts with bacterial DNA. That the lethal effect of metronidazole also depends on an interaction with DNA is suggested by the finding that some E. coli DNA repair mutants are more sensitive to metronidazole and related compounds than their parent strains (Jenkins and Bennett, 1976; Goldstein et al., 1977).

A kinetic analysis of the bactericidal effect of metronidazole in relation to the formation of acetamide in DNA repair mutants of E. coli argues for the importance of the postulated intermediate M^* and helps to explain why it was found that not all $RecA^-$ and $UvrB^-$ mutants have an enhanced sensitivity to metronidazole. Thus, $RecA^-$ and $UvrB^-$ mutants that all also lack full capacity to reduce metronidazole are relatively resistant to metronidazole. According to the proposed model their resistance is the result of their inability to convert metronidazole to its lethal form. Nevertheless, when analyzed in terms of acetamide formation, which indicates exposure to M^*, all $RecA^-$ and $UvrB^-$ mutants are found to be more sensitive to metronidazole than the parent strain, as would be expected if M^* is the actual mediator of metronidazole's action (Yeung et al., 1984).

Is the model advanced for bacteria applicable to the fate of metronidazole in mammalian tissues? The answer to this question is important because it might provide a starting point for understanding why metronidazole is tumorigenic for certain rodents (Goldman, 1980). To test whether the findings in bacteria are applicable to mammalian tissues one must first determine whether metronidazole is reduced in mammalian tissues to an extent that leads to the formation of acetamide.

Determining whether nitro group reduction occurs in mammalian tissues is not a simple problem. The flora of conventional laboratory animals reduce the nitro group of metronidazole (Koch and Goldman, 1979), and therefore metabolism at this site may obscure some metabolism of the drug that might occur in the tissues. Although the use of tissue cultures or perfused organs enable the effects of the flora to be removed, the metabolism of nitroheterocyclic compounds in such systems, depending as it does on the experimentally imposed oxygen tension (Jonen, 1980), is not a reliable guide to what may occur physiologically. With germfree animals, on the other hand, one may determine the reductive capacity of animal tissues under physiological conditions. Therefore, we administered to germfree rats metronidazole of the highest specific activity and sought some evidence for the formation of reduced metabolites as a means of providing the most sensitive indication of whether or not this conversion might occur in mammalian tissues. By such methods we were unable to find metabolites in the urine that would indicate the formation of their biologically active precursors in rat tissues (Yeung et al., 1983). Of course, this negative finding does not exclude the possibility that metronidazole is reduced to some extent in the tissues. It is possible that the amount of acetamide formed is not sufficient to detect. Alternatively, the biologically active form, M^*, might escape detection because it is reoxidized to restore the parent drug (Goldman, 1982).

ACKNOWLEDGEMENT

Much of the experimental work reported in this paper was supported by U.S. Public Health Service Research Grant CA 34957 from the National Cancer Institute.

REFERENCES

Chrystal EJT, Koch RL, McLafferty MA, Goldman P (1980) The relationship between metronidazole metabolism and bactericidal activity. Antimicrob Agents Chemotherap 18: 566.

Goldman P (1980) Drug Therapy: Metronidazole. N Engl J Med 303: 1212.

Goldman P (1982) The development of 5-nitroimidazoles for the treatment and prophylaxis of anaerobic bacterial infections. J Antimicrob Chemotherap 10 (Suppl. A): 23.

Goldstein BP, Nielsen E, Berti M, Bolzoni G, Silvestri LG (1977). The mechanism of action of nitroheterocyclic antimicrobial drugs. Primary target of 1-methyl-2-nitro-5-vinylimidazole is DNA. J Gen Microbiol 100: 271.

Jenkins ST, Bennet PM (1976). Effects of mutations in deoxyribonucleic acid repair pathways on the sensitivity of Escherichia coli K-12 strains to nitrofurantoin. J Bacteriol 125: 1214.

Jonen HG (1980). Naunyn-Schmiedeberg's Archs Pharmac 315: 167.

Koch RL, Goldman P (1979) The anaerobic metabolism of metronidazole forms N-(2-hydroxyethyl)-oxamic acid. J Pharm Exp Therap 208: 406.

Koch RL, Chrystal EJT, Beaulieu BB, Goldman P (1979) Acetamide: A metabolite of metronidazole formed by the intestinal flora. Biochem Pharmacol 28: 3611.

Rosenkranz HS, Speck WT (1975). Mutagenicity of metronidazole: activation by mammalian liver microsomes. Biochem Biophys Res Commun 66: 520.

Voogd CE, Van Der Siel JJ, Obs JJAA (1974). The mutagenic action of nitroimidazoles. I. Metronidazole, nimorazole, dimetridazole, and ronidazole. Mutat Res 26: 483.

Yeung TC, Sudlow G, Koch RL, Goldman P (1983). Reduction of nitroheterocyclic compounds by mammalian tissues in vivo. Biochem Pharmacol 32: 2249.

Yeung TC, Beaulieu BB, McLafferty MA, Goldman P (1984) Interaction of metronidazole with DNA repair mutants of Escherichia coli. Antimicrob Agents Chemotherap 25: 65.

Germfree Research: Microflora Control and Its Application to the Biomedical Sciences, pages 133–137

METABOLISM OF 1-NITROPYRENE AND 6-NITROBENZO(a)PYRENE BY INTESTINAL MICROFLORA

CARL E. CERNIGLIA

NATIONAL CENTER FOR TOXICOLOGICAL RESEARCH
FOOD AND DRUG ADMINISTRATION
JEFFERSON, ARKANSAS 72079

There has been considerable interest in the environmental occurrence of nitropolycyclic aromatic hydrocarbons because of their high mutagenicity toward *Salmonella typhimurium* (Lofroth et al. 1980; Rosenkranz et al. 1980). Nitropolycyclic aromatic hydrocarbons have also been shown to be carcinogenic and genotoxic in various mammalian test systems (El-Bayoumy et al. 1982; Marshall et al. 1982; Ohgaki et al. 1982). These compounds are formed by nitration of polycyclic aromatic hydrocarbons in airborne particulates and are present in diesel particulate matter emitted from automobile engines (Pitts et al. 1978; Wang et al. 1980; Rosenkranz 1982). The postulated mode of the biological activity of this class of compounds is due to the enzymatic reduction of the nitro group via nitroso-intermediates to an hydroxamate with further conversion to an amino-derivative. The hydroxamate can be esterified and form an electrophile capable of reacting with DNA. This metabolic activation pathway is illustrated in Figure 1. Bacterial and mammalian nitroreductases have been implicated in the metabolic activation of nitropolycyclic aromatic hydrocarbons (El-Bayoumy et al. 1983; Howard et al. 1983; Kinouchi et al. 1982; Messier et al. 1981; Nachtman and Wei 1982). This paper describes the metabolism of 1-nitropyrene and 6-nitrobenzo(a)pyrene by anaerobic bacterial suspensions from human and rat feces and intestinal contents and in pure cultures of anaerobic bacteria.

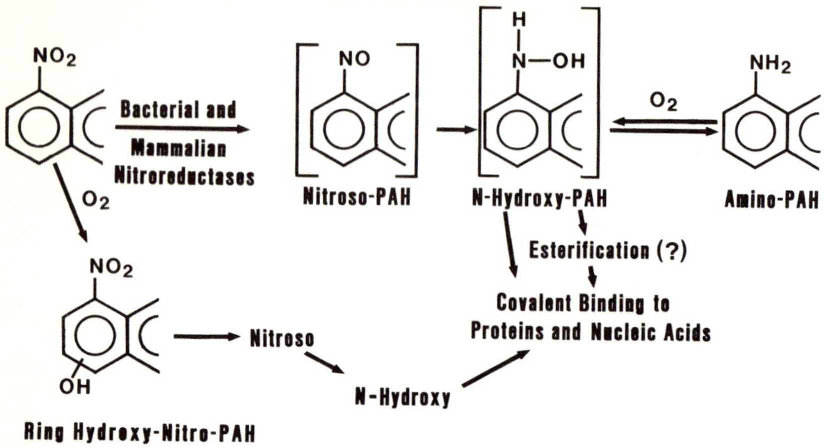

Fig. 1. Metabolic activation pathways of nitropolycyclic aromatic hydrocarbons.

MATERIALS AND METHODS

Microorganisms and Nitropolycyclic Aromatic Hydrocarbon Biotransformation Conditions

All the media used in this study were prereduced and transfers were made anaerobically under 100% CO_2 with an oxygen-free cannula. Cultures were suspended in 0.5 ml of chopped meat glucose medium and a 0.4 ml aliquot from each culture was used to inoculate 5 ml of brain heart infusion supplemented broth. The cultures were incubated at 37°C for 24 hr. Clostridium perfringens ATCC 3626 was incubated at 45°C. After 24 hours of incubation, a 10-ml aliquot (1 X 10^9 cells/ml) was aseptically removed from each incubation bottle and transferred anaerobically to 45 ml of brain heart infusion broth [4,5,9,10-^{3}H]-1-Nitropyrene (3.7 uM) or [G-^{3}H]-6nitrobenzo(a)pyrene (2.6 uM) was then added into the media and the cultures were incubated as described above. After 8 hr, the cultures were centrifuged at 10,000 x g for 20 min and the supernatants extracted with ethyl acetate and analyzed by hplc.

1-Nitropyrene and 1-aminopyrene were purchased from Aldrich Chemical Co., Milwaukee, WI. [G^{3}H]6-Nitrobenzo(a) pyrene (sp. act. 93.2 mCi/mmol radiochemical purity 99%),6-

nitrobenzo(a)pyrene and 6-aminobenz(a)pyrene were synthesized and purified as described previously (Fieser et al. 1939; Fu et al. 1984).

RESULTS AND DISCUSSION

Various anaerobic bacteria and human and rat intestinal microflora cultures were surveyed for their ability to metabolize [^{3}H]-nitropyrene and [^{3}H]6-nitrobenzo(a)pyrene. All of the cultures metabolized these nitropolycyclic aromatic hydrocarbons to the corresponding amines (Table 1).

Metabolism of [^{3}H]1-Nitropyrene and [^{3}H]6-Nitrobenzo[a]pyrene to 1-Aminopyrene and 6-Aminobenzo[a]pyrene by Anaerobic Bacteria[a]

Organism (ATCC)		1-Aminopyrene	6-Amino-benzo[a]pyrene
Rat Intestinal Microflora		+++[b]	+++
Human Intestinal Microflora		+++	+++
Bacteroides thetaiomicron	29148	+++	+++
Bifidobacterium infantis	15697	++	++
Citrobacter sp	25405	++	++
Clostridium perfringens	3626	+++	+++
Lactobacillus acidophilus	332	+	+
Peptococcus anaerobius	14955	+++	+++
Clostridium sp.		+++	+++
Peptostreptococcus productus	27340	+++	+++
Escherichia coli	25990	+	+

[a] Microorganisms were incubated with [^{3}H]1-nitropyrene and [^{3}H]6-nitrobenzo[a]pyrene for 8 hr under anaerobic conditions. The contents in the incubation bottles were extracted with ethyl acetate and the metabolites separated by hplc and quantified as described in Materials and Methods.

[b]
+ = 1 to 30% formation
++ = 31 to 50% formation
+++ = 51 to 100% formation

To confirm the identification of the nitroreduction products as 1-aminopyrene and 6-aminobenzo(a)pyrene respectively, batch experiments were conducted in which suspensions of rat and human intestinal microflora were incubated in the presence of 1-nitropyrene and 6-nitrobenzo(a)pyrene and the metabolites isolated and identified by high performance liquid chromatography and mass spectrometry. The mass spectral fragmentation patterns for 1-aminopyrene [M^+ at m/z 217, m/z 202 (M^+-15; NH loss), m/z 189 (M^+-28, HCNH loss)] and 6-aminobenzo(a)pyrene [M^+ at m/z 267, m/z

239 (M^+-28, HCNH loss] formed from 1-nitropyrene and 6-nitrobenzo(a)pyrene by rat and human intestinal microflora were identical to those of authentic standards. There is a rapid and quantitative conversion of these nitropolycyclic aromatic hydrocarbons to amino derivatives. Within 1 hr of incubation 85 to 95% of 1-nitropyrene and 6-nitrobenzo(a)pyrene were converted to 1-aminopyrene and 6-aminobenzo(a)pyrene.

The toxicological significance of this nitroreduction reaction should be that if nitropolycyclic aromatic hydrocarbons do pass through the gastrointestinal tract and interact with the intestinal microflora, the aminopolycyclic aromatic hydrocarbon formed may be returned to the liver and subsequently undergo further metabolism by mammalian or oxidative enzymes to form arylhydroxylamines. (See Figure 1)

ACKNOWLEDGEMENTS

The author would like to thank Dr. Peter P. Fu and Dr. Paul C. Howard for helpful discussions throughout this study, Wirt Franklin for technical assistance and Virginia Calaway for helping prepare this manuscript.

El-Bayoumy K, Hecht SS, Wynder EL (1982). Tumor-initiating activity and in vitro metabolism of 6-nitrochrysene and 1-nitropyrene. Proc Am Assoc Cancer Res 23:84.

El-Bayoumy K, Sharma C, Louis YM, Reddy B, Hecht SS (1983). The role of intestinal microflora in the metabolic reduction of 1-aminopyrene in conventional and germfree rats and in humans. Cancer Lett 19:311.

Fieser LF, Hershberg EB (1939). The orientation of 3,4-benzpyrene in substitution reactions. J Am Chem Soc 61: 1565.

Fu PP, Chou MW, Unruh LE, Freeman JP, Miller DW, Evans FE, Roth R (1984). Synthesis of 6-substituted 3-deuterobenzo(a)pyrene. Org Prep Proc Int (In press).

Howard PC, Beland FA, Cerniglia CE (1983). Reduction of the carcinogen 1-nitropyrene to 1-aminopyrene by rat intestinal bacteria. Carcinogenesis 4:985.

Kinouchi T, Manabe Y, Wakisaka K, Ohnishi Y (1982). Biotransformation of 1-nitropyrene in intestinal anaerobic bacteria. Microbiol Immunol 26:992.

Lofroth G, Hefner E, Alfheim I, Moller M (1980). Mutagenic activity in photocopies. Science 209:1037.

Marshall TC, Royer RE, Li AP, Kusewitt DF, Brooks AL (1982). Acute and genetic toxicity of 1-nitropyrene and its fate after single oral doses to rats. J Toxicol Environ Health 10:373.

Messier F, Lu C, Andrews P, McCarry BE, Quilliam MA, McCalla DR (1981). Metabolism of 1-nitropyrene and formation of DNA adducts in Salmonella typhimurium. Carcinogenesis 2:100.

Nachtman JP, Wei ET (1982). Evidence for enzymatic reduction of 1-nitropyrene by rat liver fractions. Experientia 38:837.

Ohgaki H, Matsukura N, Morino K, Kawachi T, Sugimura T, Morita K, Tokiwa H, Hirota T (1982). Carcinogenicity in rats of the mutagenic compounds 1-nitropyrene and 3-nitrofluoranthene. Cancer Lett 15:1.

Pitts Jr JN, Van Cauwenberghe D, Grosjean JP, Schmid JP, Fitz DR, Belser Jr WL, Knudson GB, Hynds PM (1978). Atmospheric reactions of polycyclic aromatic hydrocarbons; facile formation of mutagenic nitro derivatives. Science 202:515.

Rosenkranz HS (1982). Direct-acting mutagens in diesel exhausts: magnitude of the problem. Mutat Res 101:1.

Rosenkranz HS, McCoy EC, Sanders DR, Butler M, Kiriazides DK, Mermelstein R (1980). Nitropyrenes: isolation, identification and reduction of mutagenic impurities in carbon black and toners. Science 209:1039.

Wang CY, Lee MS, King CM, Warner PO (1980). Evidence for nitroaromatics as direct-acting mutagens of airborne particulates. Chemosphere 9:83.

Germfree Research: Microflora Control and Its Application to the Biomedical Sciences, pages 139–143

Hematopoiesis in Conventional Mice After Wound Trauma[1]

G.D. Ledney, L.K. Steel, H.M. Gelston, Jr., W.E. Jackson, III, and E.D. Exum, Departments of Experimental Hematology and Biochemistry, Armed Forces Radiobiology Research Institute, Bethesda, Maryland 20814-5145

Macrophages, granulocytes, and platelets serve in wound debridement, bacterial neutralization, and homeostasis. Replacement of such cells may involve perturbations in the clonogenic cell populations responsible for replenishing specific adult cell populations. For example, the human blood granulocyte progenitor cell compartment is increased following abdominal hysterectomy (Philip et al, 1980). In mice, skin wound trauma produced changes in the proliferative cell compartments of the hematopoietic tissues 24 hrs after injury (Ledney et al, 1980). In this study we report (1) some of our findings on the hematopoietic clonogenic cell changes in conventional mice after trauma and (2) circulating substances which may mediate these changes.

Materials and Methods

Animals: Groups of 10-20 week old female B6CBF1 Cum BR mice were wounded under Metafane anaesthesia between the hours of 10AM and 2PM. A non-lethal 2.5 cm^2 circular wound (4% skin surface) was cut in the anterior-dorsal skin fold

1. Supported by the Armed Forces Radiobiology Research Institute, Defense Nuclear Agency (work unit 4421-00129). Views presented are those of the authors, and no endorsement by the Defense Nuclear Agency has been given nor should be inferred. Research was conducted according to the principles enunciated in the "Guide for the Care and Use of Laboratory Animals" prepared by the Institute of Laboratory Animal Resources, National Research Council.

and underlying panniculus carnosus muscle (between the shoulder blades) with a steel punch. The punch was cleaned after the wounding of each animal by immersion in 70% ethanol. Wounds were untreated and left open to the environment. After injury, all mice were placed in sanitized cages with autoclaved hardwood chip bedding. Control mice were anaesthetized and handled in the same manner as the wounded animals except for the wounding procedure.

Clonogenic Cell Assays: Colony forming unit-spleen assays (CFU-s) were used to determine the pluripotent cell numbers in hematopoietic tissues. Experimental groups (6-8 mice) were exposed to 10 Gy ^{60}Co radiation (dose rate 0.4 Gy/min). Within 4 hrs of irradiation, each mouse was injected i.v. with either 25 x 10^4 spleen or 25 x 10^3 marrow cells derived from control (C) or wounded (W; 3,7,10 or 14 days post-wounding) donors. Additional irradiated mice received 1,2, or 5 x 10^5 isolated peripheral blood cells from C or W (day 3) donors. Spleens were removed at 8 days, fixed in Bouin's solution (2-4 hr) and surface colonies counted.

Endogenous CFU (E-CFU) studies were performed on groups of mice wounded 24 hr prior to radiation (7 or 9 Gy) exposure. Spleen colonies were counted at 5,7,10 and 12 days after irradiation.

Tissues containing granulocyte-macrophage colony forming cells (GM-CFC) and macrophage colony forming cells (M-CFC), were taken from C or W mice (days 3,7,10 or 14 post-wound). 10^6 spleen, 25 x 10^3 marrow, and 0.5, 1 or 2 x 10^5 nucleated peripheral blood cells were grown in triplicate double-layer agar cultures with 5% growth stimulator (pregnant mouse uterine extract). Colonies (>50 cells) were counted at day 10 (GM-CFC) and day 25 (M-CFC).

All clonogenic data are quantitated on the basis of 10^6 nucleated cells.

Circulating Mediators: Serum from C or W mice (days 2 & 3) was examined for colony stimulating activity (CSA) via M-CFC and GM-CFC assays. C-reactive protein (C-RP) was measured nephelometrically using human C-RP antisera and standards. Plasma prostaglandin E_2 (PGE_2) levels were determined by ^{125}I-radioimmunoassay of samples previously subjected to organic extraction and purification on C-18 columns. C-RP and PGE_2 samples were obtained from C and W mice days 1-7,9,11,14,17 and 21.

Results

Clonogenic Cells After Wound Trauma: Splenic CFU-s in W mice were increased 2-fold on day 3 after injury. Also at this time, peripheral blood of W mice contained 4-fold more CFU-s (71±4) than C mice (16±4). No significant changes were seen in femoral marrow CFU-s.

The splenic GM-CFC concentration was increased 5-fold 3 days after trauma (80±4 vs 16±2). Thereafter, a 1.5-2-fold increase in splenic GM-CFC was observed during the remainder of the 2-week wound healing period. In the femoral marrow, the GM-CFC concentration was increased only on day 3 after trauma (1870±139 vs 1328±116). Also on day 3, an 8-fold increase in the GM-CFC concentration in the peripheral blood (48±11 vs 6±2) was seen.

The splenic M-CFC compartment of W mice was reduced 30-70% from that of C mice (153±9) during the 2-week healing period. After injury, the femoral marrow concentration of M-CFC was increased on day 14 only (3224±403 vs 2136±157). The peripheral blood concentration of M-CFC was similar in both W and C mice (130-160) on day 3.

Wounding 24 hr prior to 7 Gy resulted in 3-fold more E-CFU at 5,7, and 10 days after irradiation than that formed after 7 Gy only (control E-CFU were 0.8, 1.7, and 6.4 respectively). A 10-fold increase in E-CFU at 5, 7, and 10 days was seen when wounding preceded 9 Gy by 24 hr (control E-CFU were 0.2, 0.4, and 2.7 respectively. Thus wounding had a greater effect at 9 Gy ($p<.05$) than at 7 Gy. The doubling time for E-CFU was 1.2 days and was independent of the radiation dose or the combined injury.

Circulating Mediators: CSA was detected in the serum of mice 2-3 days after trauma. About 250 GM-CFC and 20 M-CFC normal marrow cells cloned in 0.1 ml of serum from W mice. No colonies were formed in the serum from C mice while approx. 1200 and 200 GM-CFC and M-CFC respectively, cloned in enhancing medium.

Serum C-RP concentration increased 50-300% 2-4 days after wound trauma. One week after wounding the C-RP level returned to the control level of 4.0 ug/ml.

Plasma PGE_2 concentration increased immediately after trauma and was increased 10-20-fold 2-3 days after injury. By one week, PGE_2 returned to control values of about 50 pg/ml.

Discussion

In conventional mice, the pathophysiological stimulus of skin wounding and subsequent healing results in significant perturbation in clonogenic cell populations and circulating substances that may mediate hematopoiesis. Our findings suggest that tissue injury provokes cellular and humoral responses in the host's attempt to maintain homeostasis. The utilization of granulocytes to clear cellular debris and bacteria from the wound site may result in the mobilization of proliferative cells from the bone marrow and spleen. These cells produce mature elements to assist in repair. As the mature cells are consumed, the demand for more mature cells is met by proliferation and amplification of progenitor cell compartments (GM-CFC and M-CFC), the less differentiated E-CFU compartment, and the pluripotent CFU-S compartment.

One focus of our work is to identify mediators which may regulate clonogenic cells after trauma. PGE_2 is a known modulator of hematopoiesis (Kurland and Moore, 1977) and cells in resting stage (G_0) are known to undergo proliferation when exposed to higher levels of PGE_2 (Williams, 1979) and this could explain the reduced splenic M-CFC seen after trauma. Additionally, there is *in vitro* evidence that C-RP inhibits M-CFC expression (Marcelletti et al, 1982). The finding of a serum substance (CSA) capable of promoting GM-CFC growth suggests negative feedback regulation (consumption of granulocytes) evoked by trauma. But, *in vitro* demonstration of a CSA effect does not mean that it is active *in vivo*.

The hematopoietic proliferation and attending mediator release may be the result of host responses to infection or endotoxins. However, abscesses, pus formation, lymph node involvement, and histologic organ involvement with bacteria were never seen. Also, a secondary peak in C-RP, indicative of infection (Rowe et al, 1984) was never found. However, all wounds were contaminated with one to three naturally occurring skin or alimentary tract bacteria. Wound colonization with specific bacteria may enhance healing (Levenson et al, 1983) and/or inhibit colonization with more desirable organisms (Papageorgiou et al, 1976).

Bacterial endotoxin is a known modifier of hematopoietic responses. This substance could be released by bacteria into circulation at the wound site or through disrupted intestinal cell tight junction barriers after injury (Walker and Porvaznik, 1983). Germ-free animals may provide a clue to understanding the clonogenic and mediator substances released after trauma. In germ-free rats, the intensity of the inflammatory reaction is less than that of rats with a "normal" flora (Dorati et al, 1971). Future experiments measuring hematopoietic proliferation and mediator substance release in germ-free animals are planned.

References

1. Dorati RM, Frank DW Stromberg LMR, McLaughlin MM (1971). The effect of the germ-free state on wound healing. J Surg Res 11: 163.
2. Kurland J, Moore MA (1977). Modulation of hemopoiesis by prostaglandins. Exp Hemaol 5: 357.
3. Ledney GD, Stewart DA, Exum ED (1980). Proliferative responses of lymphomyeloproliferative cells of mice after wound trauma. J Trauma 20: 141.
4. Levenson SM, Kan-Gruber D, Gruber C, Molnar J, Seifter E (1983). Wound healing accelerated by Staphylococcus aureus. Arch Surg 18: 310.
5. Marcelletti JF, Johnson CS, Mortensen RF, Furmanski P (1982). Effect of C-reactive protein on granulocyte-macropahge progenitor cells. J Lab Clin Med 100: 70.
6. Papageorgiou PS, Alqusus Z (1976). Impairment of natural defenses 1. Exogenous cause, mechanisms, and opportunistic infection. Pediatric Annals 5: 440.
7. Philip MA, Standen G, Fletcher J (1980). The effects of surgical trauma on human granulopoiesis. Brit J. Hematol 44: 263.
8. Rowe IF, Worsley AM, Donnelly P, Catovsky D, Goldman JM, Galton DAG, Pepys MB (1984). Measurement of serum C-reactive protein concentration after bone marrow transplantation for leukemia. J Clin Path 37: 263.
9. Walker RI, Porvaznik M (1983). Association of bacteria and endotoxin with post-trauma events. In: Ninneman JL (ed): "Traumatic Injury, Infection and other Immunologic Sequelae," Baltimore: U Park Pres, p. 1.
10. Williams, N (1979). Preferential inhibition of murine macrophage colony formation by prostaglandin E. Blood. 53: 1089.

Germfree Research: Microflora Control and Its Application to the Biomedical Sciences, pages 145–148

THERMAL INJURY MORTALITY IN GERMFREE AND CONVENTIONAL ANIMALS

Sol R. Rosenthal, M.D., Ph.D.
University of Illinois, Chicago
Box F, Rancho Santa Fe, CA 92067
Louise I. Lindholm
University of Notre Dame, Notre Dame, IN

ABSTRACT

Studies in germfree (GF) rats and mice demonstrated that GF animals were as sensitive if not more so to thermal injury than conventional (CV) animals. Death occurred in both groups after a similar period of time. There was no evidence of infection in either group. Using GF animals in other forms of injury such as hemorrhagic shock, others report that death occurred with equal frequency in both GF and CV animals. It is therefore postulated that sepsis cannot be the basic cause of death in severely traumatized patients. It is of utmost importance to control the toxic effect of tissue breakdown products before the vicious cycle of depressed immunological function and malnutrition ensues.

A method of neutralizing the toxic effects of thermal injury by competitin is described. Competitin is an "anti-toxin" produced *in vitro* from "toxin(s)" isolated from burned human skin. Competitins to other toxin(s) have been produced and it is postulated that competitins may also be produced from the breakdown products of all forms of injury.

INTRODUCTION

Studies were designed to compare mortality of thermal injury in GF and CV mice and rats. The GF and control CV animals came from the same breeding stock. The work was done in the Lobund Laboratory at the University of Notre Dame, Notre Dame, IN. Other forms of trauma in GF and CV animals will be cited from the literature.

MATERIAL AND METHODS

1. Mice - GF and CV Swiss mice were scalded in water (70°C) in a specially constructed basket (Rosenthal, 1961). Only the back was exposed (30 percent of body). GF and CV mice were also immersed in a 70°C water bath. The LD_{50} was determined for mice of different weights and varying exposure times. Cultures of skin of burned GF and CV animals were made at regular intervals.

2. Rats - GF and CV Wistar rats (250 to 350 grams) were burned on the back by a hot plate over 20 percent of the body (250° to 350°F) for varying periods of time (five to thirty seconds), and the LD_{50} was determined. Skin cultures were taken at regular intervals. GF animals were kept in the GF environment throughout the experiment. CV animals were kept in clean cages in the laboratory exposed to daily traffic.

RESULTS

1. Mice - One group of GF and CV mice were immersed in a 70°C water bath while the other group of mice were each placed in a basket and only the backs of these mice exposed to 70°C water bath. This showed that at every weight level it took less time to kill (LD_{50} in 24 hours) GF than CV animals. At six days, 42 of 60 animals in the GF group died (70 percent), whereas 67 of the 117 died in the CV group (57 percent). Cultures of the local burn lesions were sterile for both GF and CV animals. This was possible because the lesions were on the back and the housing of these animals was kept clean. The skin lesions healed faster in the CV animals (21 to 28 days) than in the GF mice (28 to 42 days).

2. Rats - These animals died over a period of one to 26 days in the CV rats and one to 78 days in the GF rats. The late deaths in both groups were not associated with infection since the open wounds were sterile on repeated culturing. The skin lesions healed faster in the CV animals in a time similar to the mice.

DISCUSSION

Studies dealing with the deleterious effect of tissue breakdown, whether due to trauma, burning, or surgery, are in agreement on the catabolic parameters that are associated with injured tissue (Renk, 1982; Miller, in press; Howard, 1969). Notwithstanding the fact that endotoxin is highly toxic for traumatized animals, it is not necessarily the cause of death of traumatized animals as described by Fine et al. (1959). Zweifach (1958) has shown that irreversible hemorrhagic shock can be produced in GF animals to the same extent as in CV animals. Cultures taken of the internal organs of the shocked GF animals were sterile. The CV animals showed bacterial growth in the liver and spleen.

Unquestionably, septicemia must play an important role in late deaths following burns and trauma, but it is only one of many factors. We suggest that the toxic effects of tissue breakdown play a dominant role in traumatized patients, and that it is vital to neutralize these toxic effects before the vicious cycle of depressed immunologic function and malnutrition ensues.

Rosenthal et al. described a method of controlling toxicity by an "antitoxin" produced _in vitro_ from burned normal human skin toxin(s) which he calls competitin (Rosenthal, 1982; Rosenthal, 1968). The advantages of competitin are that human tissue is used, the "antitoxin" is produced _in vitro_, and the resulting product acts immediately to neutralize the toxin(s) as well as acts as an antigen producing classical antibodies at a later date which can neutralize toxin(s). Other studies (Rosenthal, 1968) have shown that competitins can be produced from other toxin(s). On the basis of these studies, one may imply that the production of competitins to other forms of injury, including post surgery, is a possibility.

REFERENCES

Fine J, Frank ED, et al. (1959). The bacterial factor in traumatic shock. N Engl J Med 260:214.

Howard RJ (1979). Effect of burn injury, mechanical trauma and operation in immune defense. Symposium on Trauma, Surgical Clinics of North America 59:199.

Miller CL, Trunkey DD (in press). Immunobiology of sepsis.

In Flint LM, Fry DE (eds): "Surgical Infections," New York: Medical Evaluation Publishing Co.

Miller SE, Miller CL, et al. (1982). The immune consequence of trauma. Symposium of Trauma, Surgical Clinics of North America 62(1):167.

Renk CM, Long CL, et al. (1982). Comparison between in vitro lymphocyte activity and metabolic changes in trauma patients. J Trauma 22:134.

Rosenthal SR (1961). Basket technique for producing standard thermal injury in mice. J Trauma 1:560.

Rosenthal SR (1982). Burn toxin and its competitin. Burns 81:215.

Rosenthal SR, Thadani KC, et al. (1968). An inhibitor of toxic lethality: "Competitin". J Exp Med and Surg 26:124.

Zweifach BW, Gordon HA, et al. (1958). Irreversible hemorrhagic shock in germfree rats. J Exp Med 107:437.

IV MICROFLORA EFFECTS AND INTERRELATIONS

Germfree Research: Microflora Control and Its Application to the Biomedical Sciences, pages 151–154

CLEAVAGE OF LACTOSE-LYSINE, A NOVEL INDICATOR OF COLONIZATION RESISTANCE OF THE DIGESTIVE TRACT

Gjalt W. Welling and Gerda Groen

Laboratorium voor Medische Microbiologie
Rijksuniversiteit Groningen, Oostersingel 59,
9713 EZ Groningen, The Netherlands

The anaerobic bacterial flora in the gastrointestinal tract protects the host against colonization by potentially pathogenic microorganisms. This mechanism is called colonization resistance (CR) (van der Waaij, 1982). To minimize the infection risk i.e. to maintain a high level of CR in immunocompromised patients, they often receive a preventive oral treatment with antimicrobial agents which selectively eliminate the potentially pathogenic microorganisms from the gastrointestinal tract (Sleijfer et al., 1980; de Vries-Hospers et al., 1981).

To monitor the level of CR in antibiotic-treated patients an assay was developed based on the appearance of β-aspartylglycine in the feces whenever the protective bacterial flora is disturbed (Welling, Groen, 1978; Welling, 1982; Muytjens et al., 1983). During these studies, which were performed using high-voltage paper electrophoresis, it was observed that in addition to β-aspartylglycine, 2 unknown ninhydrin-positive compounds were frequently present after total decontamination (See Fig. 1). These substances were identified as ε-N-1-(1-deoxylactulosyl)-L-lysine (lactose-lysine) and ε-N-1-(1-deoxyfructosyl)-L-lysine (fructose-lysine) (Schreuder and Welling, 1983). They were identical to the compounds which were observed by Ersser et al. (1971) in the fecal contents of a germfree infant and several newborn infants.

Lactose-lysine and fructose-lysine are products of the Maillard reaction (Angrick, Rewicki, 1980). Whenever protein is heated in the presence of reducing sugars, for instance during food processing, the free ε-NH_2 group of lysines in

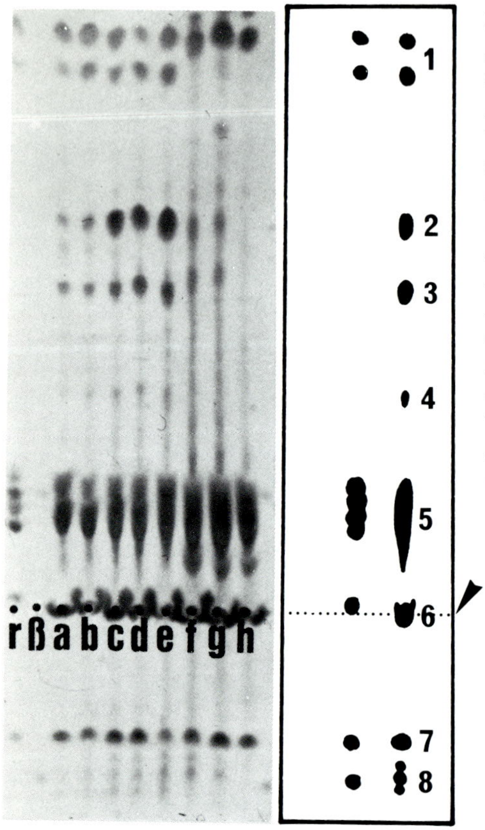

Fig. 1. High-voltage paper electrophoresis of fecal supernatants of a patient during total decontamination (a-e) and after recontamination (f-h). r, reference amino acids, β, β-aspartylglycine, 1, basic amino acids, 2, fructose-lysine, 3, lactose-lysine, 4, β-aspartyllysine, 5, neutral amino acids, 6, aspartic acid, 7, glutamic acid, 8, β-aspartylpeptides, the application line is indicated by an arrow.

a protein may react. Particularly processed milk contains a relatively large amount of lactose-lysine. Lactose-lysine is degraded by bacterial enzymes (Schreuder and Welling, 1983) and therefore it is not surprising that this compound may accumulate whenever the bacterial flora in the gastrointestinal tract is substantially reduced or not present at all i.e. in germfree infants. Since the concentration of lactose-lysine largely depends on the intake of processed heated food, it was argued that not accumulation of lactose-lysine itself but rather the presence of lactose-lysine degrading activity might be an adequate marker of CR in addition to β-aspartylglycine. In this study the lactose-lysine-degrading enzymatic activity in fecal samples of decontaminated patients and healthy human controls is determined.

METHODS

Feces were suspended in 3 volumes of water, homogenized and centrifuged for 15 min in an Eppendorf centrifuge. The supernatant can then be analyzed for the presence of lactose-lysine and β-aspartylglycine. The supernatant (800 µl) was dialyzed against water to remove substances (salts, amino acids acids, any lactose-lysine, fructose-lysine and β-aspartyl-glycine) which might interfere with the subsequent enzymatic assay. The retentate was freeze-dried and the residue was dissolved in 100 µl 10 mM sodium phosphate pH 7.0. This is the enzyme preparation. Synthetic lactose-lysine was prepared by refluxing lactose and lysine in methanol (Schreuder, Welling, 1983). It was purified by preparative high-voltage paper electrophoresis. Lactose-lysine-degrading enzymatic activity was determined by overnight incubation of 25 µl lactose-lysine solution and 20 µl enzyme solution in 10 mM sodium phosphate, pH 7.0 at 37° C. The incubation mixture was subjected to high-voltage paper electrophoresis and the extent of degradation was determined semiquantitatively from the intensity of the spots after staining with ninhydrin.

RESULTS AND DISCUSSION

Fecal supernatants from patients treated with antibiotics and those of healthy human controls were used as enzyme preparations. Preliminary studies showed that fecal enzyme preparations from healthy human controls degraded lactose-lysine from 40-100%, while fecal enzymes from totally decontaminated patients could not degrade lactose-lysine. Further studies showed that for an adequate digestion a larger quantity of enzyme was necessary. The results obtained after incubation of synthetic lactose-lysine with the amount of enzyme described in methods are shown in Table 1.

Table 1. Extent of lactose-lysine degradation by fecal supernatants

	number of persons	*number of samples*	*average % degradation*
healthy controls	*18*	*18*	*82*
antibiotic-treated patients	*6*	*17*	*0*

These results show that lactose-lysine degrading activity might be an additional marker of the level of colonization resistance in the gastrointestinal tract.

ACKNOWLEDGEMENTS

We thank Riëtte Boven, Jacqueline Magereij and Anet de Lange for their enthusiastic assistance.

REFERENCES

Angrick M, Rewicki D (1980). Die Maillard-Reaktion. Chemie Unserer Zeit 14:149.

De Vries-Hospers HG, Sleijfer DTh, Mulder NH, van der Waaij D, Nieweg HO, van Saene HA (1981). Bacteriological aspects of selective decontamination as a method of infection prevention in granulocytopenic patients. Antimicrob Agents Chemother 19:813.

Ersser RS, Seakins JWT, Gibbons ISE (1971). Effect of diet and bacteria on the fecal amino acid pattern in the newborn. Z Kinderheilk 110:276.

Muytjens HL, van Veldhuizen GL, Welling GW, van der Ros-van de Repe J, Boerema HBJ, van der Waaij D (1983). Selective decontamination of the digestive tract by pipemidic acid. Antimicrob Agents Chemother 24:902.

Sleijfer DTh, Mulder NH, de Vries-Hospers HG, Fidler V, Nieweg HO, van der Waaij D, van Saene HA (1980). Infection prevention in granulocytopenic patients by selective decontamination of the digestive tract. Eur J Cancer 16:859.

Van der Waaij D (1982). Colonization resistance of the digestive tract: clinical consequences and implications. J Antimicrob Chemother 10:263.

Welling GW, Groen G (1978). β-Aspartylglycine, a substance unique to caecal contents of germfree and antibiotic-treated mice. Biochem J 175:807.

Welling GW (1982). Comparison of methods for the determination of β-aspartylglycine in fecal supernatants of leukemic patients treated with antimicrobial agents. J Chromatogr 232:55.

Germfree Research: Microflora Control and Its Application to the Biomedical Sciences, pages 155–158

RATIONALE FOR USE OF β-ASPARTYLGLYCINE AS INDICATOR OF COLONIZATION RESISTANCE

G.W. Welling, G. Helmus, H.G. de Vries-Hospers R.H.J. Tonk, D. van der Waaij, E. Haralambie* and G. Linzenmeier*

Laboratorium voor Medische Microbiologie
Rijksuniversiteit, Oostersingel 59
9713 EZ Groningen, The Netherlands
*Institut für Medizinische Mikrobiologie
Universitätsklinikum, Essen, F.R.G.

Protein from the diet and from protein turnover in the gastrointestinal tract is primarily digested by host enzymes like pepsin from the stomach and trypsin and chymotrypsin from the pancreas. The resulting peptides are further degraded by other peptidases to amino acids.

The main target of these proteolytic enzymes is the peptide bond which connects the amino acids in a protein. However, two amino acids in the amino acid sequence of a protein i.e. asparagine and aspartic acid may cyclize with the concomitant exclusion of ammnoia and water, respectively. Decyclization occurs in solution and results in an amino acid which is no longer linked with its α-carboxyl group to the amino moiety of the next amino acid, but with its second acidic group, the β-carboxyl group. The latter form is the more stable form. This newly formed unusual peptide bond can only be cleaved by bacterial enzymes. Absence of bacteria in the gastrointestinal tract results in accumulation of β-aspartyl-containing dipeptides, especially β-apartylglycine in the feces (Welling, Groen 1978; Welling 1982).

β-Aspartylglycine was used to monitor the level of colonization resistance i.e. the completeness of the protective anaerobic bacterial flora in the gastrointestinal tract (Muytjens _et al_. 1983; Manson _et al_. 1981; van der Waaij _et al_ 1982; Koopman _et al_. 1982).

Which strains of bacteria belonging to the normal intestinal flora do produce the β-aspartylglycine-degrading enzymes? Is there a quantitative relationship between the degree of completeness of the protective intestinal flora and the concentration of β-aspartylglycine in the feces?

METHODS

80 μl of a 25% fecal supernatant is applied in portions of 5 μl to Whatman 3MM chromatography paper (46 x 57 cm) at 16 cm from the short edge of the paper. High-voltage paper electrophoresis at pH 3.5 and subsequent staining with ninhydrin is carried out as described (Welling, Groen 1978; Welling 1982). We use as a reference solution the supernatant of a 15% dry-weight fecal suspension from germfree mice. Alternatively, β-aspartylglycine from Serva (Heidelberg, GFR) or Sigma (St. Louis, USA) can be used. However, this consists of the cyclized anhydro α,β-form, which shows a yellowbrownish color after staining with ninhydrin and which migrates with the neutral amino acids. Upon incubation in water it converts to the stable β-form. A suitable standard solution of β-aspartylglycine could be a stock solution of 10 mg of commercial β-aspartylglycine which has been incubated in water for 4 weeks. Dilutions of this solution should then be kept refrigerated until use.

RESULTS AND DISCUSSION

Fecal samples from healthy human controls and from patients treated with non-absorbable antibiotics (Dankert *et al*., 1979) were analyzed for the total number of bacteria present (microscopic count). It was found (see Table 1) that the average number of bacteria in the controls was 6×10^{10} per gram of feces, while this number dropped to 6×10^{6} per gram during decontamination.

β-Aspartylglycine was not present in the feces of controls while high concentrations (1-2 μmol/gram) were found in the feces of patients undergoing decontamination. After recolonization the bacterial flora came back to normal levels within 2 to 3 weeks while β-aspartylgylcine disappeared from the feces.

In another experiment, mice were associated in a stepwise manner with an increasing number of different strains of anaerobic bacteria. Gastrointestinal colonization resistance was determined by oral contamination with a streptomycin-resistant strain of *Escherichia coli* (Welling et al. 1980). The concentration of β-aspartylglycine in fecal and cecal supernatants was determined by high-voltage paper electrophoresis. After association with 46 strains of bacteria, a 15-fold decrease in the concentration of β-aspartylglycine was found. After association with 55 different strains, β-aspartylglycine became almost undetectable. This suggests that at least a number of different strains produce β-aspartylglycine-degrading enzymes.

Table 1

	total number of bacteria	concentration (μmol/g β-aspartylglycine
healthy human controls (n=60)	100%	0
during decontamination (n=15)	0.01%	1-2

Fecal samples from patients treated with antibiotics were analyzed for the presence of anaerobic bacterial flora and the concentration of β-aspartylglycine. In more than 90% of the samples it was found that β-aspartylglycine was not present whenever a combination of different species of anaerobic bacteria: *Bacteroides sp.*, *Bifidobacterium sp.*, *Eubacterium sp.* and *Clostridium sp.*, was present at concentrations of more than 10^9 per gram. The presence of different species is required for the production of β-aspartylglycine-degrading enzymes. It was also found that the concentration of *Bacteroides sp.* had to be at least 10^{10} per gram. A number of exceptions was found. In 7 samples there was no growth of anaerobic bacteria and no β-aspartylglycine was present. In 3 of these samples, yeasts were found and in 2 *Staphylococcus albus* was present. In 2 fecal samples with an apparent normal flora, we found β-aspartylglycine. These exceptions have to be investigated further.

The results show that, in general, elimination of the anaerobic bacterial flora and thus a decrease in the level of colonization resistance, coincides with the presence of β-aspartylglycine in the feces.

ACKNOWLEDGEMENTS

We thank Pauline Schuur for her skillful assistance.

REFERENCES

Dankert J, deVries JA, Veninga TS (1979). Intestinal endotoxin and the occurrence of unexplained fever in leukemic patients during treatment with cyclophosphamide. In "Clinical and Experimental Gnotobiotes, Proceedings of the VIth International Symposium on Gnotobiology" (Fliedner T et al., eds). Gustav Fisher Verlag, Stuttgart, DBK.

Koopman JP, Mullink JWMA, Prins RA, Welling GW, Hektor MPC (1982). Association of germfree mice with intestinal micorflora obtained from normal mice. Lab Animals 16:59.

Manson WL, Welling GW, Heidt PJ (1981). The influence of cephradine on the anaerobic flora of the colon. Antoni v Leeuwenhoek 47:477.

Muytjens HL, van Veldhuizen GL, Welling GW, van der Ros-van de Repe J, Boerema HBJ (1983). Selective decontamination of the digestive tract by pipemidic acid. Antimicrob Agents Chemother 24:902.

van der Waaij D, Aberson J, Thijm HA, Welling GW (1982). The screening of four aminoglycosides in the selective decontamination of the digestive tract in mice. Infection 10:35.

Welling GW (1982). Comparison of methods for the determination of β-aspartylglycine in fecal supernatants of leukemic patients treated with antimicrobial agents. J Chromatogr 232:55.

Welling GW, Groen G (1978). β-Aspartylglycine, a substance unique to caecal contents of germfree and antibiotic-treated mice. Biochem J 175:807.

Welling GW, Groen G, Tuinte JHM, Koopman JP, Kennis HM (1980). Biochemical effects on germ-free mice of association with several strains of anaerobic bacteria. J Gen Microbiol 117:57.

Germfree Research: Microflora Control and Its Application to the Biomedical Sciences, pages 159–161

β-ASPARTYLGLYCINE DURING ANTIBIOTIC INDUCED DYSBACTERIOSIS OF INTESTINAL MICROFLORA IN THE RAT

Otar V. Chakhava, S.Z. Ruban, N.M. Shustrova

Laboratory of Gnotobiology, The N.F. Gamaleya Institute for Epidemiology and Microbiology of the USSR Academy of Medical Sciences, Moscow

Using G.W. Welling's (1979, 1982) method we detected β-aspartylglycine (BAG) in faecal supernatants of germfree (GF) and gnotobiotic (GN) E. coli monoassociated F-344 rats but not in conventional (CV) rats.

In the first experiment, 5 groups of rats were used (4 in each). Two groups received daily therapeutic doses of antibiotics during 10 days (Table 1). BAG was detected in samples of GF and GN rats. Two days after first antibiotic administration, BAG appeared in samples from all animals of 1st and 2nd groups. In these rats, first signs of dysbacteriosis (D) were detected, and they increased in 1st group: bifidobacteria, anaerobic lactobacilli, and Escherichia coli disappeared; the number of lactose negative colonies increased.

On the 21st day of the experiment (11th day without antibiotics) animals of the first three groups were given by stomach tube suspensions of Staphylococcus aureus B-243, 10^{10} cfu/ml. Two days later, S. aureus was found only in faeces of 1st and 2nd group rats. However, these bacteria were eliminated spontaneously from 2nd group but persisted in first group rats at earlier levels.

In the second experiment, 5 groups of rats were included (4 animals in each group). The antibiotics indicated in Table 2 were given in drinking water for 5 days.

BAG was first detected 2 days after the start of antibiotic "therapy", and its density on electrophoregrams increased up to 4-6 days after the end of treatment. BAG disappeared at the end of 2nd week after antibiotics were cancelled.

Animal group number	Category of animals	Antibiotic treatment	BAG before/after drugs	*Staph. aureus* cfu in 1g faeces after 2 days	7 days
1	CV	Tetracyclin p.o. Ampicillin i.m.	=/+++	10^4	10^4
2	CV	Gentamycin p.o. Kefsol i.m.	=/++	10^4	0
3	CV	None	=	0	0
4	GN	None	+++	Not infected	
5	GF	None	+++	Not infected	

Table 1. Colonization resistance of the digestive tract (CR) and BAG in faecal supernatants of rats treated with antibiotics, CV, GF, and GN associated with *E. coli* 083.
Legend: = absence of BAG; +, ++, +++ different density of BAG spot on electrophoregrams.

D lasted almost 2 weeks after drugs were cancelled in 3rd and 4th groups. On the 5th day of this period, all rats were given by stomach a suspension of *S. aureus* B-243 and *Pseudomonas aeruginosa* 93 (10^{10} cfu of each species in 1 ml) except 2nd group animals challenged only with *S. aureus*.

Most profound D, lower CR and dense spots of BAG on high voltage electrophoregrams were detected in erythromycin treated rats (Table 2).

Thus, in both experiments BAG appeared earliest detectable indicator of D and of decreased CR.

Animal group number	Category of rats	Antibiotics mg/kg body wt	BAG before/after drugs	Bacterial cfu in 1 g faeces after inoculation of	
				S. aureus	*P. aerugin.*
1	CV	None	=	10^1	10^1
2	GN	None	+++	10^3	Not infected
3	CV	Erythromycin 100.0	=/+++	10^3	10^3
4	CV	Gentamycin, 100.0	=/++	10^2	10^2
5	CV	Fradizin, 2,500	=/+	10^1	10^1

Table 2. CR and BAG in faecal supernatants of rats treated with antibiotics, and CV and GN animals infected with *S. aureus* and *P. aeruginosa*.
Legend: See Table 1 (except GN, contaminated with aerobic sporogenic bacillus).

REFERENCES

Welling GW (1979). Beta-aspartylglycine, an indicator of decreased colonization resistance? In van der Waaij D, Verhoef J (eds): "New Criteria for Antimicrobial Therapy", Amsterdam-Oxford: Excerpta Medica, p. 65.

Welling GW (1982). Comparison of methods for the determination of β-aspartylglycine in faecal supernatants of leukemic patients treated with antimicrobial agents. J Chemother 232:55.

Germfree Research: Microflora Control and Its Application to the Biomedical Sciences, pages 163–167

ELIMINATION OF SALMONELLA TYPHIMURIUM FROM THE DIGESTIVE TRACT BY THE FLORA OF CONVENTIONAL CHICKEN

Sylvie HUDAULT, Hilaire BEWA, Chantal BRIDONNEAU, and Pierre RAIBAUD
Laboratoire d'Ecologie Microbienne
I.N.R.A. - C.N.R.Z.
78350 - JOUY-EN-JOSAS (France)

Asymptomatic intestinal carriage of Salmonella in poultry flocks represents an important source of human environmental contaminations. It is now admitted that administering intestinal contents from healthy conventional fowl to newly hatched chicks decreases their susceptibility to infection (Snoyenbos, 1978). The purpose of the present work was to estimate the efficiency in axenic chickens of a microflora from a conventional chicken and to isolate the antagonistic bacteria exerting a barrier effect against Salmonella in gnotobiotic chickens and mice.

A)-Antagonistic role of complex flora towards Salmonella

When the cecal flora of a 100 day-old chicken was first established in the gut of 12 axenic chickens, a rifampicin resistant S. typhimurium strain (STMR) was completely eliminated in 100 % of the animals, while only 72% of 14 chickens eliminated STMR when this strain was first inoculated to the animals, On the other hand, we showed that from the age of 15 days, only 77 % of 30 conventional chick, harbored an antagonistic flora against STMR. Accordingly, it is advisable to inoculate chicks from hatching with a protective flora to prevent healthy carriage of Salmonella.

B)-Use of the axenic mouse as an experimental model

*Our results showed that the antagonistic activity towards STMR of a 15 day-old chicken was also expressed in axenic mice (Table 1). Moreover, the chicken flora from these*meroxenic mice was active again in axenic chickens. A comparison of the flora established in the meroxenic mice and meroxenic chickens showed that the size of the populations of strictly anaerobic bacteria and that of* enterobacteria

TABLE 1- ANTAGONISTIC EFFECT TOWARDS STMR EXERTED BY THE FLORA OF A 15 DAY-OLD CHICKEN, TRANSMITTED TO GERMFREE MICE AND THEN TO GERMFREE CHICKENS.

Host	First inoculum	Second inoculum	Mean and standard deviation of the log of the number of bacteria/g of ceca*				
			Anaerobic bacteria	STMR	Enterobacteria	Strepto-cocci	Lactobacil-li
12 germfree mice	Cecal flora from a 15 day-old chicken	STMR	10.6 ± 0.3	< 2.0 ± 0.0	8.5 ± 0.6	8.1 ± 0.8	< 6.0 ± 0.0
13 germfree chickens	Fecal flora from the 12 meroxenic mice	STMR	11.0 ± 0.1	< 2.0 ± 0.0	8.0 ± 0.6	7.3 ± 0.6	8.0 ± 0.6

*8 to 10 days after challenge with STMR.

or streptococci were similar in the ceca of both species.

C)-Antagonistic effect of various cultures of the flora from a 15-day old chicken.

To isolate bacteria responsible for the antagonistic effect in chickens we cultured fecal $1x10^{-7}$ and $1x10^{-9}$ dilutions from meroxenic mice harboring the flora of a 15 day-old chicken. Only bacteria present in the cultured $1x10^{-7}$ dilution exerted an antagonistic effect *in vivo*. The nature of the medium used did not affect this result. However, in $1x10^{-7}$ dilution, isolation of the antagonistic bacteria was not possible due to the presence of numerous colonies ;

TABLE 2

ANTAGONISTIC EFFECT TOWARDS STMR OF DILUTIONS OF THE CECAL FLORA FROM A 15 DAY-OLD CHICKEN, AFTER CULTURE.

	1.10^{-7} dilution cultured on medium		1.10^{-9} dilution cultured on medium	
	B'	F	B'	F
STMR*	2.2 ± 0.9	2.1 ± 0.5	7.7 ± 0.4	7.1 ± 0.3

*Mean and standard deviation of log of the number of STMR/g of feces of six mice, collected 5 times between 8 and 30 days after challenge with STMR.

**0,1 ml of 1.10^{-6} and 1.10^{-8} fecal dilution were deposited on plates of non selective media B' (Sacquet 1971) and F (Hudault 1982). Bacterial suspensions including all the colonies grown after 3 and 6 days of incubation at 37°C in an anaerobic chamber were obtained for each dilution (1.10^{-7} and 1.10^{-9}) and each sample.

Conversely in 1×10^{-9} dilution these bacteria were not present. It is possible that the culture media used did not allow the optimum growth and thus lowered the detection threshold.

D)-Antagostic effect of a mixture of isolated bacterial strains.

Two strains of E. coli and Streptococcus faecium as well as 14 strains of strictly anaerobic bacteria (Bacteroïdes, Eubacterium, Peptostreptococcus, Clostridium) were isolated from the dominant flora of a 15 day-old chick. A group of 10 axenic mice were challenged with all these bacteria. Before inoculating STMR, the various bacterial genera were present in the dominant flora (between $6\ 10^{8}$ and 3×10^{10} bacteria/g of feces). The association of these strains dit not eliminate STMR from the cecum of all the animals, but allowed to reduce the size of the STMR population in their ileum.

E)- Relationship between volatile fatty acid (VFA) production in vivo and the number of Salmonella.

We compared the sizes of the STMR population and the amounts of VFA in the ceca of two groups of conventional chickens fed 2 different diets. The VFA amounts were about three times smaller in chickens C_2 than in chickens C_1 (table 4). However, whatever the diet, Salmonella was no more detected in the ceca of the animals. In two groups of meroxenic mice M_1 and M_2, harboring or not an antagonistic flora towards STMR, the VFA concentration was similar to that found in the ceca of chickens C_2.

The antagonistic effect against STMR observed in chickens harboring a complex microflora is therefore speci-

TABLE 3-ANTAGONISTIC EFFECT EXERTED TOWARDS STMR BY THE ASSOCIATION OF 16 STRAINS ISOLATED FROM THE DOMINANT FLORA OF A 15 DAY-OLD CHICKEN.

		Gnotoxenic mice	Monoxenic mice
Number of Animals		10	6
Log of the number of STMR/g of*	cecum	8.0 ± 0.3	8.9 ± 0.6
	Ileum	4.9 ± 0.8	9.1 ± 0.1

*Mean and standard deviation of the log of the number of cecum or ileum, 31 days after inoculation of this strain.

TABLE 4-RELATIONSHIP BETWEEN THE VFA PRODUCIION IN SITU AND THE NUMBER OF SALMONELLA IN THE CECA OF MEROXENIC MICE AND CONVENTIONAL CHICKENS

Groups of	Log of the number of STMR/g of ceca	pH	Total VFA in mmol/kg	Respective amounts of each acid (mmol/kg)		
				C_2	C_3	C_4
C1	< 2.0	5.8 ± 0.2	53.3 ± 6	36.2 ± 2.3	3 ± 1	13.4 ± 3.6
C2	< 2.0	6.6 ± 0.0	17.0 ± 2.3	11.1 ± 1.9	2.3 ± 0.7	2.8 ± 0.6
M1	7.4	7 ± 0.0	16.3 ± 2.5	9.7 ± 2	5.7 ± 0.9	0.6 ± 0.3
M2	< 2.0	6.9 ± 0.0	18.5 ± 0.7	10.4 ± 0.5	7 ± 1.5	0.8 ± 0.0

C1 = 3 conventional chickens fed commercial diet : R 115

C2 = 3 conventional chickens fed a semi-synthetic diet.

M1 = 12 meroxenic mice harboring the culture of the 1.10^{-9} dilution cultured on medium B' and F.

M2 = 9 meroxenic mice harboring the culture of the 1.10^{-7} dilution cultured on medium B'.

fically bound to some bacteria, the growth factors of which have not yet been established. New culture media should be developed for the isolation of these antagonistic bacteria. The role of the VFA was often used to explain the antagonism exerted in vivo by the chicken flora against Salmonella (Barnes, 1979). According to our results, the mechanism of interaction between Salmonella and the bacteria of chicken microflora does not depend on the toxicity of the VFA produced in vivo.

** Meroxenic animals : germfree animals associated with a limited flora derived from conventional animals.*

Barnes E.M., Impey C.S., and Stevens, BJH, 1979. Factors affecting the incidence and anti-Salmonella activity of the anaerobic caecal flora of the young chick. J. Hyg. Camb. 82. 263-283.

Hudault S., Raibaud P., Ducluzeau R. and Bridonneau C., 1982. Effet antagoniste à l'égard de Clostridium perfringens exercé par des souches de Clostridium isolées de la microflore de souris holoxéniques dans le tube digestif de souris gnotoxéniques. Ann. Microbiol. (Inst.

Pasteur) 133A, 443-459.

Sacquet E., Raibaud P., Garnier J., 1971. Etude comparée de la microflore de l'estomac, de l'intestin grêle et du caecum de rat "holoxénique" (conventionnel) et de ses modifications à la suite de diverses interventions chirurgicales : anse aveugle jéjunale, déviations biliaires; Ann. Inst. Pasteur, 120, 501-524.

Snoyenbos G.H., Weinack O.M., and C.F. Smyser, 1978. Protecting chicks and poults from Salmonellae by oral administration of "normal" gut microflora Av. dis. 22. 273.

Germfree Research: Microflora Control and Its Application to the Biomedical Sciences, pages 169–170

ON THE MECHANISM OF CEREOBIOGEN READJUSTMENT TO DYSBIOSIS

Kang Bai M.D.
Chief, Microecologic Research
Laboratory for Normal Microbial Flora
Dalian Medical College
Dalian, People's Republic of China

We have isolated a nonvirulent strain of *Bacillus cereus* (DM423) from the soil and produced a biological preparation named "Cereobiogen", which proved to have a therapeutic effect on diarrheal diseases in infants and lambs (Wei, Kang 1981). This is the method of microecological therapy in the country of China (Wei, Kang 1981).

The preliminary results were reported in the Proceedings of the VII International Symposium on Gnotobiology, Tokyo, Japan (1981).

The present bacteriological study was made in domestic animals by quantitative anaerobic and aerobic differential analysis on contents of the upper intestinal tract. The samples for examination were collected from the intestinal lumens of lambs and piglets which had been administered "Cereobiogen" and later challenged with a pathogenic *Esherichia coli* 077 strain.

Our findings showed that the mechanisms of protection by "Cereobiogen" were as follows:

1. Desoxygen or Reduction: The growth of anaerobic bacteria was promoted by *B. cereus* in the intestinal lumens of lambs and piglets. The mechanism shown was that oxygen present in the lumen was exhausted during the growing phase of *B. cereus*.

In this experiment we found that when *B. cereus* is propagating, a state of low oxidation-reduction potential tension depresses the relative pH and En values of the surrounding fluid, creating an ideal environment for Bifidobacterium to multiply. The antagonistic action of Bifidobacterium in the intestinal lumens of lambs and piglets exerted a combating action on the invading bacteria, e.g., pathogenic E. coli and some species of Salmonella.

2. Nutrient Competition: The growth of B. cereus was found to be more rapid than other bacteria in the intestinal lumen. As a result, the nutrients in the lumen were utilized by B. cereus, thus inhibiting the growth of other bacteria, including the pathogens and potential pathogens, in the medium of the lumen.

Wei H, Kang B (1981). Bacillus cereus "Cereobiogen" as a growth promoter of Bifidobacterium in infantile diarrhea therapy. Sasaki S (ed.). Recent Advances in Germfree Research, p 475, Tokai University Press, Tokyo.

Wei H, Kang B (1981). Microflora disturbances and attempted adjustments. Chinese Medical Journal 96:79.

Germfree Research: Microflora Control and Its Application to the Biomedical Sciences, pages 171–174

"IN VIVO" AND "IN VITRO" ANTAGONISM EXERTED BY TWO ANAEROBIC STRAINS AGAINST STRAINS OF CLOSTRIDIUM PERFRINGENS

NEVZAT YURDUSEV, Monique LADIRE, Robert DUCLUZEAU and Pierre RAIBAUD
Laboratoire d'Ecologie Microbienne
I.N.R.A., C.N.R.Z.
78350 - JOUY-EN-JOSAS (France)

In a previous paper (Nicolas et al., 1981) we have studied a barrier effect produced against Clostridium perfringens in the digestive tract of gnotobiotic mice and piglet by stricly anaerobic bacteria isolated from the feces of a conventional piglet. In the present work we studied the main characteristics of a simplified barrier effect against C. perfringens in the digestive tract of mice and we attempted to reproduce the same bacterial barrier "in vitro", in order to understand its mechanisms.

"In vivo" assays :

Three groups of axenic mice were fed an irradiated diet. One group was inoculated with a strain B_2 of Bacteroides thetaiotoomicron, one group with a strain Fu of Fusobacterium necrophorum, one group simultaneously with B_2 and Fu. A target strain of Clostridium perfringens type A (Cp A), inoculated in the three groups of mice one week later, did not become established in gnotobiotic mice harboring both B_2 and Fu. But CpA was able to become established in mice monoassociated with either of strains B_2 or Fu. The antagonistic affect of the association of B_2 and Fu strains was also effective when CpA strain was previously established in axenic mice.

Another target strain of C. perfringens type C (CpC) was only weakly repressed in mice diassociated with B_2 and Fu.

"In vitro" assays :

Continuous flow culture were made in medium and in conditions described by Freter et al.(1983). The entire apparatus was maintained in an anaerobic chamber.

A bacteriostatic effect was observed neither against CpA, nor against CpC in a continuous flow culture simultaneously inoculated with both B_2 and Fu.

Feces collected from both axenic mice and mice diassociated with B_2 and Fu were diluted 1 : 4 in water and filtrated. Strains CpA and CpC were able to grow in both of the filtrates. No diffusible inhibitory substance against CpA or CpC could be evidenced from feces of mice diassociated with B_2 and Fu when tested in solid agar medium.

Table 1 shows that CpA grew in filtrate of tenfold diluted suspension of feces obtained from diassociated mice but was strongly repressed in the same non-filtrated fecal suspension (FS-(B_2 + Fu)). When the fecal suspension was incubated 24 h. at 37° before filtration, the filtrate is always able to support growth of CpA. Conversely, the growth of CpC was the same in the fecal suspension whether it was filtrated or not. Feces of mice monoassociated with either B_2 or Fu and diluted in the filtrate of the fecal suspension of the diassociated mice had no inhibitory effect. These results suggest that the in vitro bacteriostatic effect exerted by the suspension FS-(B_2+Fu) simulate the in vivo antagonistic effect exerted by the association of B_2 and Fu against CpA and CpC. After heating at 70°C for 10 pmn the inhibitory effect of FS- (B_2 + Fu) against CpA disappeared (tab. 2). After disintegration by glass beads of FS(B_2+Fu), the number of viable bacteria in the suspension decreased very strongly, but the inhibitory effect persisted against CpA, and even appeared against CpC. An inhibitory effect appeared also in disintegrated fecal suspensions from mice monoassociated with B_2 or Fu.

CONCLUSIONS :

1)-In vivo antagonistic effect against CpA exerted by two strains (B_2+Fu) implies a synergic action of strains B_2 and Fu.

2)-The antagonistic effect is not related to the production of a diffusible inhibitory substance, nor to nutrient depletion of the intestinal content of diassociated mice.

3)-The antagonistic effect exerted in vitro by fecal suspensions of diassociated mice in static culture is similar to that observed in vivo ; this effect is not observed in continuous flow culture.

4)-Insoluble substances responsible of antagonistic activity have been detected in both B_2 and Fu strains after their mechanical disintegration. These substances might be *transferred from B_2 and Fu strains to target CpA strain by an unknown mechanism in the intestinal tract of diassociated mice.*

ANTAGONISTIC ACTIVITY AGAINST CPA AND CPC EXERTED BY TENFOLD DILUTED FECAL SUSPENSIONS (FS) FROM MONOASSOCIATED (B_2 OR FU) AND DIASSOCIATED (B_2 + FU) MICE.

	FS - B_2	FS - Fu	FS-(B_2+Fu)	Filtrates of FS-(B_2-Fu) Non incubated (2 F)	Incubated (2F-I)
B_2*	9.80 ± 0.11	—	9.91 ± 0.09	—	—
Fu*	—	7.85 ± 0.14	8.18 ± 0.09	—	—
ΔLOG** CpA	3.50 ± 0.09	2.68 ± 0.07	1.40 ± 0.07	4.70 ± 0.19	3.67 ± 0.06
ΔLOG** CpC	3.83 ± 0.09	3.94 ± 0.13	4.07 ± 0.80	5.21 ± 0.10	4.35 ± 0.15

*Log_{10} N at day 0. At day 0 : log_{10} CpA = 3.38. Log_{10} CpC = 3.01

**ΔLOG = log_{10} N at day 1 - log_{10} N at day 0.

ANTAGONISTIC ACTIVITY AGAINST CPA AND CPC EXERTED BY TENFOLD DILUTED FECAL SUSPENSIONS DISINTEGRATED (FS_D) BY GLASS BEADS OR HEATED (FS_H) AT 70°C FOR 10 MIN.

	FS_D-B_2	FS_D-Fu	FS_D-(B_2+Fu)	Supernatant FS_D-(B_2+Fu)	FS_D-Axenic	FS_H-(B_2+Fu)
B_2*	8.75 ±0.11	—	8.72 ±0.08	4.20 ±0.18	—	—
Fu*	—	6.65±0.37	6.30±0.13	< 2.00	—	—
ΔLOG** CpA	0.37± 0.98	0.38±0.36	-1.07± 0.49	4.80 ±0.05	4.32±0.19	4.90±0.15
ΔLOG** CpC	2.44± 0.77	0.04± 0.88	1.58 ± 0.59	5.04 ±0.15	4.48±0.18	4.93±0.20

*Log_{10} N at day 0. At day 0 : log_{10} CpA = 3.49. Log_{10} CpC = 3.34

** LOG = log_{10} N at day 1 - Log_{10} N at day 0.

Freter R., Abrams G.D., Aranki A. (1973). Pattern of interation in gnotobiotic mice among bacteria of a synthetic "normal" intestinal flora. p. 429-433. In J.B. Henegan (ed.) Germfree research. Biological effect of gnotobiotic environments. Academic Press, Inc. New York.

Freter R., Stauffer E., Cleven D., Holdeman L.U., Moorewel (1983). Continuous flow cultures as in vitro models of the ecology of large intestinal flora. Infect. Immun. 39 : 666

Nicolas J.L., Raibaud P., Ducluzeau R., Corpet D. (1981). Antagonism against Clostridium perfringens in the digestive tract of gnotoxenic mice and piglets exerted by a small number of anaerobic bacteria isolated from the fecal flora of conventional piglet. p. 201-205. in S. Sasaki *et al. (ed.). Recent* Advances *in Germfree* Research. *Tokai University Press.*

Germfree Research: Microflora Control and Its Application to the Biomedical Sciences, pages 175–177

EXPERIMENTAL INTESTINAL MICROFLORA IN MICE WITH A CHANGEABLE POPULATION SIZE OF E. COLI

Charles Gosse, Elisabeth Faure, Patrice Ardouin,
Antoine Andremont and Cyrille Tancrède
Service d'Expérimentation Animale - Institut
Gustave-Roussy 94805 Villejuif Cedex - France

INTRODUCTION

Since 7 years S.P.F. Swiss-Nude mice (From Carl Hansen N.I.H. Bethesda) were bred in large-sized plastic isolators (Raynal et al., 1978 ; Gosse et al., 1981 ; Gosse et al., 1984). Their faecal bacterial flora (per gram of faeces) was the following :

Total anaerobic count : 5×10^9 to 1×10^{10}
Streptococcus faecalis : 4×10^5 to 5×10^6
Staphylococcus epidermidis : 2.5×10^5 to 1.6×10^6
Enterobacteriaceae : None ($<10^2$)

Over a 7 years period neither qualitative nor quantitative alterations of this defined S.P.F. nude mice flora were observed (Gosse et al., 1984). It must be specially emphasized that Enterobacteriaceae were never detected in the gut of S.P.F. nude mice reared in isolators (absolute barrier) and very unusually detected in ones maintained in other systems of protection as cages with filter cap or laminar air flow walls (partial barrier). Therefore it could be interesting to test if the bacterial flora of S.P.F. nude mice exerted a barrier effect against Enterobacteriaceae.

MATERIALS AND METHODS

Group A : 10 germfree C3H mice were fed intragastrically with the Enterobacteriaceae-free S.P.F. nude mice flora and 15 days later with a strain of E. coli (10^8/mouse).

Group B : 10 germfree C3H mice were fed intragastrically first with E. coli (10^8/mouse) and 15 days later with

Enterobacteriaceae-free nude mice flora.

- 6 weeks after establishment of E. coli and nude mice flora, mice of groups A and B received Bacitracin (1mg/ml) in sterile drinking water acidified at pH = 3.

- E. coli were counted in faeces twice a week on Drigalski agar medium. Bacterial counts/g of faeces were transformed into common logarithms.

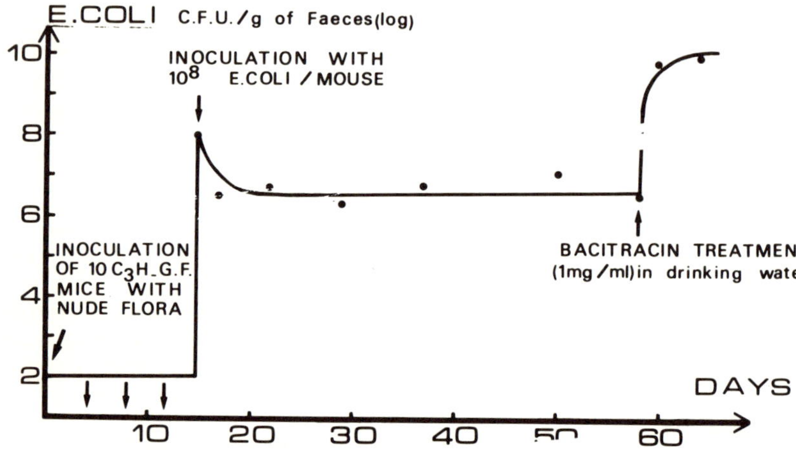

Fig. 1. Evolution of counts of E. coli in mice (Group A).

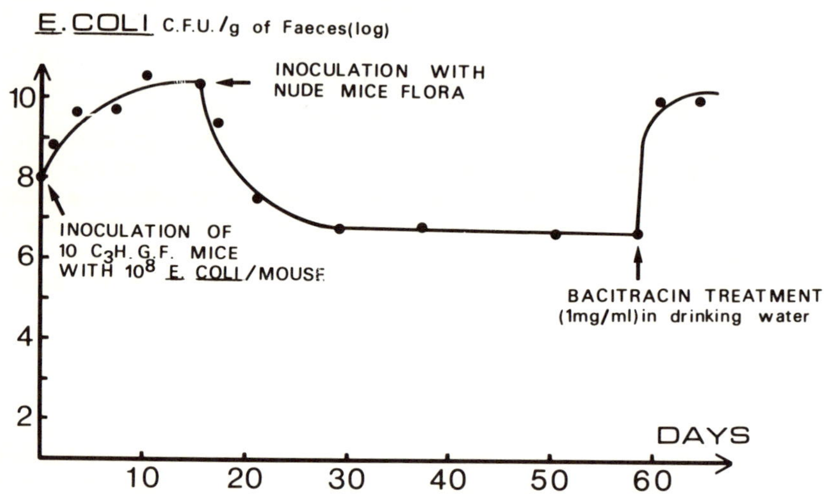

Fig. 2. Evolution of counts of E. coli in mice (Group B).

RESULTS

The evolution of counts of E. coli in groups A and B are shown on figures 1 and 2 respectively.

DISCUSSION

In group A and B the population size of E. coli was respectively 6.46 ± 0.42 and 6.69 ± 0.30 before bacitracin treatment and respectively 9.88 ± 0.21 and 9.93 ± 0.23 during bacitracin treatment i.e. the same level as in monoassociated mice with E. coli (mice of group B before inoculation of nude microflora). Therefore the S.P.F. nude mice microflora exerted : i) a permissive and preventive barrier effect against E. coli (group A) ii) a permissive and curative barrier against E. coli (group B). These partial barrier effects were destroyed by bacitracin treatment because the population size of E. coli became again the same as in monoassociated mice. Thus this model allows to study E. coli-host interactions with 3 different population sizes : i) 0 ($<10^2$) ii) between 10^6 and 10^7, iii) between 10^9 and 10^{10}.

CONCLUSION

The population size of a bacterial strain in the intestinal tract is important for bacteria-host interactions. The results of this study allowed to obtain an animal experimental microflora model with a changeable level, at will, of E. coli.

Gosse C, Dechambre RP, Ardouin P. (1981). Comparative efficiency and cost of different systems of protection against microbiological contamination of mice. In Recent Advances in Germ-free Research, Proceedings of the VIIth International Symposium on Gnotobiology, 1981 Tokyo Japan 75-78, S. Sasaki (Ed) Tokai University press.

Gosse C, Faure E, Bretou G, Ardouin P (1984). Utilisation de souris nude produites en isolateurs de grandes dimensions pour la préparation d'hybridomes et d'anticorps monoclonaux. Biosciences 3 : 3-8.

Raynal B, Chaigneau H, Eveno J (1978). Utilisation d'isolateurs, de cages à couvercle filtrant et de systèmes à flux laminaire pour la production et le maintien de souris nude. Sci. Tech. Anim. Lab. 3: 119-126.

Germfree Research: Microflora Control and Its Application to the Biomedical Sciences, pages 179–181

INFLUENCE OF ROTAVIRAL DIARRHEA ON INTESTINAL MICROFLORA ESTABLISHMENT IN CONVENTIONAL AND AXENIC MICE

CORTHIER G., MOREAU C., MULLER M.C., DUBOS F.

Laboratoire d'Ecologie Microbienne
I.N.R.A. - C.N.R.Z.
78350 - JOUY-EN-JOSAS (France)

The establishment of intestinal microflora is sequential in conventional mice. Facultative anaerobic bacteria (Streptococcus, Lactobacillus, Escherichia) were the first to colonize the gut. Then towards the age of 15 days the strictly anaerobic strains became established.

The purpose of this work was to study the consequences of an early diarrhea on the establishment of intestinal microflora. Two studies were performed on :

1) - *Conventional mice : Establishment of normal microflora bacteria*

2) - *Gnotoxenic mice : The mother was monoassociated with different Clostridium (pathogenic for man). Bacterial implantation was studied in young mice. Barrier effect of normal microflora does not permit the use of*

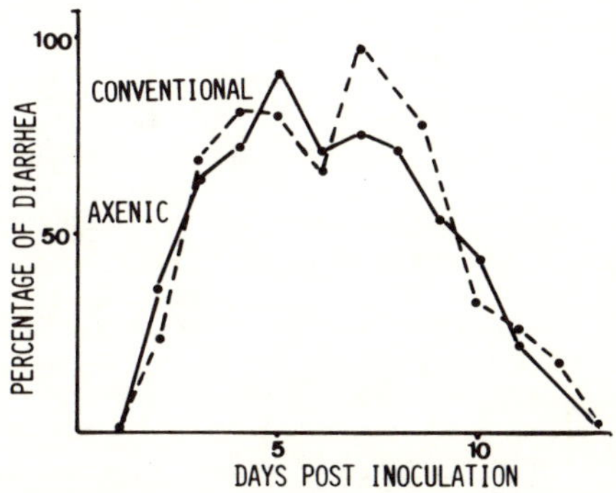

Figure 1 : *Influence of intestinal microflora* on *rotavirus diarrhea kinetics*

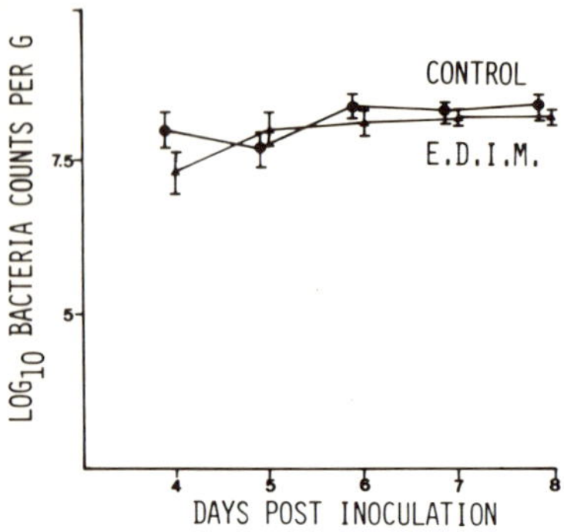

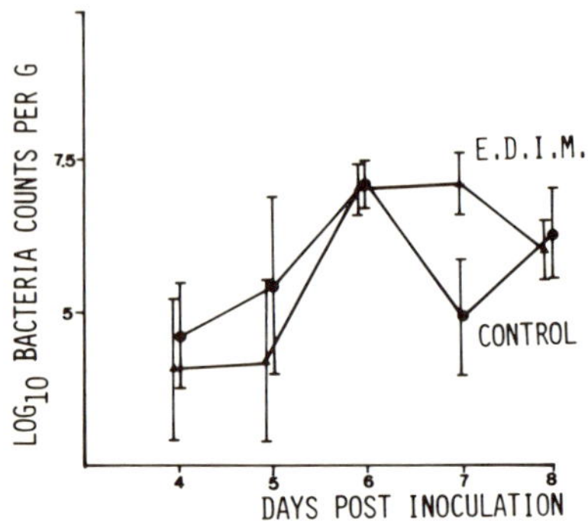

Figure 2 : Lactobacillus *Figure 3 : E. coli*
Rotaviral diarrhea in holoxenic mice ; influence on *bacterial implantation.*

conventional mice for such an experiment.

Mouse rotavirus (EDIM = Epizootic Diarrhea of Infant Mice) was used as a viral diarrhea inducer.

Rotavirus diarrhea induced in 2 day-old mice could be detected at the age of 4 days and did not persist after about 15 days. No significant differences could be detected between conventional and axenic mice on diarrhea kinetics. (Fig. 1).

In conventional mice, the establishment of E. coli and Lactobacillus sp. was studied during the acute phase of diarrhea, i.e. between 4 and 8 days. No difference was found between control and diarrheic mice, (fig. 2 and 3).

Using monoxenic models, we studied the kinetics of C. perfringens and C. difficile establishment in the feces of 8 to 25 day-old mice. For C. perfringens, a slight difference was observed between diarrheic and control mice before two weeks of age. C. difficile became established from day 11 in diarrheic mice and only towards day 16 in the controls. (Fig. 4 and 5).

Intestinal microflora did not seem to influence duration and intensity of diarrhea due to rotavirus.

In conventional mice kinetics of E. coli and Lactobacillus establishment was not modified during a period of diarrhea due to rotavirus.

In gnotobiotic mice, establishment and development of C. difficile or C. perfringens were significantly promoted by rotavirus diarrhea.

Figure 4 :

C. perfringens

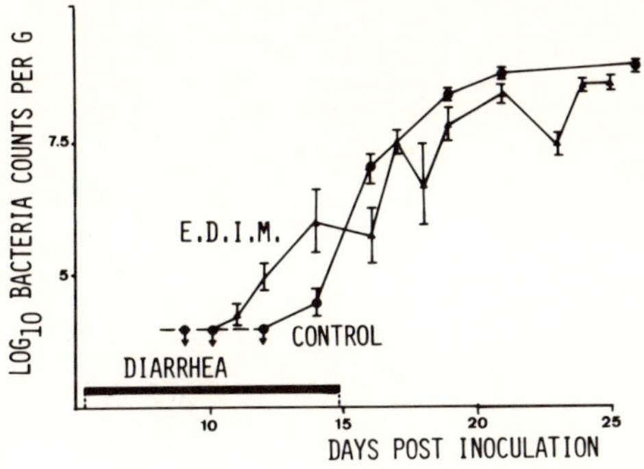

Figure 5 :

C. difficile

Rotaviral diarrhea in gnotobiotic mice: Influence on bacterial implantation.

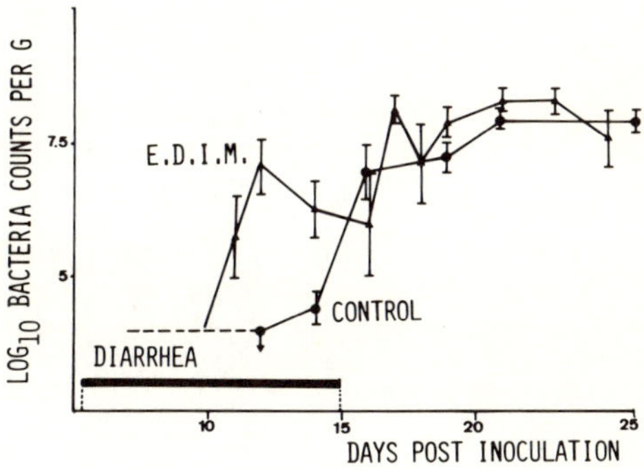

These observations could be of clinical importance. In healthy conditions (from a bacterial point of view), rotavirus induced diarrhea did not disturb the early establishment of microflora in the young. However, when some bacterial disorder occurred (antibiotic treatment which could destroy dominant intestinal flora or the presence of a small amont of pathogenic bacterial strains), rotavirus induced diarrhea could enhance the bacterial pathology.

Reference

Davis C.P., Mc Allister J.S., Savage D.C. (1973). Microbial colonization of the intestinal epithelium in suckling mice. Infect. Immun. 7 : 666.

Germfree Research: Microflora Control and Its Application to the Biomedical Sciences, pages 183–185

INFLUENCE OF ROTAVIRAL DIARRHEA ON SOME BACTERIAL FUNCTIONS IN THE INTESTINE OF CONVENTIONAL MICE

MOREAU C., CORTHIER G., MULLER M.C.

Laboratoire d'Ecologie Microbienne
I.N.R.A. - C.N.R.Z.
78350 - JOUY-en-JOSAS (France)

In conventional mice, a number of bacterial strains, namely those which are extremely oxygen sensitive (EOS), (Gordon, 1970), became established in the intestinal tract at weaning (21 days) and then represented the dominant flora of the adult. They had different functions, i.e. a permissive barrier effect towards Escherichia coli (Syed, 1970) and a stimulatory role on IgA intestinal immune system (Moreau, 1978).

The purpose of this work was to study the incidence of diarrhea induced by EDIM (epizootic diarrhea of infant mice) rotavirus on further EOS establishment. All EOS bacteria could not be enumerated, but their functions could be measured.

1) - Permissive barrier effect towards E. coli:

Fecal number of E. coli decreased from 10^8 bacteria/g of feces before weaning to 10^6 bacteria/g after weaning as a consequence of the establishment of bacterial strains responsible for this barrier effect. No difference on barrier effect was observed between rotavirus infected and noninfected mice between 8 to 48 days after birth (Fig. 1).

2) - Development of IgA intestinal immune system:

IgA plasmocyte number in intestinal mucosa increased to a maximum level at the age of 7 weeks. Previous studies have shown the determining role of some unknown bacterial strains, present at weaning, in this immune maturation process. If diarrhea induced by EDIM rotavirus disturbed the bacterial establishment of such strains at weaning, an incomplete development of the intestinal IgA immune system should be expressed at the age of 7 weeks. As a matter of fact at that age the number of IgA plasmocytes per villus was similar in control

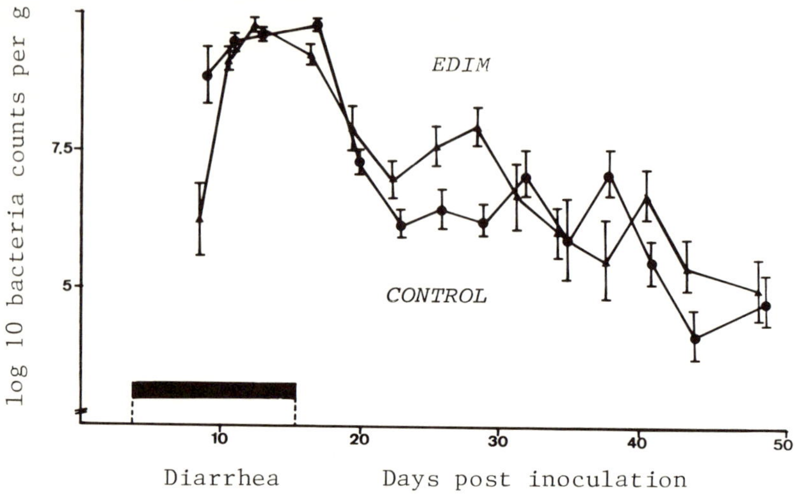

Figure 1 - Rotavirus in conventional mice: Influence of intestinal microflora involved in barrier effect against *E. coli*.-

and in diarrheic mice (Table 1).

Rotavirus-induced diarrhea between the age of 4 and 15 days did not seem to have further consequence on the establishment at weaning of the bacterial strains involved in the *E. coli* repression and of strains responsible for

IGA PLASMOCYTE NUMBER PER VILLUS	
CONTROL MICE	EDIM INFECTED MICE
40.5 + 1.07	39.5 + 3.7
N.S	

TABLE 1 : INFLUENCE OF ROTAVIRUS DIARRHEA IN CONVENTIONAL MICE: DEVELOPMENT OF IGA INTESTINAL IMMUNE SYSTEM IN 7 WEEKS-OLD MICE

the full maturation of the intestinal IgA immune system. Accordingly, it was not possible to demonstrate that rotavirus-induced diarrhea disturbed the establishment of the intestinal microflora at weaning.

REFERENCES

Gordon JH, Dubos R (1970). The anaerobic bacteria of the mouse cecum. J Exp Med 132:251.

Moreau MC, Ducluzeau R, Guy-Grand D, Muller MC (1978). Increase in the population of duodenal immunoglobulin A plasmocytes in axenic mice associated with different living or dead bacterial strains of intestinal origin. Infect Immun 21:532.

Syed SA, Abrams GD, Freter R (1970). Efficiency of various intestinal bacteria in assuming normal functions of enteric flora after association with germ free mice. Infect Immun 2:376.

Germfree Research: Microflora Control and Its Application to the Biomedical Sciences, pages 187–190

IN VIVO ESCHERICHIA COLI K88 ADHESION IN GNOTOBIOTIC PIGLETS : COMPARISON BETWEEN LARGE WHITE AND CHINESE PIGLETS

Duval-Iflah Yvonne and J.P. Chappuis

Laboratoire d'Ecologie Microbienne
I.N.R.A. - C.N.R.Z.
78350 - Jouy-en-Josas (France)

A large number of enterotoxigenic *Escherichia coli* (ETEC) strains implicated in porcine neonatal diarrhea were shown to bear a plasmid-specified proteinaceous surface K88 antigen (Söderlind and Möllby, 1979, Stirm et al., 1967). These ETEC-K88$^+$ strains specifically adhere to receptor sites located on brush borders pigs' enterocytes (Jones and Rutter, 1972). Using an *in vitro* brush border test, Sellwood et al. (1975) demonstrated the existence of 2 Large White (LW) pig phenotypes, the adhesive pigs which possess the K88 receptors and the nonadhesive ones which lack these receptors. No breed differences appear to have been so far suggested. The aim of the present work was to assess the resistance of Chinese and LW piglets by an *in vivo* attachment test using gnotobiotic colostrum deprived piglets.

Twenty-three LW piglets were from an INRA experimental herd and ten Chinese piglets from 2 Chinese breeds (Meishan and Jiaxing). Piglets were inoculated with both K88$^+$ and K88$^-$ *E.coli* strains. *In vivo* specific attachment of K88$^+$ *E.coli* to intestinal epithelium was assessed through differential bacterial counts of the two strains on intestinal washed walls and in luminal contents. The ratios between K88$^+$ and K88$^-$ *E.coli* on washed walls and in luminal contents were, respectively, $P = X/Yr$ and $p = x/y$. Since washing cannot completely remove bacteria which are not specifically attached, X is composed of Xa, which is the number of specifically attached K88$^+$ *E.coli* and of Xr, which is the number of K88$^+$ *E.coli* which come from luminal contents. Therefore Xr and Yr, which come from luminal contents are assumed to be at the same ratio as x and y enumerated in the contents, i.e., $Xr/Yr = x/y = p$. Thus the number of specifically attached bacteria was determined by means of direct bacterial differential enumerations through the equation $Xa = Yr (P - p)$. Adhesion was assessed through ratio between Xa and nonspecifically attached K88$^+$ and K88$^-$ E.coli on the washed walls, i.e., $\Gamma = Xa/Xr+Yr=P-p/p+1$. Three ratios were calculated for each piglet at the jejunum (S1), at the ileum (S3) and at half way of these 2 ends (S2), and ΓM was the mean value of the 3 values for each piglet. Since bacterial counts follow a Poisson distribution, a confidence interval $\Gamma \pm \Delta\Gamma$ was calculated for each value.

In a first experiment, ratios were calculated in 8 LW piglets inoculated with 2 different K88$^-$ E.coli strains in order to determine the limits of the ΓM, $\Gamma - \Delta\Gamma$ and $\Gamma + \Delta\Gamma$, which are to be expected in the absence of any attachment. These were respectively < 3, < 3 and < 8 (Table 1). In a second experiment, 15 LW piglets were inoculated with both ETEC-K88$^+$ and non ETEC-K88$^-$ strains. According to ΓM and $\Gamma \pm \Delta\Gamma$ they were distributed into 2 groups (Table 1). In one group of 8 animals (group n° 2) the ΓM values were always > 6, $\Gamma - \Delta\Gamma$ and $\Gamma + \Delta\Gamma$ values were respectively >6 and > 24. In the second group composed of 7 animals (group n° 3), the ΓM, $\Gamma - \Delta\Gamma$ and $\Gamma + \Delta\Gamma$ were < 2, < 2 and < 5, respectively, and similar to those found in piglets associated with two K88$^-$ strains (group n° 1). In a third experiment, 10 chinese piglets (group 4) were inoculated with both K88$^+$ and K88$^-$ strains. Table 1 shows that group 4 is homogeneous according to the ΓM, $\Delta - \Delta\Gamma$ and $\Gamma + \Delta\Gamma$ values which were < 4, < 3 and <9, respectively. These results indicated therefore that animals of group 2 belong to adhesive phenotype and those of groups 3 and 4 to nonadhesive phenotype.

Table 2 summarized the mean $\log_{10}$ CFU/g (colony forming units) of the inoculated strains in the contents and on walls of all the $S^1 + S^2 + S^3$ in each group of piglets. These data were submitted to statistical analysis (one way analysis of variance and t-test). They indicated that the wall population levels of ETEC-K88$^+$ strain in the adhesive piglets were $>10^8$ CFU/g and significantly higher than those of the same strain and of K88$^-$ strains in piglets of groups 1, 3 and 4. The ETEC-K88+ strain was at comparative levels in contents and walls in adhesive piglets. Conversely, both K88$^+$ and K88- strains were 10 to 100 fold lower on walls than in contents of piglets of groups 1, 3 and 4. It was also shown that adhesive piglets were the

TABLE 1 - DISTRIBUTION OF THE PIGLETS ACCORDING TO ΓM, $\Gamma - \Delta\Gamma$ AND $\Gamma + \Delta\Gamma$

EXPERIMENT (N° OF ANIMALS)	BREED	INOCULATED STRAINS	GROUP (N° OF ANIMALS)	Γ M	$\Gamma - \Delta\Gamma$	$\Gamma + \Delta\Gamma$
1 (8)	LARGE WHITE	ETEC-K88⁻ NON ETEC-K88⁻	1 (8)	< 3	< 3	< 8
2 (15)	LARGE WHITE	ETEC-K88⁺ NON ETEC-K88⁻	2 (8) 3 (7)	> 6 < 2	> 6 < 2	> 24 < 5
3 (10)	CHINESE	ETEC-K88⁺ NON ETEC-K88⁻	4 (10)	< 4	< 3	< 9

Γ = RATIO BETWEEN NUMBER OF SPECIFICALLY ATTACHED BACTERIA AND TOTAL NUMBER OF BACTERIA ISSUED FROM LUMINAL RESIDUES

ΓM IS THE MEAN VALUE OF 3 Γ CALCULATED FOR EACH PIGLET ;
$\Gamma \pm \Delta\Gamma$ ARE CONFIDENCE INTERVALS.

GROUP (N° PIGLETS) PHENOTYPE	E.COLI STRAINS	MEAN $\log_{10}$ CFU/G OF S1 + S2 + S3 CONTENTS	WALLS
1 (8)	ETEC-K88⁻	8,3	7,6
NOT DETERMINED	NON ETEC-K88⁻	8,2	6,6
2 (8)	ETEC-K88⁺	8.8	8.6
ADHESIVE	NON ETEC-K88⁻	7.6	6.2
3 (7)	ETEC-K88⁺	7.5	5.9
NON ADHESIVE	NON ETEC-K88⁻	7.7	6.2
4 (10)	ETEC-K88⁺	7.0	5.0
NON ADHESIVE	NON ETEC-K88⁻	8.2	6.0

TABLE 2 - POPULATION LEVELS OF K88⁺ AND K88⁻ E.COLI STRAINS IN INTESTINAL CONTENTS AND WALLS OF ADHESIVE AND NONADHESIVE PIGLETS.

most affected by the association with ETEC-K88+ strain since 7/8 were dead or diarrheic within 18 h. The nonadhesive LW piglets were not affected within 18 h of association, time at which they were sacrificed when healthy. Work is in progress to determine whether they will behave like chinese piglets, which were left to survive after inoculation with the K88+ and K88- strains. Two of these latter died at 5 days of association and 1 at 7 days. The remaining ones were healthy when sacrificed at 1, 2, 4, 7 and 9 days, respectively. This result is consistent with the fact that they belong to nonadhesive phenotype, otherwise they would have died within 18 h of association. In conclusion, the use of K88- E. coli strain as an inner marker was useful for determining specifically attached K88+ E.coli. The ГM ratios were far from zero for adhesive piglets and close to zero for nonadhesive ones. The wall population levels of K88+ E.coli were always $>10^8$ CFU/g in adhesive piglets, and $< 10^8$ CFU/g in nonadhesive piglets, and 10 to 100 fold lower than in the luminal contents. The population levels of both K88+ and K88− in the lumen contents were similar in all the LW piglets, whatever the group they belong to, indicating that adhesion does not significantly increase the population levels of K88+ strain in luminal contents. The fact that all the Chinese piglets belong to the nonadhesive phenotype suggests the existence of a resistance to attachment among some Chinese breeds. However, it would be unsafe to draw firm conclusions as to the resistant status of the Chinese breeds investigated because of the limited sample size.

Jones G.W., Rutter J.M., 1972. Role of the K88 antigen in the pathogenesis of neonatal diarrhea caused by Escherichia coli in piglets. Infect. Immun. 6, 918-927.
Sellwood R., Gibbons R.A., Jones G.W., Rutter J.M., 1975. Adhesion of enteropathogenic Escherichia coli to pig intestinal brush borders : the existence of two pig phenotypes. J. Med. Microbiol. 8, 405-411.
Söderlind O., Möllby R., 1979. Enterotoxins, O-groups and K88 antigen in Escherichia coli from neonatal pigs with and without diarrhea. Infect. Immun. 24, 611-616.
Stirm S., Ørskov F, Ørskov I, Birch-Andersen A., 1967. Episome-carried surface antigen K88 of Escherichia coli. III. Morphology. J. Bacteriol. 93, 740-748.

Germfree Research: Microflora Control and Its Application to the Biomedical Sciences, pages 191–194

ESTABLISHMENT OF GNOTOBIOTIC STATE IN GUINEA PIGS AND THEIR RESPONSE TO SHIGELIA INFECTION

Jing Shaoliang, Wang Anjing, Wu Juyuan, Chu Guangda, Zhang Xufan*, Li Zhongyuan*.
Depart of Gnotobiology, MLAC, CAMS. Beijing, China. *Lab. of Microbiology, Institute of Epidemiology and Microbiology, CNCPM, Beijing, China. 9 Dong Dan San Tiao, Beijing, China

Introduction

The GN animal is a very useful tool in the research concerning interactions between enteric pathogens and normal flora (Gordon and Pesti 1971; Maier and Hentges 1972). The purpose of our research is to achieve GN guinea pigs and study their response to Shigella infection.

Materials and Methods

Animals

Short-haired albino guinea pigs have been used.

Obtaining of GF baby guinea pigs

Guinea pigs were delivered by hysterectomy and reared in aseptic isolators. The young animals were fed with milk diet during the first few days after the operation Five days later they were fed with a combination of milk and solid diet L-477 (Pleasants et al. 1967).

Establishment of The Gnotobiotic State

At 7-10 days old, the following intestinal bacteria obtained from conventional guinea pigs were inoculated twice orally (interval: 3 days) with live cultures, namely, Bifidobacterium animalis, Streptococcus faecalis, Clostridium felsineum, Bacteroides ruminicola and B. asaccharolitica. The administered dose of each microbe was 10^9.

Bacteriological Monitoring

Fresh faeces were collected from each individual animal and cultivated onto BL, BDa and CM to culture.

Challenge

At 8 and 11 months old, the challenge was given separately to the GN and control animals (monoassociated with Peptococcus indolicus) with two doses of virulent Shigella flexneri II by mouth. The groups of the animals and their challenged dose can be seen in table 1.

Table 1. The experiment groups and challenged dose of the animals

Group No.	Number of animal		Age (month)	Dose
	Experiment	Control		
1	2	2	8	10^6
2	2	2	11	5×10^5

Result

Both GF and GN guinea pigs had been growing well, their growth curves are shown as follows (Fig. 1).

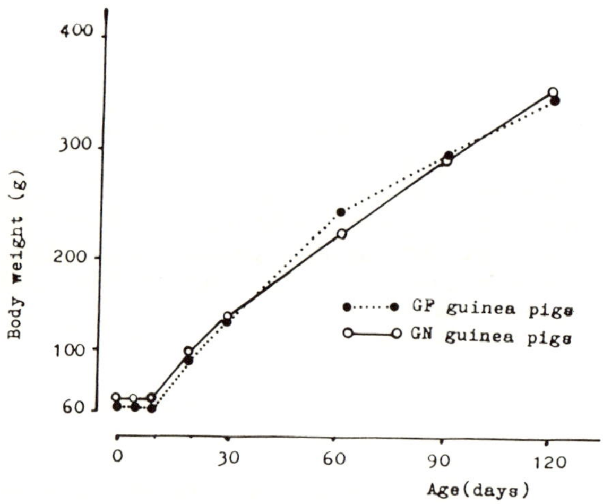

Fig.1 Growth curves of germfree and gnotobiotic pigs

There was no significant difference in body weight growth between them. However, there was a more obvious development of megacaecum in the GF guinea pigs. The microbiological monitoring showed that the above-mentioned intestinal bacteria colonized in intestinal canal of the GN animals.

After challenging, animals of the 1st group began to have a decrease in appetite and an increase in body temperature, the control animals died within 48 hrs, whereas the experimental animals restored to health in a few days. In the 2nd group, the above-mentioned symptoms began to appear on the 4th day and the control animals died on the 5th day after the challenge. The results of microbial colonization in the 2nd group of GN guinea pigs before and after challenge with Sh. flexneri IIa were shown in Fig. 2.

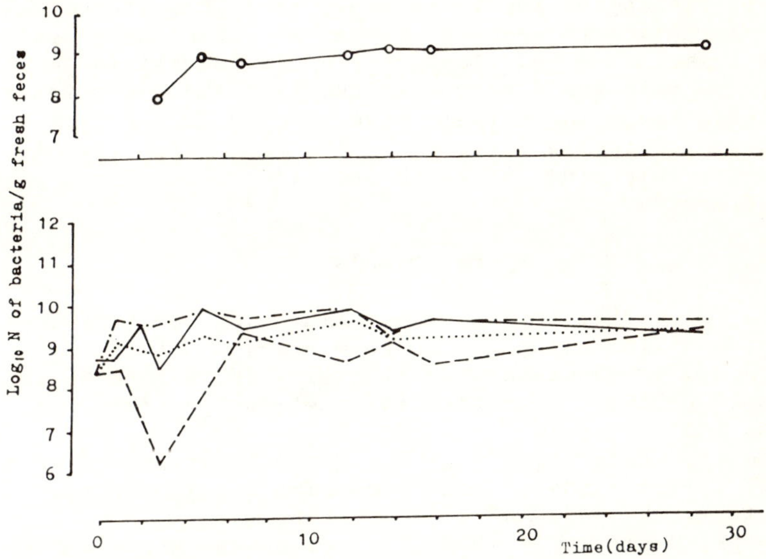

Fig. 2 Monitoring results of intestinal microbes in the gnotobiotic guinea pigs after a challenge of Sh. flexneri IIa

o—o—o Sh. flexneri IIa
-·-·-· Streptococcus faecalis
——— Bacteroides Spp.
·········· Bifidobacterium animalis
- - - - - Clostridium felsineum

Discussion

Some reports on GF animal research demonstrated that there was a development of megacaecum in these animals (Coates 1975; Gordon 1966). According to our observation, the caecums of GF guinea pigs were bigger than GN animals. It was shown that the colonization of the above-mentioned 5 species of intestinal bacteria can prevent cecal enlargement.

From our experiments, it seems evident that certain bacterial components of the lower intestinal canal of GN guinea pigs exerted antagonistic action on the Shigella population. As a result, the experimented animals survived for one month and appeared well, whereas the control animals died of Shigella infection in a few days. However, the exotic virulent Shigella flexneri IIa organism was not cleared of the intestinal tract indicating that the interaction between the normal microbial components of GN animals and Sh. flexneri IIa did not eliminate in total the pathogenic microoganisms from the lower intestinal tract and the animals have been suffering from inapparent Shigella infection. The protection effect of normal intestinal flora on the life of GN guinea pigs was apparent.

References

Coates ME (1975). Gnotobiotic animals in research: Their uses and limitations. Lab Anim 9: 275.

Gordon HA, Bruckner-kardoss E, Staley TE, Wagner M, Wostman BS (1966). Characteristics of the germfree rat. Acta Anat 64: 367.

Gordon HA, Pesti L (1971). The gnotobiotic animal as a tool in the study of host microbial relationships. Bact Rev 35: 390.

Maier BR, Hentges DJ (1972). Experimental Shigella infections in laboratory animals. 1 Antagonism by normal flora components in gnotobiotic mice. Infect Immun 6: 168.

Pleasants JR, Reddy BS, Zimerman DR, Bruckner-Kardoss E, Wostmann BS (1967). Growth, reproduction and morphology of naturally born, normally suckled germfree guinea pigs. Zschr Versuchstierk Bd 9:S 195.

Germfree Research: Microflora Control and Its Application to the Biomedical Sciences, pages 195–198

INHIBITION OF BACTERIAL TRANSLOCATION FROM THE GASTROINTESTINAL TRACT BY NONSPECIFIC IMMUNOSTIMULATION

Katrina G. Fuller and Rodney D. Berg, Ph.D.

Department of Microbiology and Immunology
Louisiana State University Medical Center
School of Medicine, Shreveport, LA 71130

INTRODUCTION

Bacterial translocation is defined as the passage of viable bacteria from the gastrointestinal (GI) tract through the epithelial mucosa into the lamina propria and then to the mesenteric lymph nodes (MLN) and other organs. Bacterial translocation does not occur in healthy, specific pathogen-free (SPF) mice. However, it does occur when: (a) the host is immunosuppressed, (b) certain indigenous bacteria reach abnormally high population levels in the intestines, or (c) the mucosal barrier is physically disrupted.

Berg and Garlington (1979) reported that viable indigenous bacteria are not cultured from the mesenteric lymph nodes, spleen, or liver of adult mice. However, viable indigenous bacteria are present in the MLN and other organs of SPF mice receiving immunosuppressive agents (Berg 1983), genetically athymic (nu/nu) mice (Owens, Berg 1980), and neonatally-thymectomized mice (Owens, Berg 1982). Grafting thymuses to nu/nu mice deficient in T-cell mediated immunity inhibits the translocation of these indigenous bacteria from the GI tract to other organs (Owens, Berg 1980). Therefore, it is of interest to determine whether stimulation of the host immune system can prevent the translocation of opportunistic indigenous bacteria from the GI tract.

Bacterial translocation from the GI tract occurs in monoassociated gnotobiotic mice (Berg 1980; Berg, Owens 1979) or antibiotic-decontaminated mice subsequently colonized with particular organisms (Berg 1980). The abnormally high population levels of bacteria in the intestines of these monoassociated mice apparently overwhelm any immunological barrier to bacterial translocation. Immunostimulation possibly could inhibit the passage of bacteria through the GI epithelium and lamina propria and increase bacterial killing in situ in the MLN. Proprionibacterium acnes is a potent nonspecific stimulator of the reticuloendothelial system (Castro 1974); and enhances host resistance to certain bacterial infections (Fauve 1975). Consequently, antibiotic-decontaminated SPF mice and germfree mice were given injections of formalin-killed P. acnes prior to monoassociation with streptomycin-resistant Escherichia coli C25 in an attempt to inhibit bacterial translocation.

MATERIALS AND METHODS

Germfree and gnotobiotic mice (CD-1 strain, Charles River Breeding Laboratories, Wilmington, MS) were maintained in flexible plastic isolators (Germfree Supply Division, Standard Safety Equipment Co., Palatine, IL) and given food (Laboratory Animal Chow 5010; Ralston Purina Co., St. Louis, MO) and water ad libitum. SPF (CD-1) mice were maintained in plastic cages (Maryland Plastics, Baltimore, MD) and also given food and water ad libitum. Antibiotic-decontaminated SPF mice were given water containing 2 mg/ml streptomycin sulfate (Sigma Chemical Co., St. Louis, MO) and 1500 U/ml penicillin-G (Sigma) ad libitum for 1 week prior to colonization. Decontamination was confirmed by aerobic and anaerobic cultures of feces.

Mice were injected intraperitoneally with 0.2 ml (7 mg/ml) formalin-killed P. acnes CN6134, lot 776/1, (Burroughs-Wellcome Co., Research Triangle Parks, NC) suspended in 0.9% saline containing 0.01% thiomersol perservative. Mice received either a single injection 1 week prior to colonization with E. coli or two injections 1 and 2 weeks prior to colonization. Control mice received injections of 0.85% saline.

Escherichia coli C25 is a streptomycin-resistant strain obtained from Dr. Rolf Freter, University of Michigan, Ann Arbor, MI (Freter 1962). Broth cultures of E. coli C25 in brain heart infusion (BHI) (Difco Laboratories, Inc., Detroit, MI) contained 1 x 10^9 viable organisms per ml after 48 hrs incubation. Germfree or antibiotic-decontaminated SPF mice were inoculated with E. coli C25 by placing viable cultures on the food and in the drinking water.

Mice were sacrificed by cervical dislocation and the MLN, spleen, liver, and cecum aseptically removed as previously described (Berg, Garlington 1979). The organs were weighed, homogenized in sterile BHI (Difco), and plated on MacConkey agar (Difco). Serial dilutions of cecal homogenate were plated on MacConkey agar containing 2 mg/ml streptomycin sulfate. Plates were incubated aerobically at 37°C for 48 hrs.

RESULTS AND DISCUSSION

E. coli C25 established cecal population levels of 10^8 to 10^9 per gram in the monoassociated SPF and gnotobiotic mice. These population levels were sufficient to promote translocation of E. coli C25 to the MLN in 75% of the SPF mice and 100% of the gnotobiotic mice. The spleens and livers did not contain any viable bacteria. Injections of P. acnes prior to colonization with E. coli C25 did not alter the cecal population levels of E. coli C25 in either the monoassociated SPF or gnotobiotic mice.

Both SPF and gnotobiotic mice injected with P. acnes showed marked splenomegaly, as demonstrated by the splenic index (spleen weight of test animals/spleen weight of controls). The splenic index was 4.8 in P. acnes pretreated SPF mice and 7.1 in pretreated gnotobiotic mice. Splenomegaly following P. acnes injections is reportedly due to an increase in the numbers of macrophages in the spleen (Fauve 1975).

The incidence of E. coli C25 translocation from the GI tract to the MLN decreased from 75% to 41% (p=.002) in SPF mice injected with P. acnes. The mean numbers of viable E. coli C25 per gram MLN also decreased from 1,900 to 300 (p=.009). However, the incidence of E. coli C25 translocation to the MLN and the numbers of viable E. coli C25 per gram MLN did not decrease in monoassociated gnotobiotes pretreated with P. acnes. Thus, nonspecific immunostimulation with P. acnes inhibits the translocation of E. coli C25 from the GI tract to the MLN of monoassociated SPF but not gnotobiotic mice.

Our preliminary results suggest that the immunopotentiating effects of P. acnes may depend on prior exposure to cross-reacting bacterial antigens of the indigenous flora. Wells and Balish (1979) report that a complex microflora is required for an intraperitoneal injection of P. acnes to enhance the splenic plaque-forming cell (PFC) response to SRBC antigen. However, gnotobiotic rats monoassociated with live P. acnes prior to intraperitoneal inoculation with killed P. acnes exhibit a decreased splenic PFC response to SRBC. Studies are currently in progress to determine whether prior colonization with an indigenous flora will enhance the immunostimulatory properties of P. acnes and decrease bacterial translocation in gnotobiotic mice.

LITERATURE CITED

Berg, RD (1980). Inhibition of Escherichia coli translocation from the gastrointestinal tract by normal cecal flora in gnotobiotic or antibiotic-decontaminated mice. Infect Immun 29: 1073.

Berg, RD (1983). Bacterial translocation from the gastrointestinal tracts of mice receiving immunosuppressive chemotherapeutic agents. Current Microbiol 8: 285.

Berg, RD, Garlington, AW (1979). Translocation of certain indigenous bacteria from the gastrointestinal tract to the mesenteric lymph nodes and other organs in a gnotobiotic mouse model. Infect Immun 23: 403.

Berg, RD, Owens, WE (1979). Inhibition of translocation of viable Escherichia coli from the gastrointestinal tract of mice by bacterial antagonism. Infect Immun 25: 820.

Castro, JE (1974). The effect of Corynebacterium parvum on the structure and function of the lymphoid system in mice. Eurp J Cancer 10: 115.

Fauve, RM (1975). Stimulating effects of Corynebacterium parvum extract on the macrophage activities against Salmonella typhimurium and Listeria monocytogenes. In Halpern (ed): "Corynebacterium parvum: Applications in Experimental and Clinical Oncology" New York: Plenum Press, p. 77.

Freter, R (1962). In vivo anatagonism of intestinal bacteria against Shigella flexneri. J Infect Dis 110: 38.

Owens, WE, Berg, RD (1980). Inhibition by thymic grafts of bacterial translocation from the gastrointestinal tract in athymic (nu/nu) mice. Infect Immun 27: 461.

Owens, WE, Berg, RD (1982). Bacterial translocation from the gastrointestinal tracts of thymectomized mice. Current Microbiol 7: 169.

Wells, CL, Balish, E (1979). Immune response modulation by colonization of GF rats with Propionibacterium acnes. Infect Immun 26: 473.

Germfree Research: Microflora Control and Its Application to the Biomedical Sciences, pages 199–202

COLONIZATION AND PATHOGENESIS OF *CAMPYLOBACTER SPP*. IN ATHYMIC AND EUTHYMIC GERMFREE MICE

John W. Yrios, M.S. and Edward Balish, Ph.D.

Departments of Surgery and Medical Microbiology
University of Wisconsin Medical School
Madison, WI 53706.

Three of eight species of *Campylobacter* are known to be associated with humans (Blaser 198]). *Campylobacter sputorum* ssp. *sputorum* is a part of our normal oral flora. *Campylobacter jejuni* is now recognized as a common cause of diarrhea, worldwide; while *C. fetus* ssp. *fetus* is an opportunistic pathogen causing systemic disease in the immunocompromised host (Blaser 1981). Conventional animals have been used, for the most part, to study *Campylobacter* pathogenesis (Blaser 1983; Caldwell 1983). The fact that conventional animals possess a very complex microbial flora complicates the study of host-*Campylobacter* interactions. To date very little research has been carried out on the pathogenesis of *C. fetus* ssp. *fetus*. The germfree (GF) BALB/c mouse is a unique animal model that can be used to assess the colonization and pathogenic potential of *C. jejuni* and *C. fetus* ssp. *fetus*. The congenitally athymic (NU/NU) BALB/c mouse also provided us with an opportunity to assess the importance of T-cell mediated immunity in the hosts' response to *Campylobacter* spp. and to study the pathogenesis of *C. fetus* ssp. *fetus* in an immunocompromised host.

Approximately 8-week-old athymic (NU/NU) and euthymic (+/NU) BALB/c GF mice were used in these studies. They were reared, housed, and challenged with a *Campylobacter* sp. in plastic GF isolators at the Gnotobiotic Laboratory, University of Wisconsin, Madison, WI. Mice in separate isolators were initially orally inoculated with $\sim 10^8$ organisms; subsequent monoassociations (MA) were allowed to occur naturally through the fecal-oral route. The bacterial strains of C. jejuni (#45100) and C. fetus ssp. fetus (#255) were human clinical isolates, the former being a diarrheal isolate and the latter was from a blood culture. At various times post-inoculation, three NU/NU and three +/NU mice were anesthetized (ether), exsanguinated by cardiac puncture, and the internal organs (liver, spleen, kidney, mesenteries) and sections of the GI tract were removed to quantitate bacteria. Inoculated plates (5% sheep blood agar; Gibco) were incubated at 37°C for 48 hrs in a microaerophilic atmosphere. Tissue specimens from each mouse were excised and immediately placed in Holland-Bouin's fixative, embedded in plastic, sectioned (3 µm) on a Sorvall JB-4 microtome and stained using a standard azure A-eosin B stain.

Semi-quantitative enumeration of Campylobacter in feces indicated that C. jejuni reached high numbers by 24 hrs, while similar numbers were not reached by C. fetus ssp. fetus until 48 hrs after oral inoculation. The initial time sequence of Campylobacter colonization was similar for GF animals that were inoculated orally or colonized naturally by a fecal/oral route. The maximum numbers of viable C. jejuni in cecal contents were ∿10.5* CFU/gm, while viable counts for C. fetus ssp. fetus were slightly lower (9.5-10.0* CFU/gm). Distribution of C. jejuni and C. fetus ssp. fetus in the GI tract of mice that were MA for 224 days is shown in Table 1. Similar viable populations of C. jejuni (#45100) were present in NU/NU and +/NU mice. The lowest numbers of C. jejuni were recovered from the stomach and upper GI tract (∿10^6 CFU/gm) and the largest population (∿10^{10} CFU/gm) were cultured from the lower GI tract (cecum and colon). Conversely, C. fetus ssp. fetus was able to grow better in the alimentary tract of NU/NU than in +/NU mice. Data in Table 1 show that NU/NU mice had 2-3 logs more viable C. fetus ssp. fetus in the stomach and upper intestinal tract and about 1 log higher viable counts in the cecum than their +/NU littermates.

Overt signs of disease were not apparent in either group of MA mice during the initial colonization experiments. After a 10 month period of monoassociation, when young (∿8 wk) GF NU/NU and +/NU mice were again introduced into the isolators, a transient diarrhea (days 5-10), occurred after a natural colonization. The diarrhea was observed only in the NU/NU mice MA with C. jejuni (#45100). Histological examination of ileal, cecal, and colonic tissue from NU/NU animals with diarrhea on day 7 revealed an

*$\log_{10}$ of numbers

intact endothelial layer, with an edematous lamina propria, congestion, PMN-filled microabscess formation in the glandular mucosa, and a slight mononuclear infiltrate. PMN's were also visible in the luminal contents of the mice with diarrhea. These abnormal histologic findings were only present in mice (NU/NU) that had visible diarrheal symptoms. The GI histology of NU/NU mice essentially appeared normal by day 22.

Table 1. Viable Bacteria Present in the GI Tract of Mice that were MA with Campylobacter spp. for 224 Days

Tissue Cultured	BALB/c Genotype	Organism: C. jejuni	Organism: C. fetus ssp. fetus
Stomach	NU/NU	6.63*	8.66
	+/NU	6.00	6.55
Upper S.I.	NU/NU	5.92	7.21
	+/NU	5.65	3.76
Lower S.I.	NU/NU	8.23	8.09
	+/NU	8.07	5.85
Cecum	NU/NU	9.71	9.58
	+/NU	9.69	8.14
Colon	NU/NU	10.04	9.50
	+/NU	9.56	8.40

*Log_{10} CFU's/gm dry wt - each value represents pooled data from three mice.

Neither Campylobacter species could be recovered from blood specimens taken at various times after inoculation (data not shown). Viable C. jejuni (10^4-10^5 CFU/gm) were consistently obtained from the mesenteries of MA NU/NU and +/NU mice over a 36 day experimental period. Internal organs cultured for C. jejuni and C. fetus ssp. fetus revealed a sporadic low recovery (10^2-10^5 CFU/gm) of organisms from pools of livers, spleens, or kidneys from NU/NU and +/NU mice over the 42 day period after oral challenge. At 42 days C. jejuni could still be cultured from the liver, spleen, and kidneys of NU/NU but not +/NU mice. Conversely, C. fetus ssp. fetus was not present in the internal organs of NU/NU or +/NU mice at 42 days. The GI colonization of BALB/c athymic and euthymic GF mice with either C. jejuni (#45100) or C. fetus ssp. fetus occurs quickly and relatively large populations (9.5-10.5* CFU/gm) are reached in the cecum and colon of the MA mice. Even with the rapid growth and large numbers of organisms in the intestinal lumen of these mice, they showed very little evidence of disease on initial exposure. The organisms were capable of penetrating the gut tissue, reaching the mesenteric lymph nodes, and disseminating into the internal organs of NU/NU and +/NU mice. A change in the pathogenesis occurred with C. jejuni (#45100) as it remained mouse monoassociated for several months. The mouse-adapted strain was consistently able to induce a transient diarrhea (day 5-10) in NU/NU mice, but not in their +/NU littermates. The symptoms, the timing, and histological picture somewhat resemble the Campylobacter diarrhea observed in humans. Even though C. fetus ssp. fetus was able to grow to very high numbers in the GI tract of MA mice and was able to invade the mesenteries and internal organs early in the colonization, neither morbidity nor mortality was observed in NU/NU or +/NU mice after prolonged periods of MA (~12 months).

Our model, using a pure culture of C. jejuni in GF mice, has demonstrated that C. jejuni can cause transient diarrhea in the MA NU/NU mouse. Our data also indicated that mice with T-cell mediated immunity were resistant to the diarrhea. The T-deficient animal (NU/NU), however, did recover from disease and remained outwardly healthy for the duration of the experiment (12 months). Further studies are needed to elucidate the mechanism(s) of the hosts' immune response and to assess the relevance of bacterial strain variation in Campylobacter pathogenesis.

Blaser MJ, Duncan DJ, Warren GH, Wang WL (1983). Experimental Campylobacter jejuni infection of adult mice. Infect Immun 39(2):908.

Blaser MJ, Reller LB (1981). Campylobacter enteritis. N Engl J Med 305:1444.

Caldwell MB, Walker RI, Stewart SD, Rogers JE (1983). Simple adult rabbit model for Campylobacter jejuni enteritis. Infect Immun 42(3):1176.

* $\log_{10}$ of numbers

Germfree Research: Microflora Control and Its Application to the Biomedical Sciences, pages 203–206

EXPERIMENTAL MODELS OF CLOSTRIDIUM DIFFICILE ENTEROCOLITIS IN GNOTOBIOTIC MICE

Takiko Sugiyama, Makio Mukai,
Ryoko Yamashita & Keisuke Sunakawa
Department of Microbiology, Pathology & Pediatrics
School of Medicine, Keio University
Tokyo, Japan 160

Germfree ICR mice intragastrically infected with *C. difficile* develops pseudomembranous colitis (PMC) without any antibiotic pretreatment.

The non-severe model using a low vilulent strain of *C. difficile* develops remarkable pseudomembrane at cecum and proximal colon and negligible mortality rate. So it is suitable for a chronological study of pathogenic activities and immune responces on whole stages of the PMC. The pseudomembrane formation was observed 2 to 4 days after infection with histopathological examinations showing that this was acute colitis with the predominantly polymorphnuclear cell infiltration. In the mice sacrificed 6 to 21 days after challenge, there was ulcer formation in the affected intestinal area while acute colitis of the predominantly mononuclear (lymphoid) cell infiltration was observed. At 4 weeks of the infection, the mucous membrane had recovered completely while there was a predominantly plasma cell infiltration in the submucosal area and these were still remained at 6 weeks after the infection (Table 1).

For the studies of antibiotic therapy and of antagonistic bacteria in the hosts, we made severe model of PMC with high mortality rate using a high vilulent strain. In the lethal model the gross and histopathological changes were observed in whole intestine. The most remarkable inflammatory change with hemorrhage was observed in four varied types; such as jejunum, ileum, cecum and colon type. Especially in jejunum type there was remarkable fluid accumulation. At the cecum and proximal colon obvious pseudomembrane formation

Table 1 Relationship among microbiological, clinical and pathological diagnoses on non-lethal *Clostridium difficile* colitis in gnotobiotic mice

Diagnosis	Days after infection: 2	3	4	6	14	21	30	44
Microbiological	*Clostridium difficile* colitis					Fecal carrier		
Histopathological	Acute colitis, pseudo-membrane-forming stage: polymorphonuclear cell infiltration			Acute colitis, ulcer-forming stage: mononuclear cell infiltration		Post-acute colitis reactions, recovery stage: plasma cell infiltration		
Clinical (equivalent to human colitis)	Pseudomembranous colitis			Nonspecific acute colitis		No sign		

was observed in all types (Table 2). On the investigation of antagonistic bacteria to *C. difficile* *in vivo*, two of 20 strains act protectively to the hosts: a strain of *Streptococcus faecalis* and *Bifidobacterium breve*. When both strains were administered individually before the challenge, these experimental groups of germfree (GF) mice did not develop enterocolitis. When both were administered at the same time with challenge, they developed enterocolitis partially but no mortal case was observed. There is a report that strains of bacteria in six of the genera ingibited the multiplication of *C. difficile* in study of bacterial interference between *C. difficile* and normal fecal flora *in vitro* (Rolfe 1981). Our data *in vivo* agree with this data. The maximum dose for human use of Cefoperazon, Cefmetazol, and Lincomycin were injected intramusclarly and Tetracycline and Vancomycin were administered perorally after developed symptom of the enterocolitis. There was no significant difference on the mortality rates between therapeutic groups and the non-therapeutic groups. Moreover some models of the therapeutic groups died earlier than the non-therapeutic groups. It was reported that exotoxin production of *C. difficile* increases under the condition with some kinds of antibiotics *in vitro*. The earlier death of the mice in therapeutic groups would be suggested that same phenomenon develops *in vivo*. The first experiment of combination with *C. difficile* and GF mice reported that the ceca in albino ND-2 monoassociated GF mice were considerably smaller than control GF mice. In non-severe PMC model of CD-1 GF mice the cecal toxin titers decreased to an undetectable level but *C. difficile* can persist in large numbers during Vancomycin therapy (Onderdonk 1980). It can be pointed out that it is also requested to use severe model of PMC for study of the antibiotic therapy.

Onderdonk AB, Cisneros RL, Bartlett JG (1980). *Clostridium difficile* in gnotobiotic mice. Infect Immun 28:277.

Rolfe RD, Helebian S, Finegold SM (1981). Bacterial interference between *Clostridium difficile* and normal fecal flora. J Infect Dis 143:470.

Skelly BJ, Trexler PC, Tanami J (1962). Effect of a Clostridium species upon cecal size of gnotobiotic mice. Proc Soc Exp Biol Med 110:455.

We thank Dr. T. Mitsuoka (Rikagaku Institute), Dr. Y. Tagaya (Yakuruto Institute) and Mr. H. Uchida (Keio University) for kindly supplying the organisms used in this study.

Table 2 Histopathological findings on lethal *Clostridium difficile* entero-colitis in gnotobiotic mice

	Main portion of histopathological changes			
	Jejunum type	Ileum type	Cecum type	Colon type
Hemorragia	jejunum	ileum	cecum	colon
Fluid accumulation	jejunum	ileum	.	.
Degenerative inflamation	jejunum	ileum	cecum	colon
Pseudomembrane formation	cecum	cecum	cecum	colon

Germfree Research: Microflora Control and Its Application to the Biomedical Sciences, pages 207–209

KINETICS OF CYTOTOXIN PRODUCTION IN FECES OF GNOTOBIOTIC MICE AND RATS ASSOCIATED WITH TWO STRAINS OF CLOSTRIDIUM DIFFICILE

Dubos F., Corthier G., Rapine P. and Raibaud P.

Laboratoire D'Ecologie Microbienne
I.N.R.A. - C.N.R.Z.
78350 - JOUY-en-JOSAS (France)

In adult human, C. difficile is involved in pseudomembranous colitis and in some antibiotic associated diarrheae. In young animals, C. difficile is involved in lethal diarrhea of young hares. C. difficile strains produce in vitro and in vivo two distinct toxins = an enterotoxin and a cytotoxin (Sullivan 1982, Lyerly, 1982). The purpose of this work was to determine the kinetics of cytotoxin production in vitro and in vivo. Two strains, a human strain and a hare strain were used. To avoid any antagonistic effect exerted by the microflora of holoxenic animals (Wilson, 1981), gnotobiotic rats and mice were used to determine the cytotoxin production in vivo.

Bacterials strains used were C. difficile VPI strain 10463 isolated from human feces (Sullivan 1982) and C. difficile FD strain isolated from a diarrheic young hare (Dabard 1979). Cytotoxin production in vitro was performed in brain heart infusion. Animals were axenic Fisher 344 rats and axenic $C3^H$/He mice. The cytotoxicity assay was performed in microtiter plates on chinese hamster ovary cells ($CHOK_1$)

Human strain produced very high titers 24 hours after inoculation. During the whole period of culture (5 days) this titer remained unchanged. However, the hare strain cultivated in the same broth, produced low cytotoxin titers. In both cases, the number of vegetative cells and spores was similar. In vivo bacterial implantation was similar in rats and mice monoassociated with human or hare strain. The number of C. difficile exceeded 10^8/g of feces two days after inoculation and population levels remained stable

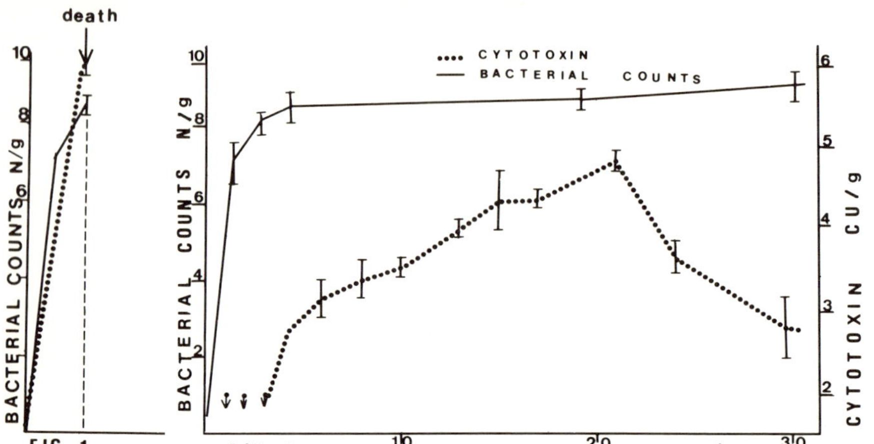

FIG. 1 FIG. 2

kinetics of cytotoxin production by the human strain (fig. 1) or the hare strain (fig. 2) in monoxenic mice.

whatever the strain or the animal host. Cytotoxin production was very high in the digestive tract of animals monoassociated with the human strain. 89 % of rats and 100 % of mice died within two to three days after inoculation. These animals were diarrheic. Mice and rats monoassociated with the hare strain remained healthy. No mortality or pathological effect were observed during three months. In rat and mouse feces cytotoxin production increased in the first three weeks and reached a maximal level. Then the production decreased. During the whole period cytotoxin production still remained lower in mice than in rats. At three weeks cytotoxin titers in rat feces were similar to those observed in rats monoassociated for 3 days with the human strain. However no diarrhea or mortality was observed at that moment. Molecular weights of cytotoxin produced in vivo were similar whatever the origin of the strain or animal host (410.000 to 440.000 daltons).

Interaction between two strains was studied in gnotoxenic rats and mice. If the hare strain was inoculated 4 days before the human one, gnotoxenic rats and mice were protected and remained healthy. A partial barrier effect against the human strain was observed. Cytotoxin titers in diassociated mice were lower than cytotoxin titer observed in monoassociated mice (human or hare strain).

Cytotoxin produced in vitro and in vivo in the diges-

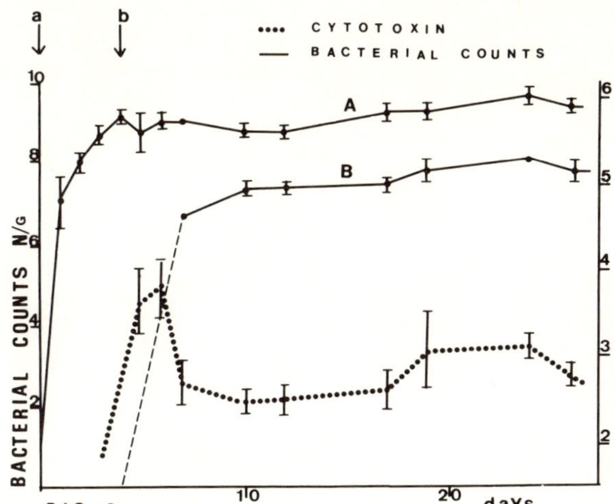

FIG. 3

Interaction between human strain (B) and hare strain (A) in diassociated mice with C.difficile.
a)-inoculation of hare strain
b)-inoculation of human strain.

tive tract of gnotoxenic animals could be different according to the origin of the C. difficile strain. Pathological effects after inoculation in gnotoxenic animals were very different. Animals monoassociated with C. difficile could be a useful tool for studying the pathogenic effect of some C. difficile strains of different origins. Furthermore, the interaction between C. difficile strains could be studied in gnotobiotic animals.

Dabard J., Dubos F., Martinet L., Ducluzeau R. Experimental reproduction of neonatal diarrhea in young gnotobiotic hares sumultaneously associated with Clostridium difficile and other Clostridium strains. Infect. Immun. 24 : 7-11, 1979.

Lyerly D.M., Lockwood D.E., Richardson S.H., Wilkins T.D. Biological activities of toxins A and B of Clostridium difficile. Infect. Immun. 35: 1147-1250, 1982.

Sullivan N.M., Pellett S., Wilkins T.D. Purification and characterization of toxins A and B of Clostridium difficile. Infect. Immun. 35: 1147-1140, 1982.

Wilson K.H., Silva J., Fekety F.R. Suppression of Clostridium difficile by normal hamster cecal flora and prevention of antibiotic associated cecitis. Infect. Immun. 34: 626-628.

Germfree Research: Microflora Control and Its Application to the Biomedical Sciences, pages 211–215

VIRULENCE FACTORS OF STREPTOCOCCUS MUTANS

Sam Rosen, Irving Shklair, E.X Beck and F.M. Beck

Ohio State University (Columbus, OH) and Naval Dental Research Institute (Great Lakes, IL)

INTRODUCTION

Streptococcus mutans is believed to be the prime etiologic agent of coronal caries in both humans (Gibbons, Van Houte 1975. Dental caries) and animals (Fitzgerald 1968). It has been suggested that the cariogenicity (virulence) of S. mutans is due to the ability of the organism to adhere to the tooth surface, then colonize or aggregate by synthesizing water insoluble glucans, and produce lactic acid by catabolizing fermentable carbohydrates to demineralize the enamel of teeth. The concept that the initiation of dental caries is associated with the development of sticky (insoluble) glucans has been proposed (de Stoppelaar, Konig, Plassecheart, Van der Hoeven 1971). They reported that a mutant of S. mutans, that was unable to synthesize insoluble glucans, was no longer cariogenic in germfree rats and that caries activity was greatly reduced in hamsters. The importance of the glucans in the etiology of dental caries has been reviewed by a number of authors (Newbrun 1972; Gibbons, Van Houte 1975. Bacterial adherence). There is little doubt that the insoluble glucan synthesized from sucrose by S. mutans plays a significant role in caries activity.

A second virulence factor characteristic of S. mutans is its ability to produce lactic acid. Some investigators (Jordan 1965; Drucker, Melville 1968) found no significant differences between cariogenic and non-cariogenic streptococci regarding either the amount of lactic acid or other types of fermentation acids produced. However, others (Hillman 1978; Mao, Rosen 1980) isolated several mutants of S.

mutans that made less lactic acid than the wild type strains and lower caries activity in test animals (Mao, Rosen 1980; Johnson, Gorss, Hillman 1978). The above data supports the importance of lactic acid in the etiology of dental caries.

In previous studies dealing with the virulence of S. mutans, either glucan synthesis or lactic acid production has been evaluated alone relative to caries activity. The objective of this study was to determine the relationship between insoluble glucan synthesis and acid formation, from the same organism, and their effect on caries development.

METHODS

Quadruplicate cultures of each organism were grown for 48 hours in 5 ml of a chemically defined media containing 5% sucrose (Osborne, Lamberts, Meyer, Roush 1976). Glucan was determined by using the total carbohydrate, phenol-sulfuric acid procedure. Lactic acid was assayed by using a gas chromatograph. DNA was determined by the diphenylamine procedure (Ashwell 1957). The amounts of lactic acid and insoluble glucan were expressed as μ moles of lactic acid per mg of DNA and mg of glucose equivalents per mg of DNA, respectively.

Strains of S. mutans used in this study were isolated from dental plaque of naval recruits. Half the isolates came from individuals with no caries and the remainder from individuals with rampant caries. Sterile dental floss was used to collect plaque from interproximal areas of posterior teeth. Serotypes of S. mutans were determined by the biotyping method of Shklair and Keene (Shklair, Keene 1974). The main criteria for selection of the isolates of S. mutans used in this investigation was that they produce varying amounts of lactic acid and insoluble glucan. Distribution by serotypes was a secondary consideration.

Germfree nursing rats of known age were purchased from Charles River Breeding Laboratories, Wilmington, MA. At 22 days of age they were placed in sterile flexible plastic isolators. The rats were distributed into 4 isolators, 10 rats per isolator. They were given drinking water (double distilled), sterilized by autoclave. Diet 2000, fortified with 1% Gustafsson's vitamin mix (Teklad Mills, Madison, Wisc.), was sterilized by exposure to 5 M rads of Cobalt 60 irradiation (Neutron Products, Dickerson, Md.). Water and diet were given *ad libitum*. The rats were infected by swabbing the molar teeth with a fresh culture and placing the remainder of the culture into the drinking water. This was done for two consecutive days. Sterility checks were performed on blood agar incubated at 37°C in anaerobic and aerobic atmospheres every two weeks. Rats were removed at the end of 5 weeks and caries scores were determined according to the method of Keyes (Keyes 1958).

RESULTS

A total of eight strains of *S. mutans* were evaluated in two experiments. The results are given in Table 1.

Table 1
Insoluble Glucan, Lactic Acid and Caries Activity of Selected Strains of *Streptococcus mutans*

Strain	Insoluble Glucan (mg/mg DNA)		Lactic Acid (μ moles/mcg DNA)		Mean Caries Scores ± S.E. in Ten Gnotobiotic Rats	
	Pre*	Post*	Pre	Post	Buccal-Lingual	Total Severity
Dowd	70.7	160	2.95	2.99	8.1±1.6	** 104±3.7
Stacey	12.2	8.3	2.72	2.21	7.6±1.6	107±6.0
Ward	19.3	33.6	1.98	1.81	1.2±0.47	96.3±5.5
Gunn	3.95	18.5	2.89	2.23	2.6±0.98	104±3.8
107B	5.42	3.24	1.72	1.14	8.8±2.1	77.4±5.4
OMZ 175	46.3	25.4	2.97	2.27	7.0±3.2	94.2±4.6
Thresher	6.52	3.39	3.79	2.25	12.3±3.3	98.3±6.4
HS-6	25.1	25.7	2.04	1.59	11.4±1.7	102.8±7.7

*Pre and post refer to strains before and after animal passage.
**Values within lines are not significantly different according to ANOVA and Newman-Keuls analysis.

The data indicated that the strains of *S. mutans* used in this study are virulent regardless of the amounts of insoluble glucan and lactic acid produced by these strains.

DISCUSSION

Possible explanations for lack of correlation between levels of insoluble glucan and buccal-lingual caries are as follows: (1) not enough insoluble glucan was produced; (2) tenacity of organisms to the teeth could have been weak; (3) in vitro may not parallel in vivo conditions; (4) the rat may not be a suitable model (Other animal model systems should be evaluated to determine the correlation of insoluble glucan with caries.); and (5) numbers of microorganisms may not have reached a critical level on the buccal-lingual surfaces of the teeth.

REFERENCES

Ashwell G (1957). Colorimetric analysis of sugars. In Colowick SP, Kaplan NO (eds): "Enzymology III," New York: Academic Press, p. 99.

De Stoppelaar JJ, Konig KG, Plassecheart ASM, Van der Hoeven JS (1971). Decreased cariogenicity of streptococcus mutans. Arch Oral Biol 16:971-975.

Drucker DB, Melville TH (1968). Fermentation end-products of cariogenic and non-cariogenic streptococci. Arch Oral Biol 13:563-570.

Fitzgerald RJ (1968). Dental caries research in gnotobiotic animals. Caries Res 2:139-146.

Gibbons R, Van Houte J (1975). Bacterial adherence in oral microbial ecology. Am Rev Microbiol 29:19-44.

Gibbons R, Van Houte J (1975). Dental caries. Ann Rev Med 26:121-136.

Hillman JD (1978). Lactate dehydrogenase mutants of streptococcus mutans: isolation and preliminary characterization. Infect and Immun 21:206-212.

Johnson CP, Gorss S, Hillman JD (1978). Cariogenic properties of LDH deficient mutants of streptococcus mutans. J Dent Res 57, Special Issue A, IADR/AADR Abstract No. 784: 270.

Jordan HV (1965). Bacteriological aspects of experimental dental caries. Am NY Acad Sci 131:905-912.

Keyes PH (1958). Dental caries in the molar teeth of rats. II. A method for diagnosing and scoring several types of lesions simultaneously. J Dent Res 37:1088-1099.

Mao M, Rosen S (1980). Cariogenicity of a "low acid" mutant of streptococcus mutans. J Dent Res 57:1620-1626.

Newbrun E (1972). Extracellular polysaccharides synthesis by glucosyltransferases of oral streptococci. Caries Res 6:132-147.

Osborne RM, Lamberts BL, Meyer TS, Roush AH (1976). Acrylamide gel electrophoretic studies of extracellular sucrose metabolizing enzymes of streptococcus mutans. J Dent Res 55:77-84.

Shklair IL, Keene HJ (1974). A biochemical scheme for the separation of the five varieties of streptococcus mutans. Arch Oral Biol 19:1079-1081.

Germfree Research: Microflora Control and Its Application to the Biomedical Sciences, pages 217–219

CHARACTERIZATION OF FUSIFORM BACTERIA AND CLOSTRIDIA ISOLATED FROM FECES OF LIMITED FLORA MICE AND THEIR ASSOCIATION TO GERMFREE MICE

ITOH, K.[1] and MITSUOKA, T.[1,2]

[1] Animal Physiology Laboratory, The Institute of Physical and Chemical Research, Wako-shi, Saitama, 351 ; [2] Department of Biomedical Science, Faculty of Agriculture, The University of Tokyo, Bunkyo-ku, Tokyo, 113, Japan.

In earlier studies, we reported that the administration of germfree mice with chloroform treated feces of conventional mice reduced the cecal size to the normal state, and only fusiform bacteria and clostridia were detectable in feces of these ex-germfree mice (Itoh, Mitsuoka 1980). There are several reports concerning characterization of fusiform bacteria harbored in the mouse cecum (Gordon, Dubos 1970; Hazenberg 1970; Koopman, Kennis 1977, 1979, 1983; Koopman 1983; Roach, Tannock 1980; Wensinck, Ruseler-van Embden 1971). but the taxonomic position of fusiform bacteria is not sufficiently reliable. Association with fusiform bacteria strains did not completely normalize germfree mice (Hazenberg 1977; Koopman, Kennis 1980; Koopman 1981). In the present study, we characterized bacteria isolated from feces of limited-flora mice produced by inoculation of chloroform treated feces with conventional biological and biochemical tests, and inoculated germfree mice with these isolates to study their effect on cecal size.

One hundred and fifteen strains of fusiform bacteria and clostridia were isolated three times from the feces of limited flora (LF) mice which were produced by inoculation with chloroform-treated (Chf) feces of conventional mice to germfree (Gf) mice previously administered with <u>Escherichia coli</u>. These strains were divided in 36 types on the basis of the colony form, gram strain, cell morphology, acid production from carbohydrates, biochemical characterics and fermentation products. Even though all these strains were isolated from one LF mouse, the bacterial types isolated varied according to the isolation time.

Forty-six strains of 115 strains were divided into three groups and each strain was inoculated to one or two GF mice. After comfirmation of their colonization in intestine, monoassociated mice were housed together in one isolator per group. Three groups of polyassociated (PA) mice were made. As the result, mean values of cecal size (cecal weight/body weight %) of mice of each group were 2.8, 2.8 and 3.2% respectively. Three weeks later, three or four PA mice of each group were housed together in one isolator, cecal size of such mice reduced to 2.5 (2.1-3.1)% at first and 2.3 (2.1-2.7)% later. These values of cecal size were similar to that 2.1(1.6-2.5)%) of GF mice inoculated with Chf-feces of LF mice.

These results suggested that ecosystem of fusiform bacteria and clostridia harboring in cecum and feces of mice was very complex and the composition of these bacteria and physiology of mouse intestine might affect each other.

Table 3. Cecal size of each mouse group.

Groups	No. of mice	Cecal size[a]
PA-(1)	13	2.8 ± 0.33
PA-(2)	8	3.2 ± 0.50
PA-(3)	8	2.8 ± 0.18
PA-(1+2+3)	10	2.5 ± 0.33
PA-(1+2+3)	8	2.3 ± 0.16
Chf-feces of a LF mouse	14	2.1 ± 0.29
CV mice	14	1.6 ± 0.26
GF mice	9	6.2 ± 1.09

[a] Mean ± S.D.

Conclusions

1. A great number of bacterial species harbored in the cecum of mice and changed their composition.

2. Fusiform bacteria isolated from feces of a LF mouse were considered to belong to the genus *Clostridium*, but it

was difficult to comfirm their spore *in vitro.*

3. Combinations of great number of fusiform bacteria are required for the normalization of the GF mouse cecal size.

References

Gordon JH, Dubos R(1970). The anaerobic bacterial flora of the mouse cecum. J. Exp. Med. 132: 251.

Hazenberg MP, Custers-van Lieshout LMC, Engels W, Kock-van Dalen AC(1977). The clostridial flora of conventional mice. Z. Versuchstierk. 19: 167.

Itoh K, Mitsuoka T(1980). Production of gnotobiotic mice with normal physiological functions. I . Selection of useful bacteria from feces of conventional mice. Z. Versuchstierk. 22: 173.

Koopman JP, Kennis HM(1977). Differentiation of bacteria isolated from mouse ceca. Z. Versuchstierk. 19: 174.

Koopman JP, Kennis HM(1979). Characterization of anaerobic cecal bacteria of mice. Z. Versuchstierk. 21: 185.

Koopman JP, Kennis HM(1980). Influence of normal mouse intestinal bacteria on cecal weight in mice. Z. Versuchstierk. 22: 224.

Koopman JP, Kennis HM(1983). Characterization of anaerobic fusiform-shaped bacteria isolated from a mouse cecum with API ZYM. Z. Versuchstierk. 25: 311.

Koopman JP, Welling GW, Huybregts AWM, Mullink JWMA, Prins RA(1981). Association of germ-free mice with intestinal microfloras. Z. Versuchstierk. 23: 145.

Koopman JP, Welling GE, Kennis HM, Groen G(1983) Characterization of the intestinal bacteria of mice by API ZYM and SDS-polyacrylamide gel electrophoresis. Z. Versuchstierk. 25: 185.

Roach S, Tannock GW(1980). Anaerobic fusiform-shaped bacteria isolated from the caecum of conventional mice. J. Appl Bacteriol. 48: 115.

Wensinck F, Ruseler-van Embden JGH(1971). The intestinal flora of colonization-resistant mice. J. Hyg. Camb. 69: 413.

Germfree Research: Microflora Control and Its Application to the Biomedical Sciences, pages 221–223

ENTEROCOCCI IN LAMBS

Yen S. Liao, B. Sc, M.S., Ph. D.

Director, Laboratory of Microbiology in Zoonoses
Jiangxi Academy of Sciences
Nanchang, Jiangxi 330029
People's Republic of China

In a few years study of newborn lamb diarrhea, short chain streptococci, frequently in diplococcus form, had been isolated quite often from dead lambs. The streptococci could be easily isolated from rectums of all diseased and clinically healthy lambs. They were enterococci. Therefore internal organs of diseased and healthy lambs were cultured immediately or within 12 hours at a temperature below 10 C. Results were shown by Fig. 1. Weisser(1979) had enterococci also from emergency slaughtered cattle and calf.

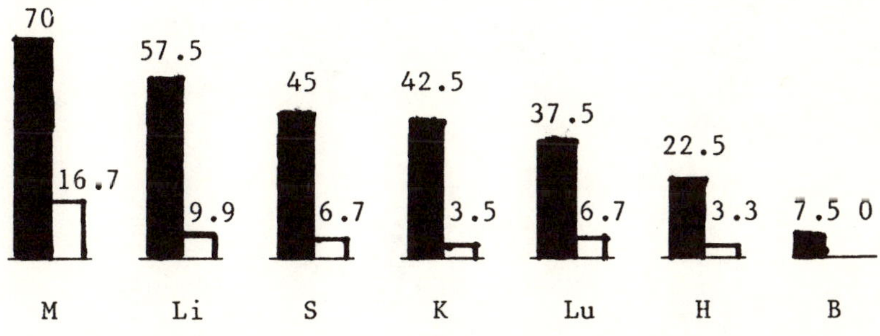

Fig. 1. Percentage of enterococci isolated from various organs of lambs. Black column....diarrheal lamb; White column....clinically healthy lamb. M, mesenteric lymph gland/ Li, liver/ S, spleen/ K, kidney/ Lu, lung/, H, heart/ B, brain.

Co-workers were S.M. Chang, S.Y. Chao, B.Y. Chang of Kansu Agricultural University and Y.W. Chu of Chenghai Animal Husbandry and Veterinary Medical College.

In organ cultures, the streptococci usually mixed with E. coli and sometimes with Clostridium welchii. Streptococci could be found in the colon of newborn lambs as early as two hours after birth.

Of the streptococci isolated, 94 strains were purified and identified. They were all gram positive spherical cocci, non-motile, mostly arranged in two and some in short chains, facultative anaerobic and very easily grown in nutrient broth, nutrient agar, cooked meat broth or thioglycollate broth. Stained by methods of Muir (Sheng et al, 1957) and Jasmin (1945) did not show capsule. Most strains could grow at 10 or 45 C, in pH 9.6, in 6.5% NaCl, in 0.1% methylene blue. They all resisted heat(60 C 30 minutes). They all fermented glucose, maltose and lactose but four. A few fermented arabinose, raffinose, inulin. Reactions to mannitol, glycerol, sucrose, sorbitol were variable. Five strains liquified gelatine. They can survive for 16 months in cooked meat broth at room temperature. Eighty-four strains identified as Streptococcus faecalis and five as Streptococcus faecalis var. liquifaciens.

Twenty four hours' broth culture of S. faecalis injected to mice (0.5 ml) or guinea pigs (0.1 ml) subcutaneously, intraperitoneally or intramuscularly and to pigeons (0.1 ml) intramuscularly did not cause any illness. The same broth culture given orally to lambs of two days of age, half (8/16) showed diarrhea and four recovered, four died.

A Clostridium welchii isolated from lamb identified as type B was highly toxic to pigeon. Pigeons injected with Clostridium welchii plus isolated Streptococcus faecalis showed almost the same death rate as injected with Clostridium welchii alone (Fig. 2).

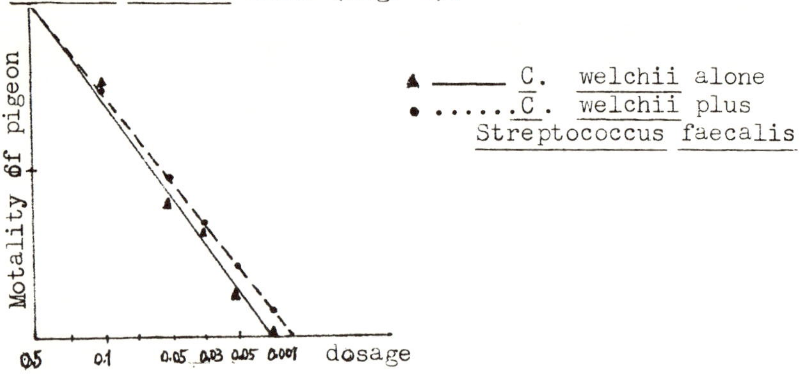

Fig. 2. Effect of Streptococcus faecalis on the toxicity of C. welchii type B.

Streptococcus faecalis used as a normal flora in control of newborn lambs diarrhea failed in this experiment. However, some authors reported that they could use some bacteria to control certain newborn animals diarrhea of E. coli origin.

Newborn animal	Bacteria used	Authors
lamb	Bacillus cereus	Kang(1984)
gnotobiotic pigs	Streptococcus faecium	Underdahi(1982)
Calf	Streptococcus faecalis and Lactobacillus	Flicoleaus et al (1978)
pig	non-pathogenic E. coli	Fong(1982)

Literature

Flicoleaux, JR, Ostre, L, Vaast, R, Cance, M (1978). Barrier effect of the digestive flora in the prevention of disease complex of newborn calves. Bulletin Mensuel de le Societe Veterinaire Pratique de France 62:563

Fong, TY(1982). Control pig diarrhea by non-pathogenic E. coli. J Inf Dis Anim 3:21

Jasmin, AM(1945). An improved staining method demonstrating bacterial capsules with particular reference to Pasteurella. J Bact 50:361

Kang, B(1984). On the mechanism of "cereobiogen" readjustment to dysbiosis (to be published).

Sheng, TS, Chu, CP, Liao, YS, Chang, BY, Chao, SY, Chu, YM (1957). A laboratory manual of veterinary microbiology. p. 147, Liberary Publisher of Animal Husbandry and Veterinary medicine, Nanking.

Underdahe, NR, Torres-Medina, A, Dester, AR(1982). Effect of Streptococcus faecium C-68 in control of E. coli-induced diarrhea in gnotobiotic pigs. Am J Vet Res 43:2227

Weisser, W (1979). Streptokokkenisolierungen aus Organen von not- und krankgeschlachteten Rindern und Kälbern im Rahmen der bakteriologischen Fleischuntersuchung. Archiv für Lebensmittelhygiene. 30:137

Germfree Research: Microflora Control and Its Application to the Biomedical Sciences, pages 225–229

THE INSTABILITY OF INTESTINAL CLOSTRIDIAL COLONISATION IN LIVING BEINGS

Elsa Haralambie and G. Linzenmeier

Inst.f.Med.Mikrobiologie,Univ.Klinik Essen

43 ESSEN, Hufelandstr. 55 FRG

Gnotobiotic surveillance cultures of fecal samples from patients with colon cancer revealed a dysbiosis in more than 90% of these patients, indicated by an increasing number of aerobes but also higher colony forming units (cfu)/g feces of various clostridial species (Schneiders 1976). Cholelithiasis and colon carcinoma are believed to have the same pathogenic cause: nutrition, intake or formation of lithogenic or cancerogenic substances (Peters, H. 1979). Experimental studies of cholecystectomy as a promoter effect in colon carcinogenesis are contradictory. Surgeons and internists are likewise interested in these problems and since bacteria seem to play a role, the clinical studies were attended by our microbiological investigations with special regard to clostridia.

<u>Surgical-Microbiological Study:</u>

The objective was to study the influence of cholecystectomy and 1,2 dimethylhydrazine (DMH) on the induction of colorectal tumor and on the intestinal bacteria of SPF-NMRI-mice. Bacteriological investigations were done monthly for 6 months and weekly thereafter. NMRI-mice were assigned to 3 groups. Group I: mice with laparotomy n=17; Group II: mice with laparotomy and 25 mg DMH/kg corp n=20, Group III: mice with cholecystectomy and 25 mg DMH/kg corp n=20. DMH administration began 3 weeks after surgery, once/week, for 12 weeks under s.c. ethylnarcotic. Forty-eight weeks after surgery, the animals were sacrificed.

Result: Tabl.I

Time and number of animals with new bacteria coming up after treatment with DMH.

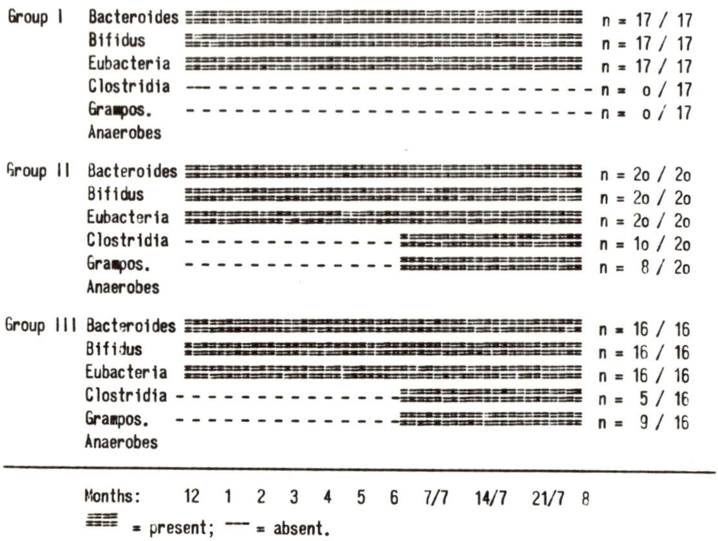

Tabl.II

Clostridia coming up in SPF mice after treatment with DMH.

Clostridia Species	Mice Group II n / animal n	Mice Group III n / animal n
C.innocuum	9 / 2o	2 / 16
C.ramosum	2 / 2o	1 / 16
C.perfringens	1 / 2o	1 / 16
C.paraputrificum	o / 2o	1 / 16
C.difficile	0 / 2o	1 / 16

Cholecystectomy and laparotomy didn't influence the incidence of tumors. Animals of groups II and III receiving a total dose of 300 mg DMH/kg corp developed tumors and the bacterial colonisation of their feces changed. In 35% of the animals, clostridia came up, at the same time the animals became colonised with gram-positive anaerobes which couldn't be identified.

Nutritive-Microbiological Study:

This study includes a hamster-study, a chimpanzee-study and the 7-Day Adventist study.

The objective of the Hamster-study was to study the role

of a lithogenic diet, the influence on the intestinal bacteria and the effect of lactulose-supplemented lithogenic diet. The animals were sacrificed after 6 weeks of feeding, colon contents were cultured under aerobic and anaerobic conditions for colony-counts/g feces. We observed 3 groups of animals:

Group I a. control group, normal chow, straw allowed n=10
b. control group, normal chow, straw not allowed n=10

Group II a. hamster with lithogenic diet, straw allowed n=17
b. hamster with lithogenic diet, no straw n=17

Group III Hamsters under lactulose-supplemented lithogenic diet, straw not allowed n=19

Result :

Tabl.3.

Hamster Bacteria	Group Ia -Ib present n /animal n		Group IIa - IIb present n/animal n		Group III present n/animal n
Lactobacilli	1o/1o	1o/1o	17/17	17/17	19/19
Gramneg.Aerobes	1o/1o	1o/1o	13/17	16/17	18/19
Eubacteria	1o/1o	1o/1o	17/17	17/17	19/19
Bifidus	1o/1o	1o/1o	17/17	17/17	19/19
Bacteroides	1o/1o	1o/1o	17/17	17/17	19/19
Grampos.anaerobes	1o/1o	1o/1o	12/17	1o/17	19/19
Clostridia	o/1o	o/1o	14/17	6/17	o/19
Peptococci	o/1o	9/1o	3/17	6/17	18/19
Megasphaera	o/1o	o/1o	7/17	5/17	6/19
Veillonella	1o/1o	1o/1o	17/17	15/17	19/19

Tabl.4.

Intestinal clostridia isolated from hamsters under lithogenic diet.

Clostridia	clostridia n/animal group IIa;	group IIb.
C.innocuum	12 / 17	2 / 17
C.difficile	1 / 17	4 / 17
C.clostridiiforme	3 / 17	-

Animals fed normal chow didn't harbour detectable amounts of clostridia and megasphaera in their fecal samples. The feeding of a lithogenic diet modified the bacterial colonisation of the gut. The modifications were stronger in animals with straw allowed. The addition of lactulose to the lithogenic diet normalised the clostridial colonisation of the gut.

Chimpanzee-study: Bacteriological aerobic and anaerobic investigations of 60 fecal samples from 3 adult chimpanzees were done fortnightly during 6 months and after a year, (20 samples/animal) to study the clostridial intestinal colonisation of chimpanzees fed their normal ovo-vegetable diet.

Result: Tabl.5.

Clostridia species	isolation-rate n / fecal sample n.
C.putrefaciens	6o / 6o
C.sporosphaeroides	3 / 6o
C.carnis	1 / 6o

The intake of a well defined diet caused no noteworthy change in the intestinal clostridial colonisation of chimpanzees. C. putrefaciens proved to be a resident intestinal clostridium of chimpanzees.

Seven Day Adventist-study: Isolated clostridia from 14 persons fed a lacto-ovo-vegetable diet were differentiated and compared to clostridia isolated from 32 fecal samples of 15 persons under "normal western diet", to study the intestinal clostridia under a lacto-ovo-vegetarian diet.

A: 35 fecal samples of 14 persons under a lacto-ovo-vegetarian diet were differentiated and compared to clostridia isolated from

B: 32 fecal samples of 15 persons under " normal western diet ".

Result: Tabl.6.

Clostridia Species	Group A	Group B
	Isolation-rate clostridia n / samples n.	
Toxin producing		
C.perfringens	2o / 35	32 / 32
C.botulinum	4 / 35	-
C.difficile	2 / 35	-
NDH clostridia		
C.paraputrificum	3 / 35	2 / 32
C.innocuum	7 / 35	1 / 32
Saccharolytic clostridia		
C.bifermentans	13 / 35	-
C.butyricum	12 / 35	1 / 32
C.tyrobutyricum	1 / 35	-
C.beijerinckii	-	2 / 32
C.ramosum	-	1 / 32
Asaccharolytic clostridia		
C.putrefaciens	31 / 35	9 / 32
C.sporosphaeroides	1 / 35	-
C.cochlearium	3 / 35	1 / 32
C.limosum	-	2 / 32
C.aminovalericum	-	1 / 32
C.sporogenes	-	2 / 32
C.subterminale	-	5 / 32

As suggested by the chimpanzee-study, we would expect a lower number of intestinal clostridia species in persons under a lacto-ovo-vegetable diet. This cannot be verified. Only the higher isolation rate of C. putrefaciens in vegetarians was found to be significant.

In a preceding study, the higher content of NDC (steroid nucleus dehydrogenating clostridia) in patients with colon cancer was due to the increased number of *C. innocuum* and *C. paraputrificum* (Haralambie 1981) but *in vitro*, these clostridia are not able to form cancerogenic metabolites of bile acids. (Edenharder 1982). Every species only performs a specific step of transformation in the bile acid metabolism. If bacteria are involved in the cancerogenic process, the degradation to cancerogenic metabolites may be performed by succeeding activities of a number of different bacterial species.

We gratefully thank Ms. P. Jager and U. Hotzel for excellent technical assistance.

REFERENCES

Edenharder HR (1982). Lecture, Workshop of Deutsche Gesellschaft fur Hygiene und Mikrobiologie, Mainz.

Haralambie E, Linzenmeier G (1981). Clostridia in the human gut. In Sasaki S *et al*. (eds): "Recent Advances in Germfree Research," Tokai University Press, p 267-211.

Peters H, Keimes AM (1979). Die Cholecystektomie als pradisponierender Faktor in der Genese des Kolorektalen Karzinoms DMW 45:1581.

Schneiders H, Haralambie E, Towfigh H, Eigler FW, Linzenmeier G (1976). Uber die Wirksankeit der Antibiotikavorbereitung in der Colonchirurgie. Chirurg 47:33.

Germfree Research: Microflora Control and Its Application to the Biomedical Sciences, pages 231–234

ATTEMPTS TO INFECT GERMFREE NU/NU MICE WITH ENTAMOEBA HISTOLYTICA

Gerd D. Burchard, M.D., Stefan Engelhorn, M.D., Manfred Dietrich, M.D.
Clinical Department, Bernhard-Nocht-Institut für Schiffs- und Tropenkrankheiten, Hamburg, Federal Republic of Germany

The pathogenicity of Entamoeba histolytica raises a number of questions which are still unsolved in human amebiasis. Few animal models are available to test the hypothesis of increasing the pathogenicity of Entamoeba histolytica by association of certain bacteria. Very few experiments have been done with germfree animals. In contrast to hamsters and guinea pigs, mice are less susceptible to an infection with E. histolytica. Recent experiments failed in oral inoculation of nu/nu mice, whereas intrahepatic infection was achieved only after treatment with anti-mouse lymphocyte serum.

So far a suitable animal model for experiments to elucidate the function of the T-cell and B-cell system as well as the phagocytic capacity in the host-parasite relationship is not available. Therefore, we tried to establish the association of Entamoeba histolytica in germfree nu/nu mice with a defective T-cell system and eventually with simultaneous suppression of the phagocytic capacity as well as the B-cell response by cytostatics and silica.

Material and methods.

Animals.

The experiments were carried out in nu/nu and nu/+ mice of the NMRI strain. The animals were kept germfree in plastic isolators. Age: at least 8 weeks. One group of animals was treated by cyclophosphamide 400 mg/kg one day before inoculation of E. histolytica. Another group of animals was treated with silica, 30 mg/animal IP, one day before inoculation of E. histolytica. A further group of animals was treated simultaneously by cyclophosphamide and silica. A fourth group of animals served as controls.

Strain of Entamoeba histolytica.

Entamoeba histolytica A3, cultivated with crithidia, was used. The crithidia were kept in TTY-medium, E. histolytica was grown in TP-S-1 monophasic medium (Diamond, 1978).

Infection.

E. histolytica was inoculated either intraperitoneally or by oral route. Between 10^5 and 10^6 trophozoites were given in 0.3 ml medium. The animals were sacrificed 1-4 weeks after inoculation.

Results.

Control animals.

The control animals (13 nu/nu mice and 28 nu/+ mice) had no gross organ lesions when sacrificed one week after IP inoculation of E. histolytica. E. histolytica could not be found in the intestinal mucosa by light microscopy. However, in 4 nu/nu mice and in 6 nu/+ controls, E. histolytica could be cultured from the cecal contents.

Cyclophosphamide group.

Only in 2 of 21 IP inoculated nu/+ mice E. histolytica could be grown from the cecal contents one week after inoculation, in no case was this possible in 8 nu/nu mice. In all of these IP infected mice as well as in 5 nu/nu and 8 nu/+ orally infected mice no histologically visible lesions could be found.

Silica group.

Pretreatment with silica also did not cause visible lesions at autopsy or histological signs of invasion in 5 nu/nu mice and in 8 nu/+ mice inoculated IP as well as in 3 nu/+ mice inoculated by oral route.

Cyclophosphamide-silica group.

Eight nu/+ mice were pretreated with silica and cyclophosphamide simultaneously and inoculated IP; 10 nu/+ mice were inoculated by oral route. In none of these cases were macroscopically or histologically visible lesions found. The results concerning cecal colonization after oral and IP inoculation of E. histolytica are summarized in Table 1 as follows:

Table 1

Cecal colonization with E. histolytica in germfree mice one week after inoculation

Animals	Treatment	Route of inoculation	Cecal colonization
nu/nu	-	i.p.	4/13
	-	oral	0/5
nu/+	-	i.p.	6/28
	-	oral	0/8
nu/nu	Cy	i.p.	0/8
	Cy	oral	0/5
nu/+	Cy	i.p.	2/21
		oral	0/8
nu/nu	Si	i.p.	0/5
nu/+	Si	i.p.	2/13
		oral	0/3
nu/+	Si + Cy	i.p.	2/8
	Si + Cy	oral	3/10

Discussion.

According to our results, infection of NMRI nu/nu mice with E. histolytica is neither possible by oral infection nor by intraperitoneal infection. Viable E. histolytica can be found only in some cases in the cecal contents after one week's time. It cannot be excluded definitely that during the intraperitoneal injection accidental injuries of the intestine occurred which allowed the ameba to enter the intestinal lumen. On the other hand, we made experiments in conventional mice with a safe intraperintoneal inoculation by laparotomy and found ameba in the cecal contents one week later, too. Obviously, E. histolytica is capable of crossing the small bowl wall from the peritoneal cavity to the lumen.

Our experiments were carried out with a culture strain of E. histolytica. This strain is pathogenic in the hamster liver, but it is generally accepted that E. histolytica gradually loses its virulence during long-term culture. Moreover, we did not try an intrahepatic inoculation of E. histolytica which generally is more successful than the intraperitoneal application. Bearing in mind these two limitations, our experiments show that the immune response of the host and the phagocytic activity are less important in the infection due to E. histolytica than species-specific factors and possibly a bacterial association.

Literature.

Diamond LS et al. (1978). A new medium for the axenic cultivation of E. histolytica and other Entamoeba. Trans Roy Soc Trop Med Hyg 72:431

Owen DG (1984). Attempt at oral infection of rats and mice with trophozoites of Entamoeba histolytica. Trans Roy Soc Trop Med Hyg 78:160

Phillips BP et al. (1955). Studies on the ameba-bacteria relationship in amebiasis. Comparative results of the intracecal inoculation of germfree, monocontaminated, and conventional guinea pigs with E. histolytica. Am J Trop Med Hyg 4:675

Phillips BP et al. (1972). Results of intracecal inoculation of germfree rats with axenically cultivated E. histolytica. J Protozool 19:498

Stern JJ et al. (1984). Murine amebiasis: The role of the macrophage in host defense. Am J Trop Med Hyg 33:372

Wijesundera M de S (1980). Hepatic amoebiasis in immunodepressed mice. Trans Roy Soc Trop Med Hyg 74:216

Germfree Research: Microflora Control and Its Application to the Biomedical Sciences, pages 235–238

ALTERATIONS IN CELL NUMBER, SIZE AND COLLAGEN CONTENT IN LIVERS OF CONVENTIONAL AND GERMFREE MICE BEARING SCHISTOSOMA MANSONI

L.Q. Vieira, T. Moraes-Santos, and E.C. Vieira

Dept. Bioquimica-Immunologia, ICB, UFMG
CP 2486, 30000 Belo Horizonte, MG
Brasil

Schistosomiasis is an endemic disease in the third world including Brazil (Machado 1982). It is caused by the blood fluke S. mansoni which lays thousands of eggs in the liver portal system. Approximately 50% of the eggs remain in the host with the major portion being trapped in the liver and intestines (Warren 1972). The eggs elicit granulomatous inflammation with infiltration by macrophages, T and B lymphocytes, eosinophils, fibroblasts, plasmocytes, neutrophils and mastocytes (Smith 1977; Bentley et al 1982). The granuloma results from a delayed type hypersensitivity (Warren et al 1967) and culminates with the cicatrization of the area due to the synthesis of collagen by fibroblasts (Wyler 1983). Mice are partially resistant to infection with S. mansoni cercariae. Only small numbers of adult worms are recovered from mice after the cercariae have penetrated the host skin (Smithers, Gammage 1980). Since resistance may be enhanced by the host's microflora (Dean 1983) and germfree mice have a milder inflammatory response to schistosomiasis (Vieira et al 1981), we assessed hypertrophy, hyperplasia and collagen accumulation in the liver of conventional (CV) and germfree (GF) mice infected with S. mansoni.

GF snails (Vieira 1967) were infected as previously described (Vieira et al 1981). The axenic cercariae were ampouled, introduced into an isolator and used to subcutaneously infect GF and CV LOB:(CFW) mice of the same age. After 7 weeks the animals were injected i.p. with 10 µCi of ^{3}H-thymidine (New England Nuclear) and sacrificed 24 hours later. Worms and eggs in the liver and intestine were counted as described by Pellegrino, Siqueira (1956) and Cheever (1964).

Liver protein (Hartree 1972), DNA (Prasad et al 1972), thymidine incorporation into DNA, and hydroxyproline (Stegmann 1958) were determined.

Vieira et al (1981) reported less accentuated hepatomegaly in _S. mansoni_ infected GF than in CV mice. In this study the recovery of adult worms as a percentage of cercariae inoculated into 13 GF mice was 25.4%, but only 13.2% were recovered from 16 CV mice. The results of the liver assays are shown in the following table.

Liver response to _S. mansoni_ eggs trapped in conventional (CV) and germfree (GF) mice. (Mean & standard deviation)[1]

	DNA/Liver	Protein/DNA	^{3}H-Thymidine incorporation[2]	OHpro/egg[3]
	(mg)	(mg/mg)	%	(ng)
CV	4.9+1.2 (15)	45.9+12.9 (15)	77.9+9.6 (16)	69.2+33.6 (9)
GF	5.2+1.3 (12)	46.4+13.7 (12)	87.8+9.8* (8)	37.8+22.3* (10)

[1]Asterisks indicate statistical difference from CV ($P<0.05$). Numbers in parenthesis indicate sample size.
[2]Radioactivity in total DNA, related to radioactivity in the liver.
[3]Liver hydroxyproline was divided by the number of eggs in the liver.

GF mice harbored about twice as many worms as CV mice. A CV host normally kills the worms during the migration of the schistosomula larva from the skin through the lungs to the liver, but the GF mice are less effective in removing the worms. The mechanisms that kill the worms are not known; however, macrophages (White 1976) and lymphocytes (Ellner et al 1982) are nonspecifically activated in this process.

The total content of DNA in liver and the ratio of Protein/DNA are the same in both groups; therefore, the average cell number and cell size do not seem to be responsible for the less accentuated hepatomegaly seen in GF mice (Vieira et al 1981). On the other hand, it has been reported that eggs or factors from eggs induce replication of cells involved in granuloma formation (Borojevic et al 1981; Bentley et al 1982; Wyler 1983). The table shows that GF mice incorporate more ^{3}H-thymidine into their liver DNA than do their CV counterparts. However, the number of eggs found in the liver of GF mice (1200 eggs) was twice that found in CV mice (660 eggs), while the increase in the percentage of ^{3}H-Thymidine incorporation was only 15% higher. In order to show the influence of eggs or egg factors on collagen production, which is proportional to the number of eggs trapped in the liver (Cheever et al 1983), we divided the amount of collagen in liver by the number of eggs found in the liver. It can be seen that GF mice incorporate less ^{3}H-thymidine and produce less collagen than CV ones, when related to the number of eggs in the liver. Our results suggest that GF mice are less responsive to S. mansoni infection in addition to allowing a better worm recovery.

REFERENCES

Bentley AG, Doughty BL, Phillips SM (1982). Ultrastructural analysis of the cellular response to Schistosoma mansoni. III. The in vitro granuloma. Am J Trop Med Hyg 31:1168.

Borojevic R, Stocker S, Grimaud JA (1981). Hepatic eosinophil granulocytopoiesis in murine experimental schistosomiasis mansoni. Br J Exp Pathol 62:480.

Cheever AW (1964). Condition affecting the accuracy of potassium hydroxide digestion techniques for counting Schistosoma mansoni eggs in tissues. Bull WHO 39:328.

Cheever AW, Dunn MA, Dean DA, Duvall RH (1983). Differences in hepatic fibrosis in ICR, C3H and C57BL/6 mice infected with Schistosoma mansoni. Am J Trop Med Hyg 32:1364.

Dean DA (1983). A review: Schistosoma and related general acquired resistance in mice. Exp Parasitol 55:1.

Ellner JJ, Alds R, Lee CW, Kleinhenz ME (1982). Destruction of the multicelluar parasite Schistosoma mansoni by lymphocytes T. J Clin Invest 70:369.

Hartree EF (1972). Determination of protein: a modification of the Lowry method that gives a linear photometric response. Anal Biochem 48:422.

Machado PA (1982). The brazilian program of schistosomiasis control, 1975-1979. Am J Top Med Hyg 31:76.

Pellegrino J, Siqueira AF (1956). Tecnica de perfusao para colheita de Schistosoma mansoni em cobaias experimentalmente infestadas. Rev Bras Malariol Doencas Trop 8:589.

Prasad AS, DuMouchelli E, Koniuch D, Oberleas D (1972). A simple fluorometric method for the determination of RNA

and DNA in tissues. J Lab Clin Med 80:598.
Smith MD (1977). The ultrastructural development of the schistosomal egg granuloma in mice. Parasitology 75:119.
Smithers SR, Gammage K (1980). Recovery of Schistosoma mansoni from the skins, lungs and hepatic portal system of naive mice and mice previously exposed to Schistosoma mansoni: evidence for two phases of parasites attrition in immune mice. Parasitology 80:289.
Stegemann H (1958). Mikrobestimmung von hydroxyprolin mit chloramin-T und p-dimethylaminobenzaldehyd. Hoppe-Seyler's Z Physiol Chem 311:41.
Vieira EC (1967). Influence of vitamin E on reproduction of Biomphalaria glabrata under axenic conditions. Am J Trop Med Hyg 16:792.
Vieira EC, Bezerra M, Nicoli JR, Pleasants JR, Coelho PMZ, Bambirra EA (1981). Host-parasite relationship in Schistosomiasis mansoni in the germfree mouse. In Sasaki S, Ozana A, Hashimoto K (eds): "Recent Advances in Germfree Research", Tokyo: Tokai Univ Press, p 433.
Warren KS (1972). The immunopathogenesis of schistosomiasis: a multidisciplinary approach. Trans Royal Soc Trop Med Hyg 65:417.
Warren KS, Domingo EO, Cowan RBT (1967). Granuloma formation around schistosome eggs as a manifestation of delayed hypersensitivity. Am J Pathol 51:735.
White RG (1976). The adjuvant effect of microbial products in the immune response. Ann Rev Microbiol 30:579.
Wyler DJ (1983). Regulation of fibroblast functions by products of schistosomal egg granulomas: potential role in the pathogenesis of hepatic fibrosis. In "Cytopathology of Parasitic Disease", (Ciba Foundation Symposium 99), London: Pitman Books, p 190.

This work was supported by FINEP and CNPq-Brasil. LQV and TMS are fellows of CNPq.

V ANTIBIOTICS: MANIPULATION OF THE MICROFLORA

Germfree Research: Microflora Control and Its Application to the Biomedical Sciences, pages 241–244

THE INFLUENCE OF ANTIBIOTICS UPON MICROFLORA-ASSOCIATED CHARACTERISTICS IN MAN AND MAMMALS

Tore Midtvedt, PH.D., M.D.

Chief Department of Germfree Research
Karolinska Institutet, Stockholm, Sweden

Arvid Bjørneklett, National Hosp. Oslo,Norway
Bodil Carlstedt-Duke, Karolinska Inst.,Sweden
Bengt Erik Gustafsson,Karolinska Inst.,Sweden
Torgeir Høverstad, National Hosp. Oslo,Norway
Egil Lingaas, National Hospital, Oslo,Norway
Karin Elisabeth Norin,Karolinska Inst.,Sweden
Henri Saxerholt, Karolinska Institutet,Sweden
Martin Steinbakk, National Hosp. Oslo,Norway

It is a well established fact that antimicrobial agents - when given to a macroorganism - may interfer with the gastrointestinal microflora, and a variety of aims are possible when studying alterations of the flora caused by administration of such drugs. In the past, most attention has been paid to determinations of quantitative or qualitative changes in the groups of microbes present as well as of determinations of emerge and colonization of antibiotic-resistant microbes.

However, it is now more and more realized that normally many structures and functions in the macroorganism are influenced by the microbial flora. For studying these complex interactions, introduction of two new terms - MAC and GAC - might be of some value.

MAC: a microflora-associated characteristic - can be defined as the recording of any anatomical structure, biochemical or physiological function in a macroorganism which has been influenced by the microflora.

GAC: germfree animal characteristic - when microbes are absent, as in germfree animals, the recording of a microflora-associated characteristic (MAC) can be defined as a germfree animal characteristic (GAC).

In Table 1, some biochemical MAC's known to be influenced by antimicrobial agents are listed.

Table 1. MAC's known to be influenced by antimicrobial agent

Conversion of cholesterol to coprostanol
Deconjugation of bile acid conjugates
Dehydroxylation of bile acids
Deconjugation of bilirubin conjugates
Conversion of bilirubin to urobilinogens
Transformation of mucin
Inactivation of enzymes (trypsin etc)
Production of gases (H_2, CH_4, etc, measured in respir. air)
Production of short chain fatty acids (SCFA's)
Absence of dipeptides (ß-aspartylglycine etc)

Previously, we have studied the influence of some antimicrobial drugs upon some of these MAC's.

Table 2. Influence of antibiotics upon MAC's

Species (reference)	Characteristic	Agent	Effect
Man (Midtvedt & al 1977)	Coprostanol	AMP,CLIN,OXI NEO/BAC PEN V,DOX	GAC pattern " " No effect
Man (Bjørneklett & al 1981)	H_2-productin	MET,PEN V DOX	Reduction No effect
	CH_4-production	MET DOX,PEN V	Reduction No effect
Rat (Gustafsson & al 1977)	Bile acids: deconjugation/ dehydroxylation	ERY,NEO/BAC NEO,PEN G MET	GAC pattern " " Minor effect
Rat (Gustafsson & al 1977)	Coprostanol/ urobilinogen/ mucin/trypsin	AMP,COL,ERY NEO/BAC,NEO PEN G MET	GAC pattern " " " " Minor effect

Species (reference)	Characteristic	Agent	Effect
Rat (Genell & al 1977)	Trypsin	AMP,CLIN DOC, PEN G	GAC pattern " "
Rat (Carlstedt-Duke & al, in press)	Coprostanol/ urobilinogen/ mucin/trypsin	CLIN (dose variation)	GAC pattern (even at very low doses)

Abbreviations: AMP = ampicillin, CLIN = clindamycin, COL = colistin, DOX = doxycycline, ERY = erythromycin, MET = metronidazole, NEO/BAC = neomycin/bacitracin, NEO = neomycin, PEN G = benzylpenicillin, PEN V = phenoxymethylpenicillin, OXI = oxitetracycline.

Generally, the effects in rats upon MAC's were found to be longlasting. In many cases, the animals did not turn from GAC to MAC values until one or more enemas with cecal content from untreated rats were given weeks or months after sessation of antibiotic administration.

In an ongoing study, clindamycin, ampicillin or metronidazole have been given for six days to healthy volunteers. The following MAC's have been studied: Conversion of cholesterol to coprostanol, deconjugation of bilirubin, formation of urobilinogens, degradation of mucin, inactivation of trypsin, production of SCFA's (up to C_6), and absence of ß-aspartylglycine (Midtvedt & al, to be published). Additionally, presence of Clostridium difficile and its toxin(s) have also been followed. Fecal samples were taken twice prior to and three times during antibiotic administration. Thereafter, the subjects were followed weekly to monthly until normalization of their MAC's.

So far, clindamycin has been found to have the most and metronidazole the least pronounced effect upon the characteristics investigated. The influence upon deconjugation of bilirubin conjugates and formation of urobilinogens is reported elsewhere (Saxerholt et al. This volume). In all students receiving clindamycin, the electrophoretic pattern of mucin turned from MAC to GAC value during administration of antibiotic. Similarly, the conversion of cholesterol to coprostanol was markedly reduced, and so were the inactivation of trypsin and production of SCFA's. No effect was found upon formation of ß-aspartylglycine. In most of the students, MAC patterns were restored 6 weeks after sessation of therapy. No attempt was made to isolate or enumerate the microbes actually involved in the biochemical interactions studied.

To conclude: Antibiotic may influence upon many microflora-associated characteristics in man and mammals, and flora-related biochemical functions can be depressed for shorter or longer periods. However, the physiological or pathophysiological importance of turning from MAC to GAC patterns remains to be settled.

Bjørneklett A, Midtvedt T (1981). Influence of three antimicrobial agents - penicillin, metronidazole, and doxycycline - on the intestinal microflora of healthy humans. Scand J Gastroent 16:473.

Carlstedt-Duke B, Gustafsson BE, Midtvedt T (1984). Clindamycin - induced alterations in intestinal microflora-associated characteristics in rats. Scand J Gastroent. In press.

Genell S, Gustafsson BE (1977). Impaired enteric degradation of pancreatic endopeptidases in antibiotic-treated rats. Scand J Gastroent 12:801.

Gustafsson BE, Gustafsson J-Å, Carlstedt-Duke B (1977). Prolonged induction of germfree bile acid pattern in conventional rats by antibiotics. Acta Med Scand 201:155.

Gustafsson BE, Norin KE (1977). Development of germfree animal characteristics in conventional rats by antibiotics. Acta Pathol Microbiol Scand, Sect B, 85:1.

Midtvedt T, Frederichsen P (1977). Influence of antibiotics on microbial intestinal transformation of cholesterol to coprostanol in man. Scand J Gastroent 12:669.

Midtvedt T, Carlstedt-Duke B, Høverstad T, Lingaas E, Norin KE, Saxerholt H, Steinbakk M (1984). Influence of peroral antibiotics upon some biochemical microflora-associated characteristics in healthy volunteers. To be published.

Germfree Research: Microflora Control and Its Application to the Biomedical Sciences, pages 245–250

SPREAD OF MULTIRESISTANT GRAM-NEGATIVE BACILLI AMONG SEVERELY IMMUNO-COMPROMISED MICE DURING PROPHYLACTIC TREATMENT WITH DIFFERENT ORAL ANTIMICROBIAL DRUGS

D. van der Waaij, J.M. van der Waaij

Professor of Medical Microbiology
University Hospital,
Oostersingel 59, 9713 EZ Groningen
The Netherlands

SUMMARY

Gram-negative infections are generally regarded as a major cause of severe infections in granulocytopenic patients. To prevent such infections, selective decontamination (SD) of the digestive tract has been introduced. It is assumed that if SD is applied on all patients in an oncologic station it may not only provide protection to patients thus treated, but it may have also epidemiologic consequences. SD may considerably reduce the number of circulating (antibiotic-resistant) strains in the ward and therewith significantly further reduce the chance of severe infection during periods of granulocytopenia.

Because the effect which different kinds of antibiotics may have on the accumulation and spread of resistant bacteria cannot be studied in patients, it was investigated experimentally. Groups of 15 conventional mice were lethally irradiated and each group was treated with one antibiotic. On the day of irradiation one mouse per cage was orally contaminated with a multi-resistant strain either of *E. coli* or *Ps. aeruginosa*. The spread of these strains among cage mates as well as their effect on mortality was investigated. It appeared that ampicillin promoted spread and enhanced mortality in comparison with a control group, whereas antibiotics currently used for SD did prevent spread of the multi-resistant strain and reduced mortality.

INTRODUCTION

If antimicrobial drugs like nalidixic acid and polymyxin are applied in sufficiently high oral doses, the susceptible Gram-negative bacilli will soon disappear from the digestive tract of a so treated individual. The anaerobic mucosa-associated flora however, which establishes colonization resistance (CR) remains during such treatment apparently unaltered (van der Waaij et al, 1971; van der Waaij, 1979). Therefore, this kind of treatment, which is presently successfully applied for infection prophylaxis in patients with severe neutropenia, is called selective decontamination of the digestive tract.

Prophylactic treatment with drugs which affect the mucosa-associated flora such as ampicillin on the other hand may cause a dramatic decrease of the CR. The consequences of maintenance of CR-associated flora in immuno-compromised individuals during antibiotic prophylaxis were subject of the present experimental study. Our investigation regarded both consequences for the development of lethal infections as well as for epidemic spread of resistant bacteria among immunocompromised individuals.

EXPERIMENTAL DESIGN

To investigate what may occur in patients during chemotherapy for remission induction in acute leukemia and who are treated with (an) antibiotic(s) which either does or does not affect the CR, the following experiments were performed in mice. Two groups of randomly bred Swiss female mice of 12 weeks of age, were x-irradiated with 700 rads and placed with 15 in a cage. Per cage the animals shared one drinking bottle and a hopper with pelleted food.

Two cages with 15 animals were not treated, two were given ampicillin (CR-decreasing) in the drinking water (0.5 mg/ml) (Thijm, van der Waaij 1979); two were treated with nalidixic acid (CR-indifferent) in the drinking water (1 mg/ml) (Wiegersma et al, 1982) and the mice in two other cages were treated with polymixin B in the drinking water (1 mg/ml) (Emmelot, van der Waaij, (1980). In four of these eight cages (one of each treatment kind) one earmarked animal had been orally contaminated the day before treatment and irradiation with 10^9 bacteria of a ampicillin and nalidixic acid resistant E. coli strain. In the remaining four cages one ear-marked mouse had been orally contaminated the day before with 10^9 bacteria of an ampicillin and nalidixic acid resistant strain of Ps. aeruginosa. In all cages individual sampling of feces was attempted every other day. In mice which died, heart blood was taken aseptically for culturing.

RESULTS AND DISCUSSION

The results of the study can be divided in epidemiologic results (Table 1) and in results concerning survival (Table 2).

Spread of both E. coli and Ps. aeruginosa was rapid in the ampicillin treated group. This was obviously correlated with the decreased CR in these animals. Spread in the control group, as well as among the nalidixic acid treated mice, was only seen in the Ps. aeruginosa contaminated groups. The fact that the contaminant strains as well as the endogenous Gram-negative bacilli were sensitive to polymyxin B, may explain why these bacteria were not cultured from the feces of polymyxin treated mice.

The influence of "overgrowth" in the GI-tract of ampicillin treated mice by the resistant potentially pathogenic bacteria (E. coli and Ps. aeruginosa) on mortality following x-irridation, is since long known (Miller et al, 1950; Gonshery et al, 1953). It is, however, perhaps not always realized that the post-irradiation treatment with CR-decreasing antibiotics may enhance mortality like in this animal model. When irradiated individuals become heavily colonized (overgrown) by resistant potentially pathogenic organisms the almost inevitable infection will become soon lethal if specific antibiotic therapy under clinical circumstances is not rapidly started. The effect of CR-decreasing treatment is more dramatic if it concerns a more pathogenic species such as Ps. aeruginosa than if the digestive tract of irradiated individuals becomes loaded with massive numbers of E. coli.

Selective decontamination (SD) with nalidixic acid or polymyxin on the other hand reduced mortality in our mouse model in comparison with the control group. SD with polymyxin B was not only effective in reducing of mortality in E. coli contaminated mice but also in pseudomonas contaminated animals (Table 2). Nalidixic acid on the other hand had no obvious effect on survival in the pseudomonas contaminated

Table 1. Epidemic spread of ampicillin resistant E. coli or ampicillin-nalidixic acid resistant Ps. aeruginosa among lethally irradiated mice (7 Gy) treated with either one of three antimicrobial drugs after oral contamination of one mouse per cage of 15 animals.

contaminant	antibiotic	days after irradiation and contamination (+ 1 day) 2	4	6	8	10	12	14	16	18	20	22	24
E.coli/Ps.aerug.	control	7*/7	7/7	7/60	7/100	7/+	7/+	0/+	+/+	+/+	+/+	+/+	+/+
E.coli/Ps.aerug.	ampicillin	21/21	79/100	19/100	100/+	100/+	100/+	+/+	+/+	+/+	+/+	+/+	+/+
E.coli/Ps.aerug.	nalidixic acid	7/7	7/7	7/7	7/21	0/41	0/+	0/+	0/+	0/+	0/+	0/+	0/+
E.coli/Ps.aerug.	polymyxin B	0/0	0/0	0/0	0/0	0/0	0/0	0/0	0/0	0/0	0/0	0/0	0/0

* figures in the body of the table show percentage of mice excreting the contaminant in the feces; + indicates no survivors.

Table 2. Mortality in mice following lethal irradiation (7 Gy) and antimicrobial treatment with either one of three antimicrobial drugs in the presence of ampicillin resistant E. coli or ampicillin-nalidixic acid resistant Ps. aeruginosa; 30 mice per antimicrobial treatment group.

contaminant	antibiotic	days after irradiation and contamination (+ 1 day) 2	4	6	8	10	12	14	16	18	20	22	24
E.coli/Ps.aerug.	control	100*/100	100/100	100/100	93/21	60/0	27/0	14/0	0/0	0/0	0/0	0/0	0/0
E.coli/Ps.aerug.	ampicillin	100/100	100/100	100/41	67/0	27/0	14/0	0/0	0/0	0/0	0/0	0/0	0/0
E.coli/Ps.aerug.	nalidixic acid	100/100	100/100	100/100	100/100	93/60	72/0	67/0	67/0	67/0	67/0	67/0	67/0
E.coli/Ps.aerug.	polymyxin B	100/100	100/100	100/100	100/100	100/100	86/79	79/48	79/48	79/48	79/48	79/48	79/48

* figures represent percentage of surviving mice

mice.

The conclusion that could be drawn from this experiment is that, if our animal model - with mice which were not individually medically treated other than with some form of antimicrobial treatment - is representative for circumstances which may occur in patients with acute leukemia, the following statements seem valid. Antibiotic treatment:

1. must not be performed with CR-decreasing antibiotics against which resistance occurs among endemically present potentially pathogenic microbes.
2. should be of SD-type in order to selectively eliminate all susceptible potentially pathogenic organisms from the GI-tract as to prevent infection and to limit spread of resistant organisms by maintaining the CR-maintaining (anaerobic) microflora (Sleijfer et al, 1980; Dekker et al, 1981; Guiot et al, 1981).
3. may reduce mortality if it is of SD-type (Sleijfer et al, 1980).
4. may reduce spread and incidence of infection of resistant Gram-negative bacteria as well as of such as Salmonella-, Shigella-, Yersinia- and Campylobacter species.
5. may - with properly selected antimicrobial drugs - also limit the incidence of acute bacterial bronchopneumonias such as staphylococcal and pneumococcal pneumonia.

REFERENCES

Dekker AW, Rozenberg-Arska M, Sixma J, Verhoef J (1981). Prevention of infection by trimethoprim-sulfamethoxazole plus amphotericin B in patients with acute nonlymphocytic leukaemia. Ann Int Med 95:555.

Emmelot CH, van der Waaij D (1980). The dose at which neomycin and polymyxin B can be applied for selective decontamination of the digestive tract in mice. J Hyg 84:331.

Gonshery L, Marston RQ, Smith WW (1953). Naturally occurring infections in untreated and streptomycin-treated x-irradiated mice. Am J Physiol 172:359.

Guiot HFL, van der Meer JWM, van Furth R (1981). Selective antimicrobial modulation of human microbial flora: Infection prevention in patients with decreased host defense mechanisms by selective elimination of potentially pathogenic bacteria. J Inf Dis 143:644.

Miller CP, Hammond CW, Tompkins M (1950). The incidence of bacteremia in mice subjected to total body x-radiation.

Science 111:540.
Sleijfer DT, Muller NH, de Vries-Hospers HG, Fidlar V, Nieweg HO, van der Waaij D, van Saene HKF (1980). Infection prevention in granulocytopenic patients by selective decontamination of the digestive tract. Eur J Cancer 16:859.
Thijm HA, van der Waaij D (1979). The effect of three frequently applied antibiotics on the colonization resistance of the digestive tract of mice. J Hyg 82:397.
van der Waaij D, Berghuis-de Vries JM, Lekkerkerker-van der Wees JEC (1971). Colonization resistance of the digestive tract in conventional and antibiotic-treated mice. J Hyg 69:405.
van der Waaij D (1979). Colonization resistance of the digestive tract as a major lead in the selection of antibiotics for therapy. In van der Waaij D, Verhoef J (eds): "New Criteria for Antimicrobial Therapy: maintenance of digestive tract colonization resistance" Excerpta Medica, Amsterdam-Oxford, p 271.
Wiegersma N, Jansen G, van der Waaij D (1982). Effect of twelve antimicrobial drugs on colonization resistance of the digestive tract of mice and on endogenous potentially pathogenic bacteria. J Hyg 88:221.

Germfree Research: Microflora Control and Its Application to the Biomedical Sciences, pages 251–253

SELECTIVE DECONTAMINATION OF THE DIGESTIVE TRACT FOR THE PREVENTION OF INFECTION IN ACUTE LEUKEMIA

A. Dekker, W. Gaus, E. Kurrle*, H. de Vries-Hospers, D. van der Waaij and F. Wendt
E.O.R.T.C. Gnotobiotic Project Group
*University of Ulm, Steinhövelstr. 9 7900 Ulm, FRG

Infections are major complications and causes of death in patients with acute leukemia during remission induction therapy. For the prevention of these infections several methods of antimicrobial decontamination have been developed within the last decade. The methods which are currently in use are mostly based on the principles of selective decontamination which aims to eliminate potentially pathogenic microorganisms from the digestive tract without affecting the anaerobic flora. A series of studies have documented that selective decontamination significantly reduces the incidence of infections in patients with acute leukemia (Hughes 1977; Gurwith 1979; Sleijfer 1980; Dekker 1981; Guiot 1983; Kauffman 1983). The antimicrobial drugs which were applied in these studies were cotrimoxazole, polymyxin B, colistin and nalidixic acid given either as single substances or in various combinations, mostly combined with antifungal drugs. There is only limited information available which of the drugs or drug combinations should be given preference in respect to efficacy, tolerance and side effects (Watson 1982; Kurrle 1983; Wade 1983). Therefore a prospective randomized study was performed in which a combination of non-absorbable antimicrobial drugs (regimen A) was compared with a combination of absorbable and non-absorbable drugs (regimen B).

A total of 140 patients were randomized into the two treatment groups A and B. Group A was given polymyxin B (800 mg/day), neomycin (1000 mg/day) and amphotericin B (2000 mg/day). Group B received a combination of cotrimoxazole (trimethoprim 480 mg/day and sulfamethoxazole 2400 mg/day), polymyxin B (800 mg/day) and amphotericin B (2000 mg/day).

The incidence of febrile days with temperature above 38°C was 32% in group A and 26% in group B, a small but significant difference ($p < 0.01$). The reduction of febrile days resulted in a decrease in the use of parenteral antibiotics in group B which were applied on 28% of the study days in this group as compared to 41% in group A ($p < 0.01$).

The number of patients who acquired major infections with positive blood cultures (14 in group A, 6 in group B) and without positive blood cultures (13 in group A, 10 in group B) was higher in group A. In group B more patients remained free of acquired infections (18 in group A, 25 in group B) or had only minor infections (13 in group A, 18 in group B) or fever episodes without proven infection (2 in group A, 4 in group B). The overall differences in the severity of infections were significant ($p = 0.036$).

The most striking differences between the two groups were found in the occurrence of septicemias without localized infections (15 in group A, 1 in group B) and of urinary tract infections (8 in group A, 2 in group B). The incidence of other localizations of infections was similar and the death rate was the same in both groups.

The spectrum of microorganisms causing infections did not differ between the two groups as far as Enterobacteriaceae, Pseudomonas spp. and Candida spp. were concerned. However, gram-positive cocci caused substantially more infections in group A as compared to group B (16 vs. 5).

The mean duration of granulocytopenia with a neutrophil count less than 500/µl was 26.8 days in group A and 32.1 days in group B, a difference which was not significant ($p > 0.05$). Allergic reactions occurred in 5 patients of group A and in 19 patients of group B.

The data of this study indicate that regimen B was more effective than regimen A, especially in the prevention of severe infections. This finding is in agreement with previous studies which reported that cotrimoxazole or combinations with cotrimoxazole were superior to other prophylactic regimen (Watson 1982; Kurrle 1983; Wade 1983). On the other hand, more allergic side effects which had to be attributed to contrimoxazole occurred in group B. The duration of severe granulocytopenia was slightly but not statistically significantly longer in group B. Other studies have described a significant prolongation of neutropenia in patients receiving cotrimoxazole, a possible side effect of this drug which should be considered in the final assessment of regimen B (Dekker 1981; Wade 1983).

Dekker AW, Rozenberg-Arska M, Sixma J, Verhoef J (1981). Prevention of infection by trimethoprim-sulfamethoxazole plus amphotericin B in patients with acute nonlymphocytic leukemia. Ann Intern Med 95: 555.

Guiot HFL, van den Broek PJ, van der Meer JWM, van Furth R (1983). Selective antimicrobial modulation of the intestinal flora of patients with acute nonlymphocytic leukemia: a double-blind, placebo-controlled study. J Infect Dis 147:623.

Gurwith MJ, Brunton JL, Lank BA, Harding GKM, Ronald AR (1979). A prospective controlled investigation of prophylactic trimethoprim/sulfamethoxazole in hospitalized granulocytopenic patients. Am J Med 66: 248.

Hughes WT, Kuhn S, Chaudhary S, Feldman S, Verzosa M, Aur RJA, Pratt C, George SL (1977). Successful chemoprophylaxis for pneumocystis carinii pneumonitis. N Engl J Med 297: 1419.

Kauffman C, Liepman MK, Bergman AG, Mioduszewski J (1983) Trimethoprim/sulfamethoxazole prophylaxis in neutropenic patients. Am J Med 74: 599.

Kurrle E, Bhaduri S, Krieger D, Pflieger H, Heimpel H (1983). Antimicrobial prophylaxis in acute leukaemia: prospective randomized study comparing two methods of selective decontamination. Klin Wschr 61: 691.

Sleijfer DT, Mulder NH, de Vries-Hospers HG, Fidler V, Nieweg HO, van der Waaij D, van Saene HKF (1980). Infection prevention in granulocytopenic patients by selective decontamination of the digestive tract. Eur J Cancer 16: 859.

Wade JC, de Jongh CA, Newman KA, Crowby J, Wiernik PH, Schimpff SC (1983). Selective antimicrobial modulation as prophylaxis against infection during granulocytopenia: trimethoprim-sulfamethoxazole vs. nalidixic acid. J Infect Dis 147: 624.

Watson JG, Jameson B, Powles RL, McElwain TJ, Lawson DN, Judson J, Morgenstern GR, Lumley H, Kay HEM (1982). Co-trimoxazole versus non-absorbable antibiotics in acute leukaemia. Lancet i: 6.

Germfree Research: Microflora Control and Its Application to the Biomedical Sciences, pages 255–258

SELECTIVE DECONTAMINATION OF THE DIGESTIVE TRACT BY NORFLOXACIN. A PRELIMINARY REPORT

Sophie S. Pecquet, Antoine O. Andremont
Cyrille H. Tancrède
Service de Microbiologie Médicale
Institut Gustave-Roussy
94805 Villejuif Cedex, France

Infections in neutropenic patients are often caused by potentially pathogenic bacteria, as enterobacteria and P. aeruginosa, originating from the intestinal tract. For preventing these infections, oral antimicrobial prophylaxis is used. Potentially pathogenic bacteria are then eliminated from the digestive tract, but the anaerobic flora is expected to remain intact (Van der Waaijd and col, 1980). We have evaluated the possible use of Norfloxacin, a new carboxilic acid, to perform microbial modulation of the intestinal tract. It seems that this compound is a good candidate for this purpose, because its absorption is intermediate (40 %) (Swanson, Boppana and col, 1983), it has an higher activity than other quinolones on the strains which are susceptible to Nalidixic acid and its spectrum is extended to Nalidixic acid resistant strains, especially P. aeruginosa (Norrby, Jonsson, 1983). Like other quinolones, it does't select plasmid resistance.

Then we have tested its activity in the lumen of the intestinal tract.

Twelve adults healthy fully informed volunteers were orally treated during five days. Six of them received 200 mg twice a day, the other six 400 mg twice a day. Every day blood samples were collected two hours after the morning administration, to determine the serum concentration of Norfloxacin. Fecal samples were also collected to determine the total count of bacteria, the number of enterobacteria and enterococci, and the fecal concentration of Norfloxacin. Blood and fecal samples were tested by microbiological assay using the susceptibility of E. coli IPP 7624.

Serum Concentrations of Norfloxacin were 1.03 ± 0.09 mcg/ml and 1.73 ± 0.17 mcg/ml during treatment with 400 and 800 mg daily regimens respectively. A residual concentration of 0.43 ± 0.11 mcg/ml persisted 4 days after the end of treatment.

Fecal Concentrations of Norfloxacin gradually increased during treatment and ranged between 1 and 5 mg per gram of feces. Large sample to sample variations were observed in the two regimens. The mean fecal concentrations were 2.5 mg per gram of feces and 4 mg per gram of feces for the two dosages respectively. Four days after the end of treatment, the residual concentrations ranged from 0.1 to 1.5 mg per gram of feces.

Total counts of bacteria were performed in an anaerobic glove box on Freter medium (Aranki, Kenney, Freter, 1969).

Table 1 FECAL COUNTS OF ANAEROBES

	DOSES (mcg/day)	TOTAL COUNTS*
PRETREATMENT	400	■ 10.59 ± 0.25
	800	10.42 ± 0.38
MID TREATMENT	400	10.17 ± 0.31
	800	■ 9.86 ± 0.55
POST TREATMENT	400	■ 10.27 ± 0.28
	800	10.32 ± 0.23

* Geometric mean of 6 subjects ± SEM (cfu/g of feces (log))

■ $p > 0.05$

The results observed before treatment are in agreement with those normally found in a normal fecal flora (table 1). On day 4 of treatment, the two regimens haven't different effects. The total counts of anaerobic bacteria decrease during treatment, but this decrease was not more important in the 800 mg group.

As there were no difference between the two groups, the table 2 shows the results with the two regimens together.

Table 2 FECAL COUNTS OF ANAEROBES

PRETREATMENT	10.50 ± 0.32*		
		$p < 0.05$	
MID TREATMENT	10.01 ± 0.46		ns
		$p < 0.05$	
POST TREATMENT	10.30 ± 0.25		

* Geometric mean of 12 subjects ± SD (cfu/g feces (log))

The mean total counts decrease significantly during treatment. This decrease was observed for each of the twelve volunteers. After treatment the mean total count was not significantly lower than before treatment.

Elimination of Enterobacteria : The mean number of enterobacteria was 7.22 ± 1.21 cfu/g of feces before treatment. On day 5, no Enterobacteria could be detected from the feces of any volunteers ($<$ 2.00 cfu/g of feces). Neither Norfloxacin nor Nalidixic acid resistant enterobacteria were isolated during the experimentation. Four days after the end of treatment six out of twelve volunteers were recolonized with susceptible strains and it took more than 8 days to recolonize the other six.

The mean number of enterococci was 6.17 ± 1.57 and 6.40 ± 1.19 cfu/g of feces respectively before and during treatment. The difference is not significant.

So enterobacteria were eliminated but enterococci, with higher M.I.C. (1-32 mcg/ml) remained unchanged during treatment. Despite of high fecal concentration of Norfloxacin, it seems that only the bacteria with M.I.C less than 1 mcg/ml are eliminated. We can suppose that just a part of active Norfloxacin, found in the feces, is available.

In conclusion, we have seen that with the two regimens, enterobacteria were eliminated without emergence of resistant strains. From a functional point of view, the barrier effect against enterococci was not modified, but the total fecal bacterial counts decreased significantly during treatment. When the anaerobic study will prove there is no ecological modifications of the dominant anaerobic microflora, Norfloxacin could be used for microbial modulation of the intestinal tract.

Aranki Syed SA, Kenney EB, Freter R (1969). Isolation of anaerobic bacteria from human gingiva and mouse caecum by means of a simplified glove box procedure. App. Microbiol. 18 : 568-76.

Norrby SR, Jonsson M (1983). Antibacterial activity of Norfloxacin. Antimicrob. Agents Chemother. 23 : 15-18.

Swanson BN, Boppana VK and col (1983). Norfloxacin disposition after sequentially oral doses. Antimicrob. Agents Chemother. 23 : 284-8.

Vand der Waaij D and col(1980). Infection Prevention in granulocytopenic patients by selective decontamination of the digestive tract. Europ. J. Cancer 16 : 859-69.

Germfree Research: Microflora Control and Its Application to the Biomedical Sciences, pages 259–262

NORFLOXACIN FOR SELECTIVE DECONTAMINATION: A STUDY IN HUMAN VOLUNTEERS

De Vries-Hospers HG. Welling G.W., Van der Waaij D.

Laboratory for Medical Microbiology
Oostersingel 59, 9713 EZ Groningen
The Netherlands

INTRODUCTION

Selective decontamination of the digestive tract (SDD) aims at elimination of potentially pathogenic microorganisms (ppmo) from the gi-tract, without affecting the anaerobic microflora responsible for the Colonization Resistance (CR) (Van der Waaij 1971). A decreased CR increases the chance for overgrowth by (resistant) ppmo which subsequently can result in infection by these microorganisms. Especially in granulocytopenic patients maintenance of an intact CR is important, as the majority of the infections in these patients is caused by aerobic gram-negative bacilli. The number of such infections can be considerably reduced by application of SDD (Sleijfer et al 1980). SDD of gram-negative bacilli can performed by oral administration of nalidixic acid (100 mg/kg/day) (De Vries-Hospers et al 1981).
In the present study it was investigated whether SDD could be performed with norfloxacin, a nalidixic acid derivative which is more active against *Enterobacteriaceae* (MIC < 1 µg/ml) (Shungu 1982). Most *Pseudomonas* species, enterococci and staphylococci will be inhibited by concentrations of 10 µg/ml while anaerobes are resistant (Gadebusch 1983; Norrby 1983). Three different dosages of norfloxacin were administered to volunteers. Fecal samples and oral washings were collected for bacteriological investigations. β-Aspartylglycine (β-asp-gly) was used as a parameter for the CR (Welling 1978). This dipeptide is normally degraded by enzymes originating from anaerobic bacteria. Elimination of these bacteria will result in accumulation of β-asp-gly in the fecal samples.

VOLUNTEERS, MATERIALS AND METHODS

Ten healthy males between 21 and 40 years of age participated in this study. Liver- and kidney function tests were performed. After being informed written informed consent was obtained.
Tablets of 200 mg norfloxacin were provided by Merck, Sharp and Dohme, Haarlem, The Netherlands. Three different dose regimens of 200, 400 and 800 mg were tested. Each dose was given during five consecutive days. The daily dose was divided into two portions.
Each study period (Table I) consisted of the five treatment days plus two days before and seven days thereafter. The interval between two study periods was at least two weeks.

Table I: Norfloxacin study period.

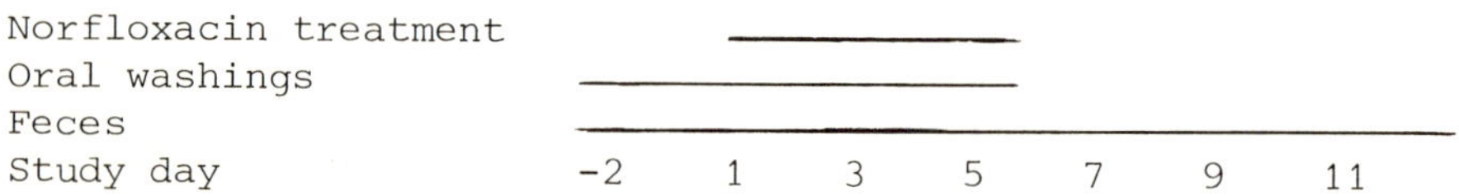

During the study period oral washings and fecal samples were collected (Table I). Oral washings were used for culturing of viridans streptococci and other ppmo. In feces the logarithmic concentration of enterococci, gram-negative bacilli and yeasts was determined. In addition a fecal supernatant was prepared for the determination of β-asp-gly by high voltage paper electrophoresis (Welling 1982).

RESULTS

All volunteers completed the study. Adverse effects were not mentioned except by two volunteers: one complained of mild diarrhea, the other of moderate obstipation during treatment with the highest oral dose of 800 mg. Liver- and kidney function tests did not change during the study.
Oral flora consisted mainly of viridans streptococci. Their concentration was not influenced by norfloxacin treatment. Gram-negative bacilli, yeasts and staphylococci were cultured incidentally in three, two and six volunteers respectively. A relation with norfloxacin treatment could not be found.
The inventory of the feces consisted of enterococci and *Enterobacteriaceae*. In one volunteer *Pseudomonas aeruginosa* was cultured in addition. All isolated gram-negative bacilli were

susceptible to norfloxacin.
The three dosages tested were almost equally effective in eliminating gram-negative bacilli from the fecal samples (Table II).

Table II: Cumulative number of volunteers free of fecal gram-negative bacilli on a certain study day.

Oral dose of norfloxacin	Study day with gram-negative-free feces 1	2	3	4	5	6	7
200 mg	1	4	7	9		10	
400 mg	1	2	8	9	10		
800 mg		5	8	9	10		

Enterococci tended to decrease. In fact enterococci could no longer be cultured from the feces of three, six and five volunteers receiving 200, 400 and 800 mg norfloxacin respectively.
β-Asp-gly was present in 7.6% of the fecal samples. Two of t these sampees were derived from one volunteer, one during the 400 mg course and another sample during the 800 mg course. All the other positive samples were derived from one volunteer who also had β-asp-gly in his pretreatment fecal samples. Overgrowth with resistant microorganisms was not seen in any of the volunteers.

DISCUSSION

The results of this study indicate that it is possible to eliminate fecal gram-negative bacilli by oral treatment with norfloxacin. In this respect even the lowest dose of 200 mg/day was effective within a reasonable period of time: fecal gram-negative bacilli were no longer culturable from the feces in most volunteers from the third treatment day on. Enterococci disappeared in some volunteers, while in others they did not. However, overgrowth with resistant enterococci or other ppmo was not seen.
β-Asp-gly was present in the fecal samples of two volunteers. The majority of these samples were derived from one volunteer who also was positive for β-asp-gly before he started oral treatment with norfloxacin. The presence of β-asp-gly in this person can therefore not be ascribed to the norfloxacin. It was concluded that daily dosages of 200 mg norfloxacin did not affect the CR in these ten volunteers.

SUMMARY

During five consecutive days three different dosages of 200, 400 and 800 mg norfloxacin were administered to ten healthy males. Cultures of feces and oral washings revealed that even the lowest dose was effective in eliminating gram-negative bacilli from the gi-tract within a reasonable period of time. At this dose signs of a decreased CR due to norfloxacin and measured by both overgrowth by resistant ppmo or the appearance of β-asp-gly in feces were not seen. Treatment with 100 mg norfloxacin twice daily seems therefore suitable for SDD, at least in volunteers. However, further studies in patients should be performed.

REFERENCES

De Vries-Hospers HG, Sleijfer DT, Mulder NH, Van der Waaij D, Nieweg HO, Van Saene HKF (1981). Bacteriological aspects of selective decontamination of the digestive tract as a method of infection prevention in granulocytopenic patients. Antimicrob Agents Chemother 19:813

Gadebusch HH, Shungu DL, Weinberg E, Chung SK (1982) . Comparison of the antimicrobial activity of norfloxacin (MK 0366, AM 715) a new organic acid, with that of other orally absorbed chemotherapeutic agents. Infection 10:41

Norrby SR, Jonsson M (1983). Antibacterial activity of norfloxacin. Antimicrob Agents Chemother 23:15

Shungu DL, Weinberg E, Gadebusch HH (1983). In vitro antibacterial activity of norfloxacin (MK-0366, AM-715) and other agents against gastrointestinal pathogens. Antimicrob Agents Chemother 23:86

Sleijfer DT, Mulder NH, De Vries-Hospers HG, Fidler V, Nieweg HO, Van der Waaij D, Van Saene HKF (1980). Infection prevention in granulocytopenic patients by selective decontamination of the digestive tract. Eur J Cancer 16:859

Van der Waaij D, Berghuis-de Vries JM, Lekkerkerk-van der Wees JEC (1971). Colonization Resistance of the digestive tract in conventional and antibiotic treated animals. J Hyg 69:405

Welling GW, Groen G (1978). β-Aspartylglycine, a substance unique to caecal contents of germfree and antibiotic treated mice. Biochem J 175:807

Welling GW (1982). Comparison of methods for the determination of β-aspartylglycine in fecal supernatants of leukemic patients treated with antimicrobial agents. J Chromatography 232:55

Germfree Research: Microflora Control and Its Application to the Biomedical Sciences, pages 263–265

A RANDOMIZED TRIAL OF ORAL VANCOMYCIN IN NEUTROPENIC PATIENTS

D. Guyotat, C. Plonton, D. Fiere

Department of Hematology, Hopital E. Herriot, 69374, Lyon, France

As infection in neutropenic patients mainly originates from intestinal tract, a variety of decontamination regimens have been designed, mostly to suppress gram-negative flora. Vancomycin, a non-absorbable antibiotic active against gram-positive flora, has been added in some regimens. Our trial has been designed to test its role in severly neutropenic patients receiving a standard decontamination by gentamycin and colistine.

PATIENTS AND METHODS

34 patients (aged 17-66 years) receiving intensive chemotherapy for acute leukemia in first perceptible phase or in relapse, were randomly allocated on admission to one of two oral antibiotics regimens. Antibiotics were given in capsules (4 a day) containing each: in regimen A (17 patients) gentamycin 20 mg, colistine sulfate 0.5 million units, nystatin 1.5 million units; in regimen B (17 patients), the same antibiotics plus vancomycin, 500 mg. Both groups were similar with regards to age, sex and stage of the disease. Patients with documented infection at presentation were excluded. Parenteral antibiotics (ticarcillin and amikacin) were started when temperature was superior to 38.5 degrees.

Patients were considered as febrile when temperature was superior or equal to 38.5 degrees. Cultures of faeces, throat and urine were taken on admission and then once a week. Blood cultures were taken on each day with pyrexia. Stool specimens were mixed with tryptic soy broth and plated onto Hektoen, cetrimide, bromocresol-purple, Sabouraud and Chapman media, and selenite broth which was plated onto Hektoen medium one day later. Anaerobic germs were not researched. Cultures were considered as pathologic when showing presence of pseudomonas, or predominance of any germ (more than 80%), and decontamination was modified accordingly.

The Mann-Whitney test and the corrected X2 test were used for statistical analysis.

RESULTS

Clinical results are given in table 1 :

Groups	A	B
Total number of days : $< 0,5 \times 10^9/l$ granulocytes $< 0,1 \times 10^9/l$ granulocytes	 277 232	 337 268
Total number of days : - of pyrexia - of I.V. antibiotics	 173 291	 139 284
Number of complete remissions	7	7
Number of deaths	2	1

The differences between both groups are not significant with regards to total number of days of granulocytes < 0.5 or $0.1\ 10^9/l$, or to number of days of pyrexia or parenteral antibiotics. However, the rate number of days of pyrexia/ number of days at risk is significantly lower in patients treated with vancomycin ($P < 0.05$).

Bacteriological results are given in table 2 :

Groups	A	B
Episode of septicaemia :	10	5
- *Staphylococcus epidermidis*	5	2
- *Staphylococcus aureus*	0	1
- *Streptococcus mitis*	3	0
- *E. Coli*	1	2
- *Enterobacter agglomerans*	1	0

Number of stool cultures : (number of patients)	54 (17)	58 (17)
- Staphylococcus epidermidis	10 (9)	0
- Streptococcus group D	2 (2)	0
- pseudomonas	5 (3)	0
- proteus	1 (1)	1 (1)
- candida albicans	1 (1)	8 (4)
- sterile	11 (11)	16 (12)
- without predominance of pathogen	23 (17)	31 (16)

Staphylococcus epidermidis was never found in the stools on admission, but was found in the stools of 9 patients on regimen A after a median duration of 16 days. Vancomycin was added to the decontamination and the cultures returned to normal in all but 1 cases. No patient treated with vancomycin acquired Staphylococcus epidermidis contamination ($p < 0{,}001$). Although the number of gram-positive septicaemias was greater in group A than in group B (8 vs 3), the difference is not significant.

CONCLUSION

Although vancomycin is active on anaerobic flora, it did not seem in our experience to enhance colonization of the faeces by gram-negative organisms, as shown in animal or human models (Van der Waaij, 1971).

The major point of our study is the fact that staph. epidermidis is frequently found in stools of patients without vancomycin. In a similar trial, Bender (1979) did not find any benefit in adding vancomycin to gentamycin and nystatin. However, as S. epidermidis seems to become a growing source of infection (Wade, 1982), we think that vancomycin can be useful for intestinal decontamination and possibly for reducing the number of systemic infections.

Bender J.F., Schimpf S.C., Young V.M. (1979). Role of vancomycin as a component of oral nonabsorbable antibiotics for microbial suppression in leukemic patients. Antimicrob. Agents Chemother 15: 455.

Van der Waai J.D.,(1971). Colonization resistance of the digestive tract in conventional and antibiotic-treated mice. J. Hyg. 69: 405.

Wade J.C., Schimpf S.C. (1982). Staphylococcus epidermidis : an increasing cause of infection in patients with granulocytopenia. Ann. Intern. Med. 97 : 503.

Germfree Research: Microflora Control and Its Application to the Biomedical Sciences, pages 267–270

PROLONGED GRANULOCYTOPENIA DUE TO SELECTIVE ANTIMICROBIAL MODULATION OF THE INTESTINAL FLORA: BIAS OR BIOLOGY?

H.F.L. Guiot, E.F. Tordoir-van der Bent

Dept. of Infectious Diseases, University Hospital Leiden; J.A. Cohen Institute, 2333 AA Leiden, The Netherlands

INTRODUCTION

Analysis of five articles on the results of large-scale controlled studies on the prophylactic effect of oral nonabsorbable antimicrobials in patients with decreased host defence shows that the patients on oral prophylaxis were granulocytopenic longer than the untreated controls (table 1). From these findings the question arises whether this prolongation is the result of bias, or whether it reflects the influence of intestinal bacteria on the haematopoiesis. Although some animal studies have been reported, showing that the endogenous microbial flora might affect the haematopoietic activity, it is not known whether this is of clinical importance in man. One of

Table 1: Analysis of five articles and the results of our own study (Guiot) with respect to granulocytopenia

	Days* granulocytes < 500/mm^3		
STUDY	PEPA	PA	CONTROL
1973 LEVINE	16 (188%)	13 (159%)	8
1973 YATES	--	23 (96%)	24
ditto	26 (118%)	--	22
1977 STORRING	--	28 (110%)	26
1977 EORTC	30 (124%)	--	24
ditto	30 (117%)	--	26
1981 MUNDO	19 (109%)	--	17
1983 GUIOT (SAM)	--	36 (123%)	28

*: mean number of days per patient with acute non-lymphocytic leukaemia during remission induction therapy; PE: protective environment; PA: nonabsorbable prophylactic antibiotics; % (between brackets): percentage of the control.

the differences between the animal experiments and the clinical situation is that the supposed effect of prophylactic oral antimicrobial drugs has to be exercised during the relatively short period of complete or selective elimination of the intestinal bacteria. Because also in patients on selective antimicrobial modulation (SAM) a delay in granulocyte recovery is suggested (table 1), we performed experiments in mice to imitate the clinical situation of patients undergoning cytotoxic therapy.

METHODS

Conventional random bred NMRI mice were given either drinking water without, or with the antimicrobial agents of SAM, i.e. 50 mg polymyxin-B, 130 mg neomycin, 130 mg amphotericin,

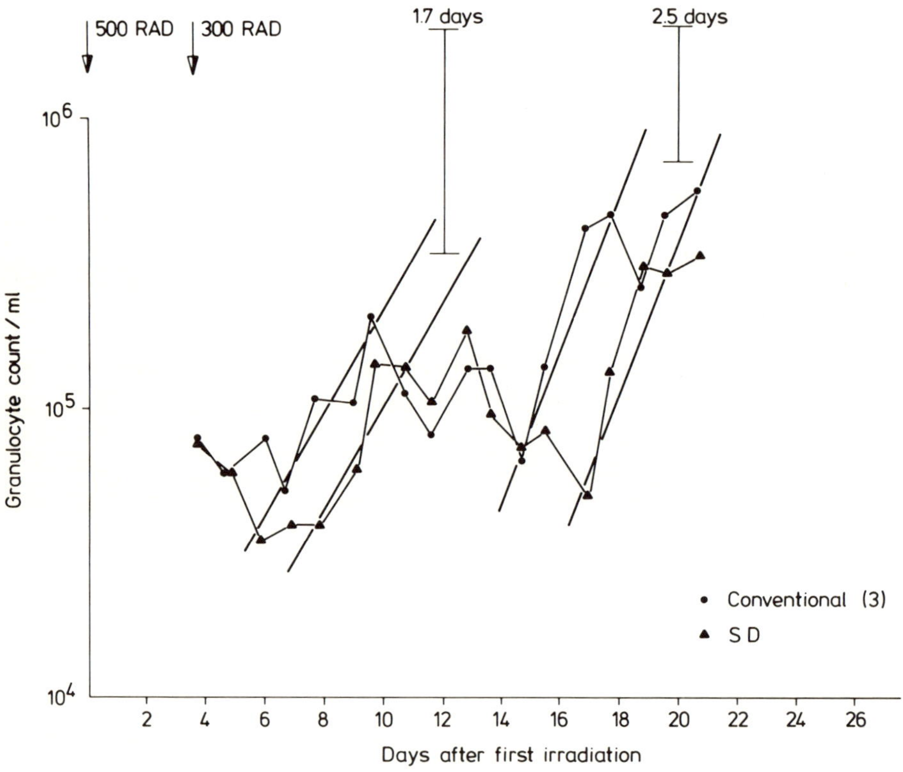

Fig. 1: Autologous recovery of peripheral granulocytes in conventional mice (●) and in selectively decontaminated mice (▲) after two dosages of irradiation (500 - 300 rad).

400 mg nalidixic acid in 100 ml drinking water. In both groups of mice, 4 g saccharose/100 ml was added to the drinking water, and pH was adjusted to pH 5.0. After 3-4 weeks, after the aerobic gram-negative rods were selectively eliminated, the mice were irradiated with a single dose of 500 rad, or one dose of 500 rad followed by 300 rad with four days interval. Each day after irradiation, the number of leukocytes and granulocytes in the peripheral blood were counted. Blood was obtained by orbital punctures once a week for each mouse, until normal values were reached. In one group of mice on SAM, the food was mixed with large numbers of *Escherichia coli, Klebsiella pneumoniae* and *Proteus mirabilis* (20 ml washed bacterial pellet of centrifuged overnight cultures, mixed with 400 g homogenized conventional food in 100 ml H_2O, dried at 45°C overnight).

RESULTS

Mice, not irradiated and given SAM during 3 months, were not different from untreated mice with peripheral granulocytes at a mean of 0.57×10^6/ml (untreated mice 0.59×10^6/ml).

After irradiation (500 rad) a transient increase to 0.9×10^6/ml was followed by a decrease with the lowest numbers of granulocytes at days 4 and 5 after irradiation. A complete autologous recovery was reached between days 10 and 12. Analysis of variance shows a delay of recovery in the mice on SAM with 1.7 days in comparison with conventional irradiated mice. In mice given SAM and food mixed with high numbers of gram-negative bacteria the results showed a complete reversal of this delay of granulocyte recovery (analysis of variance: 0.14 days delay).

In conventional mice that were irradiated twice (500 + 300 rad) a granulocytopenic episode of about 15 days was induced. The mice on SAM needed 2.5 days longer (Fig. 1).

DISCUSSION

The results show that in mice the recovery of peripheral granulocytes is delayed by selective eradication of the aerobic gram-negative rods from the gut. The results in mice given live gram-negative bacteria in their food excludes the possibility that the delay was caused by direct action of the antimicrobial drugs on the bone marrow. The delay of granulocyte recovery of 2.5 days, in mice given SAM and made granulocytopenic for about 15 days by irradiation twice (500 and 300 rad, four days interval), is of a similar proportion as the mean delay found for the reviewed studies, i.e. a 5.5 days delay in the patients on prophylactic antibiotics at a mean granulocytopenic episode of 22 days in the patients without prophylactic antibiotics. The results make it hard to resist the temptation to extrapolate the results in mice to the clinical situation in man.

REFERENCES

Dietrich M, Gaus W, Vossen JM, Waaij D van der, Wendt F. (1977). Protective isolation and antimicrobial decontamination in patients with high susceptibility to infection. A prospective cooperative study (EORTC) of gnotobiotic care in acute leukemia patients. I. Clinical results. Infection 5:107-114.

Guiot HFL, Broek PJ van den, Meer JWM van der, Furth R van (1983). Selective antimicrobial modulation of the intestinal flora of patients with acute nonlymphoblastic leukemia: A double blind, placebo-controlled study. J Infect Dis 147:615-623.

Levine AS, Siegel SE, Schreiber AD, et al. (1973). Protected environments and prophylactic antibiotics. A prospective controlled study of their utility in the therapy of acute leukemia. New Engl J Med 288:477-483.

Mundo MR, Granena A, Rozman C.(1981). Evaluation of a protective environment in the management of granulocytopenic patients: A comparative study. Cancer 48:419-424.

Storring RA, Jameson B, McElwain TJ, Wiltshaw E, Spiers ASD, Gaya H.(1977). Oral non-absorbed antibiotics prevent infection in acute non-lymphoblastic leukaemia. Lancet II:837-840.

Yates JW, Holland F. (1973). A controlled study of isolation and endogenous microbial suppression in acute myelocytic leukemia patients. Cancer 32:1490-1498.

Germfree Research: Microflora Control and Its Application to the Biomedical Sciences, pages 271–273

THE EFFECT OF SELECTIVE DECONTAMINATION OF THE GI-TRACT OF MICE ON THE SURVIVAL OF INTESTINAL MUCOSA DURING X-IRRADIATION

Toorop-Bouma AG., Van der Waaij D.
Laboratory for Medical Microbiology
Oostersingel 59, 9713 EZ Groningen
The Netherlands

INTRODUCTION:

Irreversible bone marrow and mucosal damage can be an unfortunate side effect of chemo- and radiotherapy. Lethal destruction of the bone marrow may not be an absolute dose limiting factor in many hospitals, since autologous bone marrow rescue transplantation can be performed. Autologous transplantation of intestinal crypts, however, is not possible. This makes the dose at which irreversible damage occurs to the gastro-intestinal lining more critical than that to bone marrow. Although the effect of radiation on the proliferative compartment of the intestinal mucosa of germfree and conventional mice has been studied quite intensively (Hendry et al, 1984; Von der Maase et al, 1983; Tucker et al, 1983), relatively few investigations have been done to study the influence of different fractions of the microflora on the radiosensitivity of the intestinal mucosa (Boranic et al, 1970; Mastromarino et al, 1976; Yokokura et al, 1981). We studied the effect of elimination of aerobic gram-negative bacilli from the gi-tract on the radiosensitivity of the intestinal mucosa of mice.

METHODS:

Female C57bl mice (age 12-15 weeks) were housed under aseptic conditions. Groups of selectively- (SD) and of totally decontaminated (TD) mice were obtained by suspension of non-absorbable antibiotics in the drinking water for at least 14 days prior to irradiation. We used norfloxacin (.025 g/l) for SD treatment and neomycin (2.0 g/l) combined with cephalotin (2.0 g/l) and norfloxacin (.025 g/l) for TD treatment.

Untreated mice served as controls. In all mice infection prophylaxis was achieved by total decontamination of the gi-tract (Boranic et al, 1970) 2 days before irradiation until day 30. Every group consisted of 5 subgroups of at least 10 mice. On day 0 these mice received total body irradiation with various doses of X-rays in the crypt affecting dose range (10-14 Gy). One day after irradiation the animals were transplanted with syngeneic bone marrow to obtain a fast recovery of marrow activity. Mortality was recorded three times per day for 30 days after irradiation. Mice which died between day 6 and 10 most probably died due to the intestinal syndrome and therefore provided data for the LD50-mouse intestinal syndrome survival. Feces were examined for contamination with bacteria and fungi. Organs were cultured at autopsy.

A parallel experiment was done to study intestinal crypt survival of 48 mice on day 6 after irradiation (12 and 13 Gy). Average crypt regeneration was determined by counting the number of surviving crypts per circumference of 7 μm sections of ileum and colon.

Results were compared using the log rank test, analysis of variance and linear regression analysis.

RESULTS:

Selective decontamination of the gi-tract of mice prior to X-irradiation resulted in an increase of the midlethal dose for mouse intestinal syndrome survival (± 12 Gy) as compared with the untreated group (± 10.5 Gy). Ileal and colonic crypt survival in SD-mice was in general better than in TD- and untreated mice (table 1). Infection prophylaxis prolonged the mean survival time of the untreated group from 3-4 to 7 days.

table 1: Crypt survival (%) of mouse ileal and colonic mucosa 6 days after various doses of whole body X-irradiation. Data are mean ± SEM. * $p < .001$ vs TD and untreated mice.

treatment	radiation dose ileum 12 Gy	radiation dose ileum 13 Gy	radiation dose colon 12 Gy	radiation dose colon 13 Gy
SD	31.7 (1.8)	24.7* (1.9)	66.4* (8.2)	44.0* (2.9)
TD	32.4 (2.8)	17.8 (1.1)	32.9 (5.6)	14.3 (2.3)
untreated	25.1 (2.0)	10.9 (1.5)	21.4 (3.3)	10.1 (2.4)

DISCUSSION:

These preliminary results suggest that selective decontamination of the gi-tract may increase the survival of X-irradiated mice. The increased radioresistance of these mice appears to be more than the result of prevention of post-irradiation infection. In 1983 Hendry et al. described a relationship between the gastro-intestinal syndrome and the radiosensitivity of mucosal clonogenic cells. It is believed that intestinal mucosal stem cell radiosensitivity responds to the elimination of gram-negative bacilli from the gi-tract of mice.

LITERATURE:

Boranic M, Van der Waaij d (1970). The effect of the supply of oral antibiotic on the fecal flora and mortality of mouse radiation chimeras. The J. of Inf. Dis. 122 (1):83.

Hendry JH, Moore JV, Potten CS (1984). The proliferative status of microcolony-forming cells in mouse small intestine. Cell Tissue Kinet. 17:41.

Hendry JH, Potten CS, Roberts ND (1983). The gastro-intestinal syndrome and mucosal clonogenic cells: relationship between target cell sensitivities, LD50 and cell survival, and their modification by antibiotics. Rad. Res. 96:100.

Mastromarino A, Wilson R (1976). Antibiotic radioprotection of mice exposed to supra-lethal whole-body irradiation independent of antibacterial activity. Rad. Res. 68:329.

Tucker SL, Withers HR, Mason KA, Thames HD jr (1983). A dose-surviving fraction curve for mouse colonic mucosa. Eur. J Cancer Clin. Oncol. 19 (3): 433.

Von der Maase H, Overgaard J (1983). Microcolony survival assay for jejunal crypt cells exposed to radiation alone and combined with cancer chemotherapeutic agents- methodological problems. Int. J. Radiat. Biol. 43 (1): 45.

Yokokura T, Onoue M, Uchida K, Takahashi T, Mutai M (1981). Effect of intestinal microflora on the survival time of mice exposed to lethal whole-body y-irradiation. In S. Sasaki et al (eds): "Recent Advances in Germfree Research", Tokai university press, p. 361.

Germfree Research: Microflora Control and Its Application to the Biomedical Sciences, pages 275–277

FORECAST OF ANTIBIOTIC SUSCEPTIBILITY OF TRANSLOCATING STRAINS OF ENTEROBACTERIA AND PSEUDOMONAS IN LEUKEMIC NEUTROPENIC PATIENTS

Cyrille H. Tancrède, Antoine O. Andremont

Service de Microbiologie Médicale
Institut Gustave-Roussy
94805 Villejuif Cedex - France

The choice of antibiotherapy is of primary importance in leukemic patients during the neutropenic period because of the rapidity and severity of infection in such circumstances. (Bodey, Rodriguez, Chang, Narboni 1978). The origin of infection is frequently not identified. It is, however, assumed than infection arises from the patient's indigenous microflora (Schimpff, Young, Greene, Vermeulen, Moody, Wiernik 1972), particularly from the gastrointestinal tract (Schimpff, Aisner, Wiernik 1979). The translocation of certain indigenous bacteria from the gastrointestinal tract to other organs has been studied in selected animal models (Berg, Garlington 1979). In a previous study (Tancrède, Andremont 1983), we have shown that translocation of the dominant fecal clone of enterobacteria or of Pseudomonas was the mechanism by which over 80 % (45/55) of bacteremia developped in neutropenic leukemic patients.

In order to investigate whether clinically useful information to treat these bacteremia could be obtained from the antibiotic susceptibility pattern of the aerobic gram negative fecal flora investigated 0 to 5 days before onset of febrile episodes, a prospective study was initiated. Fecal samples were regularly obtained from leukemic patients during the postchemotherapeutic neutropenic period. Freshly passed feces were weighted and serially diluted in saline from 10^{-2} to 10^{-8}. 0.1 ml of the appropriate dilutions were plated on selective media : Enterobacteria fully susceptible to antibiotics and those resistant to Ampicillin (AMP), Cefotaxime (CTX), Gentamicin (GT) and Amikacin (AMI) were counted on Drigalski agar with and without antibiotics. Total Pseudomonas and Pseudomonas resistant to Carbenicillin (CARB) were counted on Cetrimide agar (Table 1):

DETECTION AND NUMERATION OF AEROBIC GRAM NEGATIVE RODS IN FECES

SELECTIVE MEDIUM	ANTIBIOTICS	DETECTED ORGANISMS
DRIGALSKI	0	SUSCEPTIBLE
"	AMP 100 mg/1	AND
"	CTX 2 mg/1	RESISTANT
"	GT 5 mg/1	
"	AMI 5 mg/1	ENTEROBACTERIA
CETRIMIDE	0	SUSCEPTIBLE AND
"	CARB 10 mg/1	RESISTANT PSEUDOMONAS

Susceptibility pattern of dominant clones to these antibiotics was determined by their growth characteristics on these media. The data were available after an overnight culture (48 h for 20% of Pseudomonas). These results were in accordance with those obtained by standard disk susceptibility testing which were available only 24 h later.

In the 45 cases of bacteremia where translocation of the dominant fecal clone of enterobacteria or Pseudomonas was documented, the susceptibility pattern of the strain isolated from the blood culture was in accordance with the information previously obtained from the fecal sample. In practice, the antibiotic susceptibility patterns obtained from fecal samples were always available at least 24 h before the same results obtained from blood cultures. The average in 45 patients was 48 h, according to the mean interval between the fecal analysis and the first positive blood culture. The antibiotic susceptibility data obtained from fecal samples before blood cultures in the 45 patients with further documented translocation are summarised in table 2:

ANTIBIOTIC SUSCEPTIBILITY OF DOMINANT FECAL CLONES OF ENTEROBACTERIA AND PSEUDOMONAS WHICH FURTHER WERE ISOLATED FROM BLOOD CULTURE IN 45 PATIENTS.

Enterobacteria	AMP R	CTX R	GT R	AMI R
E. Coli	8/20	0	0	0
Klebsiella sp.	4/4	0	0	0
Ent. Cloacae	7/8	0	1/8	0
TOTAL	19/32	0	1/32	0
	CARB R			
Pseudomonas	3/13		4/13	0

In 49% of cases (Enterobacteria 42%, Pseudomonas 7%) AMP and/or CARB were recognized as inactive before the isolation of the gram negative rod from blood culture, and an active B-lactam antibiotic (CTX or Cefsulodin) was prescribed to the patient before obtaining antibiotic susceptibility data from blood culture.

The overall results show that serial culturing of fecal samples of neutropenic leukemic patients on selective media is a reliable method to forecast the antibiotic susceptibility pattern of the most probably translocating gram negative rod. When an episode of clinical bacteremia occurs, these results can be safely used to choose an active antibiotic treatment immediately after the drawing of blood cultures.

The results of this study suggest that, beside empiricism, this alternative mode can be used for the survey of intestinal colonization of neutropenic patients by exogenous microorganisms, and the choice of systemic antibiotic treatment when there is a clinical evidence of bacteremia. Detection and numeration of opportunistic pathogens on selective antibiotic media appears as a very relevant method to manage neutropenic patients.

Berg RD, Garlington AW (1979). Translocation of certain indigenous bacteria from the gastrointestinal tract to the mesenteric lymph nodes and other organs in gnotobiotic mouse model. Infect Immun 25:320.

Bodey GP, Rodriguez V, Chang HY, Narboni (1978). Fever and infection in leukemic patients: a study of 494 consecutive patients. Cancer 41:1610.

Schimpff SC, Young VM, Greene WH, Vermeulen GD, Moody MR, Wiernik PM (1972). Origin of infection in acute non lymphocytic leukemia: significance of hospital acquisition of potential pathogens. Ann Inter Med 77:707.

Schimpff SC, Aisner J, Wiernik PM (1979). Infection in acute non-lymphocytic leukemia: the alimentary canal as a major source of pathogens. In van der Waaij D, Verhoef J (eds): "New Criteria for Antimicrobial Therapy: Maintenance of Digestive Tract Colonization Resistance". Amsterdam: Excerpta Medica, p 12.

Tancrede C, Andremont A (1983). Translocation of aerobic gram negative rods from the intestinal tract in neutropenic patients. 21 st Annual Meeting of Association for Gnotobiotic June 26-29 1983 Baltimore, Maryland.

Germfree Research: Microflora Control and Its Application to the Biomedical Sciences, pages 279–282

IN VIVO INFLUENCE OF THREE B-LACTAM ANTIBIOTICS ON THE INTESTINAL MICROFLORA OF MAN. A PRELIMINARY STUDY IN GNOTOBIOTIC MICE.

Florence C. Léonard, Antoine O. Andremont,
Cyrille H. Tancrède
Service de Microbiologie Médicale
Institut Gustave-Roussy
94805 Villejuif, France

Cefotaxime (Ctx), Ticarcillin (Ti) and the association of Ti and Clavulanic acid (Cla) are recommended for the systemic treatment of severe gram-negative infections. The aim of this study was to compare the ecological modifications of the intestinal microflora of man induced by these treatments.

We studied the effect of the 3 drugs on the indigenous enterobacteria (EB) in treated patients and in ex-germ-free C_3H mice associated with a human intestinal microflora. This experimental model allows also to investigate the modifications of the anaerobic microflora induced by antibiotics (Andremont 1984).

Material and Methods

16 patients were treated with Ctx 60 mg/kg/24h, 3 with Ti 225 mg/kg/24h and 7 with Ti + Cla (Ti 225 mg/kg - Cla 9 mg/kg/24h).

C3H germ-free mice maintened in isolator were inoculated with a complex microflora without Ti and Ctx resistant (R) EB. They were treated subcutaneously twice daily with Ctx (75 mg/kg/24h), Ti (225 mg/kg/24h) and Ti + Cla (Cla 9 mg/kg/24h). The control group was treated with physiological saline (Van der Waaij 1982).

In man and mice, fecal counts of EB were performed on Drigalski agar. R EB were detected on the same medium added of 2 mg/l of Ctx, 100 mg/l of Ti respectively.

In mice, fecal counts of aerotolerant *Streptococci* (AS) were performed on Bile Esculin agar.

The total count of anaerobes was performed 1) on Aranki agar in an anaerobic glove box (Aranki 1969), and

2) in agar pourred in deep tubes after atmospheric contact, in such conditions Extremely Oxygen Sensitive (E.O.S) anaerobes are destroyed (Andremont 1983). Dominant anaerobic strains (> 9.00 cfu/g) were classified into genera according to their shape, respiratory type, gram stain properties, presence and position of spores.

The susceptibility to Ti, Ti + Cla and Ctx of the dominant clones of anaerobes under each treatment was investigated and compared.

Fecal concentrations of Ti, Cla and Ctx were determined in mice feces by agar diffusion assay.

Results

In man, Ctx eliminated EB (< 3.00 cfu/g of feces). Under Ti and Ti-Cla the count of EB was not affected, because Ti R EB were present before and persisted during treatment. We have not observed the action of Cla, except in one case with elimination of Ti R EB susceptible to Ti-Cla.

In gnotobiotic mice, EB were eliminated by the 3 drugs. AS were unchanged by Ctx and increased under Ti and Ti-Cla.

Table 1

Fecal counts of EB and AS in gnotobiotic mice

Treatment	EB		AS	
	Before treatment	During treatment	Before treatment	During treatment
Control	6.87±.69	7.08±.69	5.48±.42	5.38±.34
Ctx	5.80±.33	< 2.00	5.85±.80	6.20±.90
Ti	5.87±.68	< 2.00	5.51±.77	9.15±.18
Ti + Cla	5.62±.37	< 2.00	5.70±.46	9.20±.20

- Data are mean ± SEM log counts/g - Underlined : $p < 0.05$
- Five mice with each treatment

The total count of anaerobes (10.43 ± .14 cfu/g) was not modified by the 3 antibiotics, but the proportion of EOS anaerobes decreased at least by 1 cfu/g (log) under Ctx, Ti and Ti-Cla.

The gross composition of microflora was modified by the antibiotic treatments especially under Ti and Ti-Cla.

DOMINANT ANAEROBIC HUMAN FECAL FLORA IN MICE
(>10^9 cfu/g of feces)

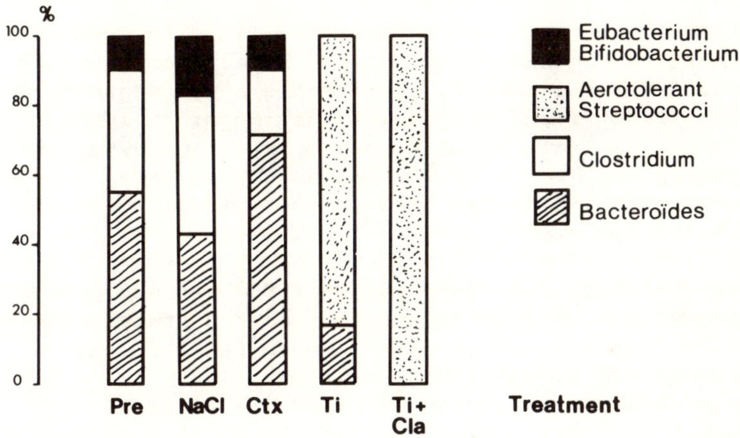

MICs of TIC on the 25 dominant clones of anaerobes ranged from <1 to 64 mg/l before and were > 128 mg/l during teatment by Ti and Ti + Cla. They were unchanged during Ctx treatment.

MICs of Ctx on the 25 dominant clones of anaerobes ranged from < 0.25 to 32 mg/l before treatment, they were > 4 mg/l under Ctx and > 32 mg/l under Ti and Ti-Cla. During Ctx treatment, the fecal concentrations were lower than 0.2 mcg/g of feces. The mean concentration of Ti were 71 and 52 mcg/g under Ti and Ti-Cla respectively and the Cla concentrations ranged from 0.2 to 0.7 mcg/g. These data are in accordance with those obtained in human feces (Ctx < 0.2 ; Ti 200 ; Cla 2 mcg/g).

Discussion

In man and gnotobiotic mice, the three treatments eliminated the susceptible EB from the fecal flora. In mice, total counts of anaerobes were unchanged but EOS strains disappeared from the dominant flora.

During Ctx treatment : The main bacterial genera of the dominant flora were unchanged and the colonization resistance to AS was maintened. The resistance of the dominant flora increased to Ctx but not to Ti and Ti-Cla.

During Ti and Ti-Cla treatment : The results emphasize a modification on colonization resistance by AS. The intestinal microflora was sharply simplified : only AS and a few Bacteroides sp. were still present. The resistance of the remaining flora increased to Ti, Ti-Cla and also Ctx.

In conclusion, Ctx, Ti and Ti-Cla eliminate susceptible EB both in man and in associated mice. The 3 drugs destroy the EOS anaerobes. The microflora was sharply simplified with Ti and Ti-Cla with an important increase of AS. The colonization resistance of the remaining microflora to various exogenous microorganisms is under investigation with the 3 antibiotics in gnotobiotic human flora associated mice.

Andremont A, Raibaud P, Tancrède C (1983). Effect of Erythromycin on microbial antagonisms : a study in gnotobiotic mice associated with a human fecal flora. J. Infect Dis 148:579.

Andremont A, Raibaud P, Tancrède C, Duval-Iflah Y, Ducluzeau R (1984). The use of germ-free mice associated with human fecal flora as an animal model to study enteric bacterial interactions. In Takeda Y, Miwatani T (eds) : "Bacterial Diarrheal Diseases" Tokyo : p 219.

Aranki A, Syed SA, Kenney EB, Freter R (1969). Isolation of anaerobic bacteria from human gingiva and mouse cecum by means of a simplified glove box procedure. Appl Microbiol 17:568.

Van der Waaij D, Hofstra H, Wiegersma N (1982). Effect of B-lactam antibiotics on the Resistance of the digestive tract of mice to colonization. J Infect Dis 146:417.

Germfree Research: Microflora Control and Its Application to the Biomedical Sciences, pages 283–285

INFLUENCE OF TETRACYCLINE ON NUMBER AND SIZE OF CANDIDAL LESIONS OF THE DORSAL RAT TONGUE

Carl M. Allen, D.D.S., M.S.D.
Frank M. Beck, D.D.S., M.A.
Fredrick A. Lurie, D.D.S.
H. Mark Pinsky, D.D.S.

The Ohio State University
College of Dentistry
Columbus, Ohio 43210

Candidal infections have generally been associated with either a defect in the host immune status or an alteration of the environment of the organism (Odds 1979). The former situation includes patients using systemic steroids or having widespread malignancy, while the latter situation frequently implies antibiotic usage with its attendant reduction in bacterial organisms, allowing overgrowth of *Candida albicans*. Some authors have implied that infection with *C. albicans* is not possible without such alterations (Odds 1981). Consequently, animal models for oral candidiasis have typically used some form of immunologic or bacterial suppression in order to establish the candidal infection (Russell,Jones 1973; Fisker, Schiott, Philipsen 1982; Sofaer, Holbrook, Southam 1982; Allen, Beck 1983).

Recently, however, differences in the pathogenicity of *C. albicans* strains for rat oral mucosa have been identified (Allen, Beck 1983). The purpose of this study was to compare the effects of a mucosally virulent strain of *C. albicans* in normal and tetracycline-treated rats with respect to oral mucosal infection.

Forty female Sprague-Dawley rats, average weight 200 gm, were randomly divided into two groups and housed five animals per cage. Group I received a 0.01% tetracycline solution as drinking water, while Group II was given double-distilled, demineralized water. Both groups were inoculated orally each week with a suspension of viable C. albicans (0.1 ml/rat; 5×10^7 CFU's/ml). A fresh inoculum was prepared each week from a culture grown on a Sabouraud's agar slant for 72 h at 25°C.

After twenty weeks, all animals were killed, and standard 35mm color transparencies of the tongues were taken from a dorsal view. The tongues were then fixed in formalin and processed for light microscopy.

Gross and histologic examination showed candidal infection in 16 of 20 animals in Group I and 17 of 20 in Group II (no significant difference, $p>0.05$). Due to the well-delineated gross appearance of the infected epithelium, comparison of the lesional areas could be performed. The standard transparencies of the dorsal tongue lesions were projected at a set distance (1 meter) and the outline of each lesion was traced on .0075 mil acetate sheets. The tracings were then cut out and weighed on an electronic balance, and their areas were calculated by comparing lesional acetate weight to standard area acetate weight. The mean lesional area of Group I was found to be significantly greater than Group II (63.0 ± 7.9 vs. 44.5 ± 4.1, $p<0.05$).

These results suggest that a tetracycline solution may increase the size of the oral mucosal lesions produced by this strain of C. albicans. More importantly, however, this study indicates that a candidal infection can be induced simply by introducing the organism into the oral cavity. No antibiotic alteration of the oral microflora or immune suppression of the host was necessary for establishment of infection.

Allen CM, Beck FM (1983). Strain-related differences in pathogenicity of Candida albicans for oral mucosa. J Infect Dis 147:1036.

Fisker AV, Schiott CR, Philipsen HP (1982). Short-term oral candidosis in rats, with special reference to the site of infection. Acta Path Microbiol Immunol Scand, Sect B 90:49.

Odds FC (1979). "Candida and Candidosis." Baltimore: University Park Press, p 82.

Odds FC (1981). The pathogenesis of candidosis. Hosp Update 7:935.

Russell C, Jones JH (1973). Effects of oral inoculation of Candida albicans in tetracycline-treated rats. J Med Microbiol 6:275.

Sofaer JA, Holbrook WP, Southam JC (1982). Experimental oral infection with the yeast Candida albicans in mice with or without inherited iron-deficiency anemia (sla). Archs Oral Biol 27:497.

VI CANCER

Germfree Research: Microflora Control and Its Application to the Biomedical Sciences, pages 289–292

PROFILE OF SPONTANEOUS TUMOR DEVELOPMENT IN GERMFREE LOBUND WISTAR RATS

Morris Pollard and Phyllis H. Luckert

Lobund Laboratory, University of Notre Dame

Notre Dame, Indiana 46556 U.S.A.

In the course of establishing germfree (GF) animals as experimental subjects, it is as important to document the profile of diseases which each strain develops spontaneously as it is important to document their susceptibilities to induced diseases. We should expect that data derived from conventional animals may be modified by elements of the microbial flora and by other factors that would be excluded from GF animals.

During the past 20 years, GF Lobund-Wistar (L-W) rats were examined for evidence of diseases that occurred "spontaneously" that might be related to their deaths, or to the phenomenon of senescence (Pollard 1971). Since 1969, GF L-W rats were fed, ad libitum, a steam-sterilized all-vegetable diet (L-485) and tap water, and maintained in steel Reyniers-type isolators on Sani-cell bedding, with 12 hour light and dark cycles. The diet (L-485) had been assessed for adequacy according to weight gain, reproduction records and survival of progeny (Kellogg and Wostmann 1969). The rats were free of microbial flora, and searches for viruses have been negative. They manifested the physical characteristics of germfree status: large thin-walled cecum, small spleen, lymph nodes and Peyer's patches, rare germinal zones, low counts of white blood cells, and relatively low levels of globulins.

It has been recorded that GF L-W rats live longer than conventional (CV) counterpart rats: the latter lived average 750 days while the GF rats lived average 1050 days (Pollard and Wostmann 1984). Current records indicate that in

GF and CV L-W rats, spontaneous cancers of the intestine and of the breast do not develop. However, GF male L-W rats develop a significantly high incidence of prostate cancer (Pollard 1973). The latter disease may be related to the extended life span of the GF group, since clinically-apparent prostate cancer generally develops in men older than 50 years. GF and CV L-W rats are resistant to the oncogenic effects of dimethylhydrazine (Asano and Pollard 1978), of N methylnitrosourea, and of methyl(acetoxymethyl)nitrosamine (Pollard and Luckert 1981). They develop a wide spectrum of malignant and benign neoplasms following whole-body exposures to non-lethal doses of x rays (Pollard and Sharon 1971).

MATERIALS AND METHODS

GF male and female L-W rats, up to the 46th generation in GF status, were examined for gross and microscopic evidence of diseases. Rats were selected for examination either arbitrarily or because they showed evidence of illness. Each rat was anesthetized with ether, exsanguinated from the exposed heart, and subjected to thorough post-mortem examination. This included microscopic examinations of tissues that had been fixed in Bouin's solution, embedded in paraffin and sections thereof were stained with hematoxylin and eosin.

RESULTS

102 male and 60 female GF L-W rats were examined for evidence of disėase. The average life span recorded for males was 1050 days. Signs and symptoms of disease were attributed to (a) hemorrhagic adenomas of the pituitary gland (ataxia and paralysis), (b) thymomas (dyspnea), (c) loss of weight and weakness (liver tumors) and (d) weakness of posterior extremities (prostate cancer). In general, most of the aging rats which were arbitrarily killed for examinations appeared active and alert.

Data on 132 rats beyond age 30 months have been reported previously (Pollard and Luckert 1979), to which subsequently accumulated data on 162 rats has now been added. Except for liver tumors, the spectrum of lesions involved endocrine glands and hormone-modulated glands (breast and prostate), as noted earlier. These were benign hyperplastic changes in endocrine glands: the highest incidence involved the medullary area of the adrenal gland and, less frequently, involving the pituitary and thymus glands. Non-metastatic liver tumors were observed in 143 rats (88%) (age range 30 to 46 months) which ranged from benign nodules to adenocarcinomas. Focal areas of gamma glutamyl transpeptidase (GGT) were observed in the livers of younger GF L-W rats by histochemical procedures (Pollard, Luckert and Adams 1982). GGT is interpreted by some investigators as preneoplastic lesions. The incidence and sizes of the GGT foci were increased in rats which were fed the tumor promoter, phenobarbital, in the drinking water. Among the male L-W rats, 11/102 (11%) had developed prostate adenocarcinomas spontaneously; and in 9 of the rats metastatic tumors had developed, usually in the lymph nodes and lungs.

DISCUSSION

GF L-W rats are unique in that they develop a significant incidence of spontaneous prostate adenocarcinomas. The incidence of disease is increased and the latent periods are shortened in L-W rats which were given implantations of testosterone in silastic depots (Pollard, Luckert and Schmidt 1982). This is exemplary of (a) hormone regulation of gene expression and (b) of genetic susceptibility to prostate cancers, since this neoplasm has not developed spontaneously among the Lobund strain of GF Sprague-Dawley rats which are maintained under the same husbandry conditions as the L-W rats. Spontaneously-occurring liver tumors in conventional rats have been observed rarely, possibly because (a) few live beyond age 30 months, (b) they must be exsanguinated for better visibility of the tumors, or (c) diet L-485 may contain hepatocarcinogens in such low levels as to require 30 months for them to grow to visibility. It is of significance to note the consistent correlation in the aging L-W rat with histological changes in endocrine glands. It is thus conceivable that the process of aging involves hormone dyshomeostasis.

It is important to establish a profile of spontaneous neoplasms among animals which are used for carcinogenesis studies, since tumors resulting from administration of test agents to them may confuse their role(s) as initiators rather than promoters of carcinogenesis. Also, it is important to know if a specific strain of test animal can actually develop the disease being investigated, since investigations on prostate cancer in Sprague-Dawley rats may yield negative results.

Asano T, Pollard M (1978). Strain susceptibility and resistance to 1,2 dimethylhydrazine-induced enteric tumors in germfree rats. Proc Soc Exp Biol and Med 158:89.

Kellogg TF, Wostmann BS (1969). Stock diet for colony production of germfree rats and mice. Lab Animal Care 19: 812.

Pollard M (1971). The germfree rat. In Ioachim HL (ed): "Pathobiology 1971," New York: Appleton-Century-Crofts, p. 83.

Pollard M (1973). Spontaneous prostate adenocarcinomas in aged germfree Wistar rats. J Natl Cancer Inst 51:1235.

Pollard M, Luckert PH (1979). Spontaneous liver tumors in aged germfree Wistar rats. Laboratory Animal Science 29: 74.

Pollard M, Luckert PH (1981). Effect of indomethacin on intestinal tumors induced in rats by the acetate derivative of dimethylnitrosamine. Science 214:558.

Pollard M, Luckert PH, Adams RA (1982). The detection of gamma glutamyl transpeptidase in the livers of germfree and conventional rats. Jr Lab Animal Science 32:147.

Pollard M, Luckert PH, Schmidt MA (1982). Induction of prostate adenocarcinomas in Lobund Wistar rats by testosterone. The Prostate 3:563.

Pollard M, Sharon N (1971). Irradiation-induced lesions in germfree rats. J Nat Cancer Inst 47:229.

Pollard M, Wostmann BS (1984). Aging in germfree rats: the relationship to the environment, diseases of endogenous origin, and to dietary modification. In Archibald, Ditchfield, Rowsell (eds): 8th ICLAS/CALAS Symposium, Stuttgart: Fischer Verlag, in press.

Germfree Research: Microflora Control and Its Application
to the Biomedical Sciences, pages 293–296

GERMFREE ANIMAL AS A TOOL TO STUDY ROLE OF GUT MICROFLORA AND NUTRITION IN CANCER.

Bandaru S. Reddy, Karam El-Bayoumy, and Yves M. Louis
American Health Foundation
Valhalla, New York 10595, USA

Agents which are genotoxic by definition interact with the gene to yield an abnormal genetic material and are considered initiating agents. It is thought that cancer causation requires the action of such agents which alter the genetic material. The second broad class of agents, which act by epigenetic mechanisms and tend to increase the development of lesions initiated by genotoxic carcinogens, include cocarcinogens and tumor promoters. As to the origin of these agents, they may be preformed in the environment, ingested in the form of a precarcinogen and metabolized to the carcinogen and/or endogenously formed and metabolized in the body.

In this context, the role of bacteria in the production of genotoxic and epigenetic agents has been investigated in several laboratories (Reddy 1981; Goldin and Gorbach 1976). In any situation where the bacteria produce or inhibit the tumorigenic agents from dietary components or from endogenous sources as a response to dietary components, there will be an association between bacteria, diet, and cancer. Another way in which bacteria can affect the carcinogenic process is by altering the tissue (for example, colon) sensitivity to initiators and promoters.

The next question that is being asked is "how can the germfree animal model contribute to the understanding of cancer causation?" Major benefits will come from the application of model tumor systems that are now available in germfree and conventional rodents which include carcinomas of the colon, breast and prostate (Pollard 1981; Reddy

(1981). These model systems can be used to seek answers to significant problems.

BACTERIAL PRODUCTION OF CARCINOGENS, CO-CARCINOGENS AND PROMOTERS

1-Nitropyrene is an important compound in the class of nitrated polynuclear aromatic hydrocarbons and found in various natural sources. Its presence in the respiratory environment, in particular diesel engine exhaust, represents a potential health hazard because of its high mutagenicity in bacterial test systems. Our recent study clearly demonstrates that intestinal microflora play a role in the metabolism of 1-nitropyrene in the rat (El-Bayoumy et al. 1983). 1-Aminopyrene and hydroxylated 1-aminopyrene were detected in the feces of conventional rats gavaged with 1-nitropyrene, but not in the feces of germfree rats. However, in germfree rats metabolites derived from ring hydroxylation of 1-nitropyrene were detected. These studies thus indicate that nitroreduction and ring hydroxylation could be involved in the activation of 1-nitropyrene. When 1-nitropyrene was incubated with human fecal contents, 1-aminopyrene was identified. Highly mutagenic intermediates such as 1-nitrosopyrene or 1-hydroxylaminopyrene are formed in the reduction of 1-nitropyrene, a process which has been shown to damage DNA.

ROLE OF GUT MICROFLORA IN THE DEVELOPMENT OF GASTROINTESTINAL AND ENDOCRINE-RELATED CANCERS

The influence of gut microflora in colon carcinogenesis was first demonstrated by Laqueur (1970) using cycasin, the glucoside of methylazoxymethanol (MAM), which is not active in germfree rats because the required hydrolysis of the glucosidic bond is mediated by bacterial glucosidase in the intestine.

In a recent study, we investigated the effect of gut microflora on 3,2'-dimethyl-4-aminobiphenyl (DMAB)-induced carcinogenesis in rats (Reddy and Watanabe 1978). The number of DMAB-induced colon, small intestinal, breast, salivary gland and ear duct tumors was higher in conventional rats than in their germfree counterparts. DMAB could be N-hydroxylated in the liver and a portion of

N-hydroxy derivatives may be conjugated with glucuronic acid or sulfate and excreted via bile into the intestine where the bacterial and mammalian enzymes hydrolyze the conjugates to active metabolites. In germfree rats, only the intestinal mucosal glucuronidase is involved in releasing the active metabolites of DMAB, whereas both intestinal and bacterial glucuronidases are involved in hydrolyzing the conjugates in conventional rats. In addition, active metabolites from the gut are also absorbed and transported in the blood stream inasmuch as the germfree rats demonstrate a low incidence of ear duct, breast and salivary gland as well as tumor in other organs, compared to that in conventional rats. This might explain the observed differences in the incidence of tumors between conventional and germfree rats.

ACKNOWLEDGMENTS

Supported in part by grants CA-17613, CA-29602 and CA-36872 from the National Cancer Institute.

REFERENCES

El-Bayoumy, K, Sharma, C, Louis, YM, Reddy, BS, Hecht, SS (1983). The role of intestinal microflora in the metabolic reduction of 1-nitropyrene to 1-aminopyrene in germfree and conventional rats and in humans. Cancer Lett 18: 311.

Goldin, BR, Gorbach, SL (1976). The relationship between diet and rat fecal bacterial enzymes implicated in colon cancer. J Natl Cancer Inst 57: 371.

Laqueur, GL (1970). Contribution of intestinal microflora to carcinogenesis. In Burdette, WJ (ed.) "Carcinoma of the Colon and Antecedent Epithelium, Springfield, Illinois: Charles C. Thomas, p. 305.

Pollard, M (1981). Development of the model systems for cancer research. In Sasaki S, Ozawa, A, Hashimoto, K (eds.) "Recent Advances in Germfree Research", Tokyo, Japan, Tokai Univ Press, p. 621.

Reddy, BS (1981). Role of gut microflora and nutritional factors in the development of large bowel and breast cancer. In Sasaki, S, Ozawa, A, Hashimoto, K (eds.) "Recent Advances in Germfree Research", Tokyo, Japan, Tokai University Press, p 627.

Reddy, BS, Watanabe, K (1978). Effect of intestinal microflora on 3,2'-dimethyl-4-aminobiphenyl-induced carcinogenesis in F344 rats. J Natl Cancer Inst 61: 1269.

Germfree Research: Microflora Control and Its Application to the Biomedical Sciences, pages 297–299

SPONTANEOUS POLYPOSIS IN THE SMALL INTESTINE OF GERMFREE AND CONVENTIONALIZED BALB/c MICE

Mizutani T[1]., D.V.M., Ph.D. and Mitsuoka T[1,2]., D.V.M.,Ph.D.
[1]Animal Physioloy Laboratory, The Institute of Physical and Chemical Research, Wako-shi, Saitama, 351 and [2]Department of Biomedical Science, Faculty of Agriculture, University of Tokyo, Bunkyo, Tokyo, 113, Japan

The development of polyposis was unexpectedly noticed in the small intestine of untreated germfree (Gf) mice in an experiment aiming to study the effect of the intestinal bacteria on the chemical carcinogenesis in the intestine of gnotobiotic BALB/c mice. The spontaneous polyposis in the small intestine of Gf and conventionalized (Cvz) BALB/c mice was studied.

Gf mice were bred in our laboratory and maintained Gf in vinyl isolator. Regular checks for bacteria and fungi gave negative results throughout the experiment. The first generation offspring of the Cvz mice derived from the Gf mice was used as Cvz animals and was placed in an ordinary animal facility. All animals were given an irradiated (5 Mrads) diet and autoclaved tap water ad libitum. When they were 12 months old, the animals were killed under CO_2 inhalation and autopsied carefully for the number and size of polyps with the aid of a dissecting microscope.

As shown in Table 1, the incidence of Polyposis was higher in the Gf mice (68% in female and 89% in male) than in the Cvz mice (37% in female and 51% in male) . The number of polyps/mouse was also higher in the Gf mice (2.3 in female and 5.7 in male) than in the Cvz mice (0.8 in female and 1.3 in male). All of the polyps were histopathologically adenomatous and developed only in the upper part (mainly duodenum) of the small intestine.

TABLE 1

INCIDENCE OF POLYPOSIS IN THE SMALL INTESTINE OF GERMFREE AND CONVENTIONALIZED BALB/c MICE

Group	Sex	No. of animals	Animals with polyp No.	(%)	No. of polyps No.	Per mouse
Germfree animals	F	40	27	68[a)]	93	2.3[e)]
	M	36	32	89[b)]	204	5.7[f)]
Convention-alized animals	F	41	15	37[c)]	32	0.8[g)]
	M	51	26	51[d)]	64	1.3[h)]

Animals were killed under CO_2 inhalation when they were 12 monts old. Significance of the data was analyzed by Student's t test and X^2test.

a) vs c): $p < 0.01$ b) vs d): $p < 0.001$ a) vs b): NS

c) vs d): NS e) vs g): $p < 0.01$ f) vs h): $p < 0.001$

e) vs f): $p < 0.001$ g) vs h): NS

The histological classification of intestinal tumors in the mouse corresponds roughly to the pathology of tumors in rats (Pozharisski, 1970). Rowlatt and Chesterman (1979) have divided polypoid tumors in the duodenum into the duodenal plaques and epithelial polyps. In the present study all visible polypoid lesions with the aid of a dissecting microscope were classified as polyps.

Present study demonstrated that the incidence of polyposis in the small intestine of BALB/c mice was reduced by the presence of the intestinal microflora. Hare and Stewart (1956) showed that decreasing the roughage in the diet would decrease the incidence of polyps in the duodenum of DBA mice. Seronde (1970) has related the duodenal plaques to pantothenate deficiency in the diet rather than to roughage. However, the present study suggested that the intestinal microflora was also related to the inhibition of polyposis in the small intestine of BALB/c mice. The inhibitory effect of the intestinal microflora on polyposis may be due to modification of the host-immune systems by the intestinal microflora, although such evidence was not obtained from the present study.

McIntire and Princler (1969) have found that the yield of plasma-cell tumors in GF BALB/c mice after prolonged adjuvant stimulation is low as compared with that in conventional animals. Bartizal et al. (1983) have also indicated that the intestinal microflora can alter murine natural cell-mediated cytotoxicity in BALB/c mice. The intestinal microflora may be involved in modification of the host-immune system in resistance to tumorigenesis.

The present study demonstrated that the development of polyposis in the small intestine of BALB/c mice was inhibited by the intestinal microflora.

REFERENCES

Bartizal KF, Salkowski C, Balish E (1983). The influence of a gastrointestinal microflora on natural killer cell activity. J Reticuloendthel Soc 33:381.

Hare WV, Stewart HL (1956) Chronic gastritis of the glandular stomach, adenomatous polyps of the duodenum, and calcareous pericarditis in strain DBA mice. J Natl Cancer Inst 16:889.

McIntire KR, Princler GL (1969) Prolonged adjuvant stimulation in germfree BALB/c mice: development of plasma cell neoplasia. Immunology 17: 481.

Pozharisski KM (1973) Tumours of the intestines. In Turusov VS, (ed): "Pathology of tumours in laboratory animals", vol 1, Lyon: International Agency for Research on Cancer, part 1, pp. 119.

Roe FJC, Grant GA (1970) Inhibition by germfree status of development of liver and lung tumors in mice exposed neonatally to 7, 12-dimethylbenz(a)anthracene: Implications in relation to tests for carcinogenicity. Int J Cancer 6: 133.

Rowlatt C, Franks L M, Sheriff MU and Chesterman, FC (1969). Naturally occurring tumors and other lesions of the digestive tract in untreated C57BL mice. J Natl Cancer Inst 43: 1353.

Seronde J Jr (1970). Focal avillous hyperplasia of the mouse duodenum. J Path 100: 245.

Germfree Research: Microflora Control and Its Application
to the Biomedical Sciences, pages 301–304

SUPPRESSIVE INFLUENCE OF METHEMOGLOBINEMIA DUE TO NITRITE ON HEPATOTOXICOSIS DUE TO DIMETHYLNITROSAMINE (DMN) IN GERM-FREE RATS

Yukiko Sumi and Masasumi Miyakawa

Laboratory of Germfree Life
Research Institute of Disease Mechanism & Control
Nagoya University School of Medicine
Nagoya, Japan

We previously reported that germ-free rats given dimethylnitrosamine (DMN) by stomach tube at a certain dose were far more susceptible to hepatic damage than were conventional rats similarly treated (Sumi, Miyakawa 1983). However, we noticed that germ-free rats given nitrite and aminopyrine, being under a condition of nitrite-methemoglobinemia, were less susceptible to the hepatic damage. This study was planned to investigate whether or not nitrite-methemoglobinemia may exert any influence on the DMN-hepatotoxicosis in germ-free rats.

Germ-free and conventional male Wistar rats weighing 350 to 390 g were used as experimental and control rats. In the first experiment, 70 mg of NaNO2 per kg body weight dissolved in 1 ml distilled water was given by stomach tube to both germ-free and conventional rats. At 1.5 hr (10 germ-free rats and 17 conventional rats), 3 hr (14 germ-free and 20 conventional), 5 hr (17 germ-free and 20 conventional), 7 hr (10 germ-free and 14 conventional) and 10 hr (10 germ-free and 10 conventional) after NaNO2 intubation, blood of each group at respective time point was collected from the abdominal aorta and analyzed for methemoglobin by the method of Evelyn-Malloy (Evelyn, Malloy, 1938). In the second experiment, 70 mg of NaNO2 per kg was intubated to 52 germ-free and 64 conventional rats. At time intervals of 3 hr or 5 hr after the NaNO2 intubation, DMN at a dose level of 8 or 10 mg per kg body weight was intubated to rats of both groups, respectively. And then 48 hr later, the blood was similarly collected, and analyzed for serum GPT and GOT by the method of Reitman-Franklin. For controls, 32 germ-free and 40 conventional rats were given 8 or 10 mg of DMN alone and similarly analyzed for serum GPT and GOT. In both experiments, all experimental and control rats were autopsied immediately after blood collection, and the liver and other organs were examined for necrosis.

In the first experiment, the results of the study of the level of methemoglobin in blood are shown in Table 1. We indicated that germ-free rats were prone to cause methemoglobinemia by NaNO2 and their recovery from the methemoglobinemia was delayed, as compared with conventional rats, so that there was a significant difference in the methemoglobin level especially at 5 hr ($P < 0.001$) or 7 hr ($P < 0.001$) after the intubation of NaNO2 between germ-free and conventional rats. From the viewpoint that a germ-free status might be one kind of pristine condition (Fitzgerald, Pollard 1967), this phenomenon might have something to do with the fact that infants are prone to cause toxic methemoglobinemia due to spinach or well water containning high concentration of nitrite or nitrate (Comly 1945; Cornblath, Hartmann 1948). In the second experiment, the results are given in Figure 1. The levels of serum GPT and GOT in germ-free experimental rats, which were treated by a dose of 8 mg of DMN at 3 hr after the pretreatment with NaNO2, were lowered to about one eighth ($P < 0.001$) and about one sixth ($P < 0.001$) of those in germ-free controls treated

Table 1. Methemoglobin concentration in the blood of germ-free and conventional rats after $NaNO_2$ (70 mg/ kg body weight) intubation.

hr after $NaNO_2$	Methemoglobin/Total hemoglobin (%) Germ-free	Methemoglobin/Total hemoglobin (%) Conventional	Significant difference
1.5	48.4 ± 6.4[a] (10)[b]	46.4 ± 4.6 (17)	NS
3	37.1 ± 3.9 (14)	33.1 ± 5.1 (20)	NS
5	28.4 ± 4.3 (17)	9.4 ± 4.3 (20)	$P < 0.001$
7	12.3 ± 4.2 (10)	0.7 ± 0.8 (14)	$P < 0.001$
10	0.2 ± 0.2 (10)	0.5 ± 0.4 (10)	NS

a Mean ± S.D. *b* Number of rats examined.

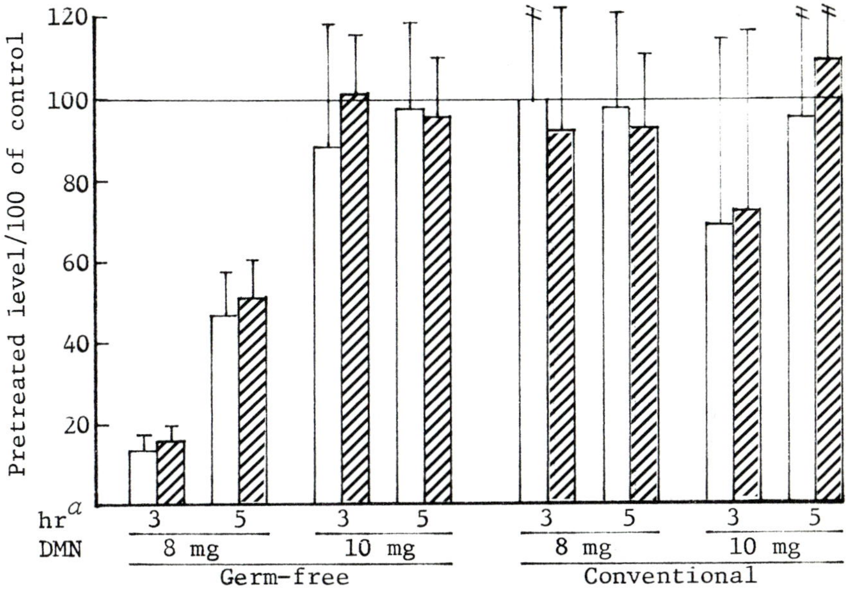

Fig. 1. Relative levels of serum GPT and GOT in germ-free and conventional rats treated with 8 or 10 mg of DMN at 3 or 5 hr after the pretreatment with 70 mg of $NaNO_2$ to those of germ-free and conventional controls treated with DMN alone (▭ GPT, ▨ GOT). *a* Hours after pretreatment.

with 8 mg of DMN alone, respectively and both levels in germ-free experimental rats at 5 hr after the pretreatment were lowered to about a half of those in germ-free controls ($P < 0.001$). This data was coincident with the morphological finding that the majority of the germ-free experimental rats showed absence of necrosis or slightly graded necrosis in the liver while the majority of the germ-free controls showed severe necrosis, suggesting that methemoglobinemia suppresses the development of hepatic necrosis. However, germ-free experimental rats treated by a dose of 10 mg of DMN at 3 hr or 5 hr after the pretreatment with NaNO2 were much susceptible to hepatotoxic effect of DMN and showed the same high levels of serum GPT and GOT as germ-free controls treated by 10 mg of DMN alone. Pathohistologically, severe necroses were observed in most of the germ-free experimental and control rats, indicating that 10 mg of DMN was too high a dose to determine the suppressive effect of the methemoglobinemia on the liver damage. On the other hand, the levels of serum GPT and GOT in conventional experimental rats, which were treated by a dose of 8 mg of DMN at 3 hr or 5 hr after pretreatment with NaNO2, remained normal, being almost similar to those in conventional controls treated with 8 mg of DMN alone. Furthermore, both levels in conventional experimental rats treated with a dose of 10 mg of DMN at 3 hr or 5 hr after the pretreatment showed no significant suppressive influence of the methemoglobinemia. Symptoms present in the period of the pretreatment of NaNO2 were those of anoxia due to the decreased oxygen-carrying capacity of blood, resulting in cyanosis in various organs. The fact that in germ-free status such a preceding methemoglobinemia suppressed the hepatotoxic effect of DMN subsequently intubated, would be valid consideration. It is well known that DMN intubated is rapidly metabolized and the liver is the main organ concerned (Magee 1956). In the process of the in vivo metabolism of DMN in the liver, the production of a hepatotoxic agent is quite possible. The production of such a toxic agent in the liver and its effect depend mainly upon the dose of DMN intubated, the rate of oxygen transfer in the liver and the status of the animal.

In summary, the results first showed that the recovery from methemoglobinemia due to NaNO2 was much delayed in germ-free rats, as compared with that in conventional rats. Secondly, the results showed that the methemoglobinemia caused by 70 mg NaNO2/kg predominantly suppressed the formation of hepatic necrosis due to 8 mg DMN/kg in germ-free rats, while such a suppressive phenomenon did not appear so distinctly in conventional rats. Therefore, it may be said that the production of the hepatotoxic agent is inhibited in germ-free rats where the rate of the oxygen transfer is not sufficient to metabolize the dose of DMN present in the liver.

Comly HH (1945). Cyanosis in infants caused by nitrates in water. J Am Med Ass 129:112.

Cornblath M, Hartmann AF (1948). Methemoglobinemia in young infants. J Pediat 33:421.

Evelyn KA, Malloy HT (1938). Microdetermination of oxyhemoglobin in a single sample of blood. J Biol Che 126:655.

Fitzgerald GR, Pollard M (1967). Interferon production by germ-free mice. Proc Soc Exp Biol Med 126:245.

Magee PN (1956). The metabolism of dimethylnitrosamine. Biochem J 64:676.

Sumi Y, Miyakawa M (1983). Susceptibility of germ-free rats of dimethylnitrosamine or dimethylamine plus sodium nitrite administered orally. Cancer Res 43:2942.

Germfree Research: Microflora Control and Its Application to the Biomedical Sciences, pages 305–309

COMPOSITION OF, AND CHOLATE TRANSFORMATION BY, THE PREDOMINANT FECAL FLORA OF PATIENTS WITH COLON OR RECTAL CANCER AND MATCHED CONTROLS

Rudolf Edenharder[+] and Rainer Hammann

[+] Institute of Hygiene, University of Mainz and Institute for Medical Microbiology, University of Bonn, [+]D-6500 Mainz, FRG

Epidemiologic studies have shown that dietary factors play a significant role in the etiology of colon cancer in humans. Diets high in total fat and low in dietary fiber are generally associated with an increased incidence of colon, but not necessarily of rectal cancer (Correa 1981). One effect of dietary fat is to increase the excretion of bile acids into the gut.

It has been hypothesized that the composition and activity of the gut microflora may be modified by bile acids, and that this flora can metabolize bile acids to tumorigenic compounds or to tumor promoters (Hill et al. 1971). While several studies compared the fecal floras or fecal bile acid composition of persons from different risk groups (Crowther et al. 1976; Finegold et al. 1975; Mower et al. 1979; Reddy, Wynder 1977; Schwann et al. 1982), it was our intention to compare the predominant fecal flora of patients with colon and rectal cancer and of matched controls with respect to both the flora composition and bile acid transforming capacities.

Anaerobic floras were isolated and cultivated within an anaerobic chamber similar to that described by Aranki et al. 1969 (modified Schaedler-agar; 10^8 and 10^9 fold dilutions). Media and fecal dilutions were prepared outside the box, using a multiple gas supply head flushed with CO_2 to strictly exclude air. For species identification, all colonies from a single plate, approximately 45 on the average, were transferred into culture tubes with Schaedler-broth, supplemented with cooked meat medium and 250 μM of cholate, and incubated until the late stationary phase. Aliquots of these cultures were used to identify the isolates by standard methods (Holdeman, Moore 1977) as well as to determine cholate metabolism. Bile acids were analyzed as described earlier (Edenharder, Knaflic 1981).

Microscopic (mc) and cultural counts (cc) of 12 patients with colon, 6 patients with rectal resp. 25 patients with colorectal cancer and of 25 controls were in the range of 10^{11} per g dry matter with no statistically significant differences. This was also true for recovery rates (rr). Example: patients with colorectal cancer: 6.3 ± 3.0 (mc); 3.2 ± 1.9 (cc); 43.1 ± 23.9 (rr); controls: 6.5 ± 2.9 (mc); 3.1 ± 2.5 (cc); 44.4 ± 24.1 (rr).

Table 1 shows the Bacteroides fraction of the predominant fecal flora of patients with colorectal cancer and of matched controls. These results were obtained from a total of 2164 pure and mixed cultures, from which about 2650 bacterial strains could be identified.

Table 1. The predominant fecal flora of 17 patients with colorectal cancer and of 24 matched controls

Species or taxon of Bacteroides	Controls	Patients	Controls	Patients	P-* value
	% of specimens harboring		Means+		
B.distasonis	63	36	1.38	0.91	-
B.eggerthii	27	47	0.93	0.39	0.06
B.fragilis	58	47	2.32	2.49	-
B.incommunis	13	18	0.12	0.09	-
B.splanchnicus	30	6	0.57	0.02	0.06
B.thetaiotaomicron	63	82	3.75	3.47	-
B.uniformis	42	35	1.07	0.63	-
B.variabilis	42	24	1.07	0.63	-
B.vulgatus	92	82	7.05	3.33	0.07
B."3452 A"	33	12	1.93	0.15	-
B."ssp a"	21	0	0.54	0	0.05
B."T 4-1"	29	12	0.64	0.10	-
B."polysacch.neg."	29	0	0.64	0	0.02
Bacteroides (not specificable)	96	88	3.29	1.71	-

+ Most probable number x 10^{10} per g dry matter;
* Wilcoxon-Mann-Whitney-test.

Each person had a typical fecal flora comprising one to four different Bacteroides species. Small, although significant differences between patients (P) and controls (C), visible for several Bacteroides species were contradictory to those found in earlier studies (Hill et al. 1971; Crowther et al. 1976). These differences were, however, no longer detectable when comparing 8 patients with colon cancer with the controls except for B.eggerthii ($p = 0.03$).

Bacterial strains from genera other than Bacteroides were infrequently isolated. Usually they are part of the accompanying flora, although they may predominate in specific individuals. Amongst our patients there was one person with a predominating Eubacterium flora and four with predominating Lactobacillus floras. The following genera or species were most frequently isolated (most probable number x 10^{10} per g dry matter): Lactobacillus: 2.95 (C); 2.56 (P); Eubacterium: 1.15 (C); 1.06 (P); E.coli: 1.93 (C); 1.57 (P). Strains of Bifidobacterium, Clostridium, Propionibacterium and anaerobic grampositive and gramnegative cocci were found to be present in approximately 10^9 counts per g dry matter.

Table 2. Transformation of cholate by human fecal isolates

Transformation product	Total number of active cultures	Number of cultures with trace activities
3α,12α	149	60
3α,12ß	4	-
3α;12-keto	8	3
24-methyl-3α,12α	127	60
3α,7α	16	4
5α-3α,7α,12α	3	1
3ß,7α,12α	28	3
3α,7ß,12α	4	1
3-keto;7α,12α	383	315
3α,12α;7-keto	972	122
3α,7α;12-keto	377	180
3ß,12α;7-keto	19	1
3ß,7α;12-keto	20	15
3,7-diketo;12α	126	110
3α;7,12-diketo	94	61
3,12-diketo;7α	55	40
3ß;7,12-diketo	7	1
3,7,12-triketo	12	4
24-methyl-3α,12α;7-keto	75	71

Number of tests analyzed: 2476; inactive: 1251 (50.5 %); active: 1225 (49.5 %)

The products of cholate transformation and their frequencies are shown in Table 2. None of these bile acids has ever been convincingly demonstrated to be a carcinogen for the large bowel. Introduction of double bonds into the steroid nucleus, side chain or further degradations were not detected. Therefore, the concept of carcinogen production from bile acids by intestinal bacteria is not confirmed by these experiments.

A comparison of the different patient groups with matched controls with respect to the frequency of cholate transformation products generated by the corresponding predominant fecal floras is shown in Table 3. The differences observed for patients with rectal cancer were even stronger when 6 further patients were added, whose carcinoma were located at the rectosigmoid junction.

Table 3. Transformation of cholate by fecal isolates of 12 patients with colon cancer (upper part of the table), 6 patients with rectal+ cancer (lower part) and of 25 matched controls

Transformation product	Controls	Patients	Controls	Patients	P-* value
	% of specimens with active strains		Mean ranks		
3α,12α	72	17	22.2	12.4	0.01
24-methyl-3α,12α	76	8	23.2	10.3	< 0.01
3-keto;7α,12α	96	83	20.3	16.3	-
3α,12α;7-keto	96	92	20.2	16.6	-
3α,7α;12-keto	96	92	19.5	17.9	-
3,7-diketo;12α	64	50	19.7	17.5	-
3α;7,12-diketo	48	33	19.8	17.3	-
3,12-diketo;7α	36	50	17.5	22.1	-
3α,12α	72	33	17.5	9.6	< 0.05
24-methyl-3α,12α	76	33	17.8	8.5	< 0.05
3-keto;7α,12α	96	50	18.2	6.8	< 0.01
3α,12α;7-keto	96	83	16.2	15.3	-
3α,7α;12-keto	96	67	17.7	9.0	< 0.05
3,7-diketo;12α	64	33	17.3	10.7	≃ 0.1
3α;7,12-diketo	48	33	16.6	13.5	-
3,12-diketo;7α	36	33	16.3	14.8	-

Only secondary bile acids produced by more than 50 cultures were considered.

* Wilcoxon-Mann-Whitney-test

\+ Rectum is understood here as the last 8 cm of the large bowel.

Oxidation of the 3α- and 12α-hydroxyl functions might be a protective mechanism with regard to rectal cancer but not to colon cancer. This points to a partly different etiology as suggested by most epidemiologic studies. Deoxycholic acid is a known colon cancer promoter in animal experiments (Weisburger et al. 1983). It is conceivable, though at the moment only speculative, that esterification abolishs this ability. Contradictory to these mechanisms is the prevalence of 7α-dehydroxylating abilities among fecal isolates of the control but not of the patient group.

Aranki A, Syed SA, Kenney EB, Freter R (1969). Isolation of anaerobic bacteria from human gingiva and mouse cecum by means of a simplified glove box procedure. Appl Microbiol 17:568.

Correa P (1981). Epidemiological correlations between diet and cancer frequency. Cancer Res 41:3685.

Crowther JS, Drasar BS, Hill MJ, MacLennan R, Magnin D, Peach S, Teoh-Chan CH (1976). Faecal steroids and bacteria and large bowel cancer in Hong Kong by socio-economic groups. Br J Cancer 34:191.

Edenharder R, Knaflic T (1981). Epimerization of chenodeoxycholic acid to ursodeoxycholic acid by human intestinal lecithinase-lipase-negative Clostridia. J Lipid Res 22:652.

Finegold SM, Flora DJ, Attebery HR, Sutter VL (1975). Fecal bacteriology of colonic polyp patients and control patients. Cancer Res 35:3407.

Hill MJ, Drasar BS, Aries V, Crowther JS, Hawksworth G, Williams RFO (1971). Bacteria and aetiology of cancer of large bowel. Lancet 1:95.

Holdeman LV, Moore WEC (1977). "Anaerobic laboratory manual". Blacksburg, VA.

Mower HF, Ray RM, Shoff R, Stemmermann GN, Nomura A, Glober GA, Kamiyama S, Shimada A, Yamakawa H (1979). Fecal bile acids in two Japanese populations with different colon cancer risks. Cancer Res 39:328.

Reddy BS, Wynder EL (1977). Metabolic epidemiology of colon cancer: Fecal bile acids and neutral sterols in colon cancer patients and patients with adenomatous polyps. Cancer 39:2533.

Schwann A, Ryden AC, Laurell G (1982). Fecal bacterial flora of four nordic population groups with diverse incidence of large bowel cancer. Nutr Cancer 4:74.

Weisburger JH, Reddy BS, Barnes, WS, Wynder EL (1983). Bile acids, but not neutral sterols, are tumor promoters in the colon in man and in rodents. Environ Health Perspect 50:101.

Germfree Research: Microflora Control and Its Application to the Biomedical Sciences, pages 311–313

PRODUCTION OF TUMOR NECROSIS FACTOR IN GERMFREE MICE

Lu Longmu

Wuhan Institute for Biological Products

Wuchang, Hubei, China

Tumor necrosis factor (TNF) is a substance that is found in serum of animals sensitized with priming agents (such as B.C.G., *Corynebacterium parvum*) and challenged later with endotoxin. The substance causes the necrosis of some tumors *in vivo*. It is a glycoprotein with molecular weight 150,000 (Ruff, Gifford 1981; Carswell et al 1975). The animals used previously for production of TNF were conventional. This paper reports the responses of GF animals to priming and eliciting agents in the course of production of TNF.

MATERIALS AND METHODS

Mice: GF AKR and CFW mice were used. They were kept in flexible plastic isolators, fed with autoclaved diet L 485 and water ad libitum. The temperature and humidity of the animal rooms were maintained at about 72°F and 58% respectively. Counterpart conventional (CV) control mice were housed in a clean conventional air-conditioned room where the temperature and humidity were also controlled.

Priming agent: *Corynebacterium parvum* was provided by Institut Mereux and Wellcome Reagents, Ltd.

Eliciting agent: Endotoxin polysaccharide (Old 1976) was prepared from *E. coli*.

Production of TNF: The mice (GF and CV) were each inoculated i.v. with 0.7 mg killed *C. parvum*. Six days later each was inoculated IV with 20 micrograms endotoxin. Ninety minutes later, the mice were bled from the orbital plexus and the serum from each mouse was separated.

Assay of TNF: (Ruff, Gifford 1980) = 96-well microtiter plates were used. 3-fold serial dilutions of serum samples were made directly into the wells. Actinomycin D (0.1 μg/ml) was added (Ostrove, Gifford 1979). Then 60,000 L-929 cells were added into each well of plates. The plates were incubated for 18 hr. and the residual cells were stained with crystal violet, washed, dried and then examined. The titer recorded was the highest dilution of serum sample which shows 50% or greater cytotoxicity on L-929 cells. A serum titer less than 1:50 was recorded as negative.

RESULTS

Germfree mice responded effectively to the priming and eliciting agents. The titers of TNF were high (Table 1). There were no significant differences in titers of TNF between GF and CV mice (Table 2).

Table 1. Assay for TNF Production by GF Mice

	Experiment Group				Control Group			
Strain	No.	Age	Sex	TNF Titer	No.	Age	Sex	TNF Titer
AKR	25	1-3 mos.	mix	1:5,478	4	1-3 mos.	mix	<1:7.5
CFW	7	2 mos.	mix	1:16,000	6	2 mos.	mix	<1:7.5

Table 2. Comparison of TNF Production by GF and CV Mice

Type of Mice	Experiment Group				Control Group			
	No.	Sex	Age	TNF Titer	No.	Sex	Age	TNF Titer
GF	32	mix	1-3 mos.	1:10,739	10	mix	1-3 mos.	<1:7.5
CV	11	mix	2 mos.	1:98,402	8	mix	1-2 mos.	<1:45

The results indicate that while the TNF titers of the uninoculated control mice were negative, the GF mice responded very well to the priming effect of C. parvum and the elicitation by endotoxin. The TNF titers of all mice in the experimental group exceeded 1:5,000.

DISCUSSION

Confirming previous observations by others, we found that AKR and CFW mice responded very well to the priming of C. parvum and elicitation of endotoxin in the course of production of TNF. Moreover, we demonstrated that GF mice responded as well as their conventional counterparts, which promises the potential use of TNF from GF mice in the treatment of some human tumors, with less chance to transmit infectious agents during treatment with TNF from GF mice. Such an oncolytic factor would be an attractive candidate for its selected destruction of transformed tumor cells.

There have been reports of some differences among strains of mice in the production of TNF. Mannel et al. (1980) reported that TNF can be produced in BCG-infected mice by endotoxin in endotoxin-sensitive mice, but not in endotoxin-insensitive mice. In the data reported here, the TNF titers in serums from CFW mice were slightly higher than in serums from AKR mice.

REFERENCES

Carswell EA, Old LJ, Kassel RL, Green S, Fiore N, Williamson B (1975). An endotoxin-induced serum factor that causes necrosis of tumors. Proc Nat'l Acad Sci USA 72(9):3666.

Mannel DN, Meltzer MS, Mergenhagen SE (1980). Generation and characterization of a lipopolysaccharide-induced and serum-derived cytotoxic factor for tumor cells. Infection and Immunity 28(1):204.

Old LJ (1976). Tumor necrosis factor. Clin Bull 6(3):118.

Ostrove JM, Gifford GE (1979). Stimulation of RNA synthesis in L-929 cells by rabbit tumor necrosis factor. Proc Soc Exp Biol and Med 160(3):354.

Ruff MR, Gifford GE (1980). Purification and physico-chemical characterization of rabbit tumor necrosis factor. The Jrnl of Immunology 125(4):1671.

Ruff MR, Gifford GE (1981). Tumor necrosis factor. In Pick E. (ed): "Lymphokines," New York: Academic Press, Vol. 2, p. 235.

Germfree Research: Microflora Control and Its Application to the Biomedical Sciences, pages 315–318

INHIBITION OF HUMAN COLON TUMOR GROWTH IN NUDE MICE BY INDOMETHACIN

James B. Heneghan, Ph.D.
Departments of Surgery and Physiology
Louisiana State University, School of Medicine
New Orleans, Louisiana 70112 U.S.A.

Introduction:

Indomethacin, a potent inhibitor of prostaglandin synthesis reduces both the incidence and the growth of intestinal tumors induced by either dimethylhydrazine, methylazoxymethanol, or methyl (acetoxymethyl) nitrosamine in rats (Pollard and Luckert, 1981). The present investigation will determine if a similar inhibitory effect of indomethacin can be observed on the growth of human colon adenocarcinoma xenografts in nude mice.

Methods:

A small breeding colony of specific pathogen free (SPF) hypothymic nude mice (Hap: (nu/nu) BR) was established and maintained with two different techniques: a) flexible-film isolators and gnotobiotic techniques and b) filter-top cages and a laminar air-flow hood. At 10 weeks of age (25 to 35 gms body weight) the nude mice were transferred into a laminar air flow hood and anesthetized with an intraperi toneal injection of sodium pentobarbital (60 mg/ kg). Two pieces of human adenocarcinoma (125 mm^3) were transplanted subcutaneously into the right and left flanks of each mouse as follows: a) Experiment 1, tumor from an untreated mouse (16 mice); b) Experiment 2, tumor from an ethanol-treated mouse (8 mice) and c) Experiment 3, tumor from an indomethacin-treated mouse (8 mice).

At the end of Experiment 1, one animal from each treatment group was used as a donor for experiments 2 and 3. When the tumor volume reached approximately 550 mm^3/ (6 weeks), half of the mice in each experiment were fed drinking water containing 2 mg/ml ethanol (control) and the other half with 2 mg/ml ethanol and indomethacin (20 mg/L). Each week tumor volume was determined by three caliper measurements: a) longest diameter (length); b) longest perpendicular diameter (width) and c) height (depth). These measurements were repeated before the sacrifice of each mouse *in vivo* and again after each tumor was removed and weighed (*in vitro*). Area and volume calculations were compared to final tumor weight. Complete necrosies were performed with emphasis on possible metastatic sites.

Results:

The data in Table 1 showed that the tumor volume measurements had the best correlation with actual tumor weight, exhibiting the smallest deviations both *in vivo* and *in vitro*.

Table 1. Human Colon Tumor Growth in Nude Mice Average Percentage Deviations from Tumor Weight

	n	In Vivo	In Vitro
Experiments 1 and 2			
Volume	38	10.4	0.5
Area	38	14.5	11.7
Experiment 3			
Volume	12	9.4	2.8
Area	10	15.5	12.7

Table 2. Human Colon Tumor Growth in Nude Mice Doubling Time in Days

	n	1st	2nd	3rd	4th
Experiment 1:	Tumor from Untreated Mouse				
Untreated	16	23	-	-	-
Indomethacin	16		45	40	-
Ethanol	16		25	8	-
Experiment 2:	Tumor from Ethanol-treated Mouse				
Indomethacin	16	12	12	18	-
Ethanol	16	6	6	14	-
Experiment 3:	Tumor from Indomethacin-treated Mouse				
Indomethacin	16	1	21	$>$ 22*	-
Ethanol	16	3	5	15	15

*Did not double; increased only 80%

The original well differentiated morphology of this

Figure 1. Human colon adenocarcinoma; H & E stain (40 x).

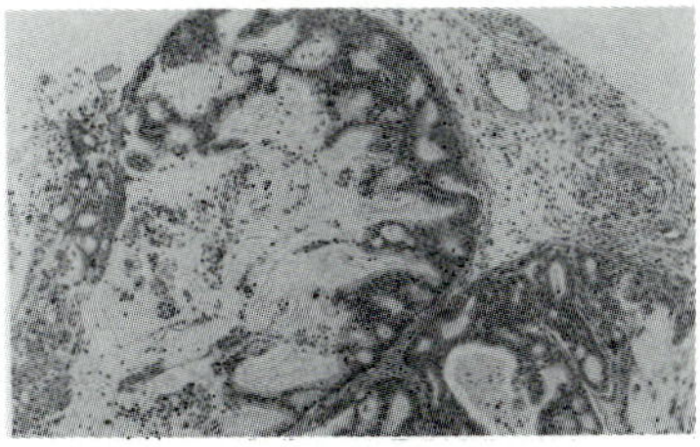

Figure 2. Growth of human colon tumor from an untreated mouse; Experiment 1.

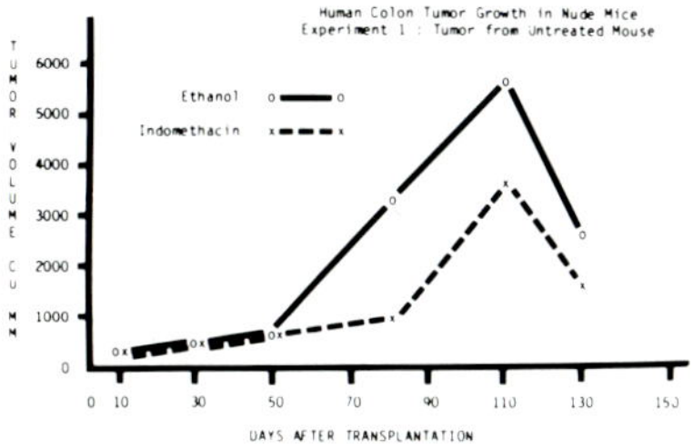

Figure 3. Growth of human colon tumor from an ethanol-treated mouse; Experiment 2.

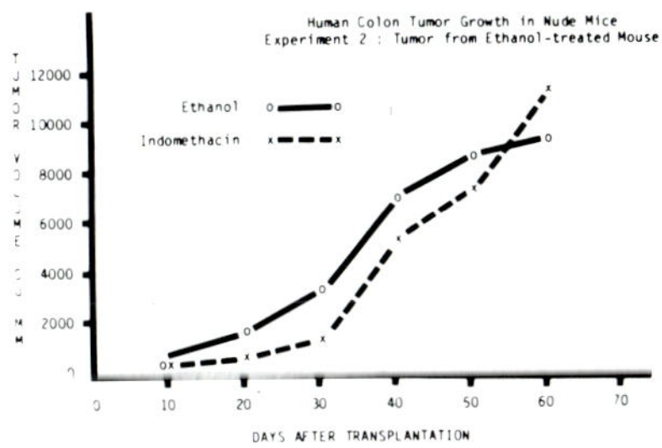

Figure 4. Growth of human colon tumor from an indomethacin-treated mouse; Experiment 3.

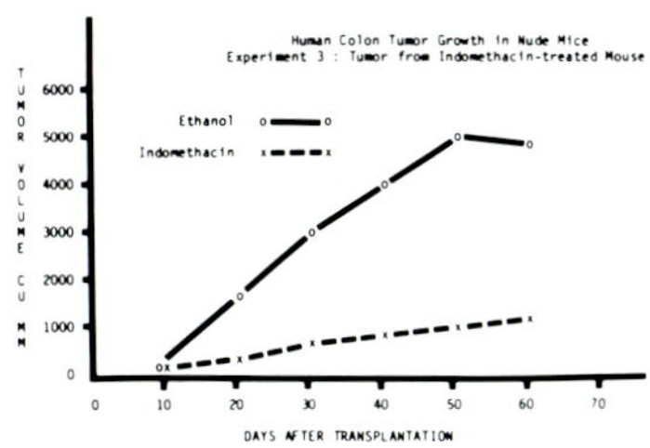

human colon adenocarcinoma as shown in Figure 1 has been retained through 10 subpassages. In Experiment 1 the growth rate of tumor taken from an untreated mouse (Figure 2) was significantly inhibited by indomethacin treatment between 50 and 75 days after initial transplantation; thereafter tumor growth rates were similar in both groups. In Experiment 2 (Figure 3) a slight reduction in growth rate of tumor from an ethanol-treated mouse, in nude mice treated with indomethacin, was seen between 10 and 30 days. Finally, in Experiment 3 (Figure 4), a significant reduction of tumor growth rate was observed between 10 and 50 days in indomethacin-treated mice with tumors from an indomethacin-treated mouse. This inhibition of tumor growth by indomethacin treatment in each experiment was confirmed when tumor doubling times were calculated (Table 2) at necropsy; no metastases were observed. In previous experiments, 5-fluouracil treatment produced a small but statistically insignificant inhibition in the growth of this same tumor (Heneghan, 1982).

Conclusions:

The original well differentiated morphology and growth characteristics of a human colon adenocarcinoma have been retained through 10 subpassages in nude mice. Tumor volume measurements both _in vitro_ and _in vivo_ had the best correlation with actual tumor weight. Consistant and significant growth inhibiting effects of indomethacin have been quantified in this nude-mouse-human-tumor model. Both tumor volume measurements and tumor doubling times confirmed these trends. Thus, the tumor inhibiting effects of indomethacin, initially observed with chemically induced tumors in rats, have been demonstrated on human colon adenocarcinoma in nude mice. (Supported by the American Cancer Society Grant PDT 223).

Bibliography:

Pollard M, Luckert PH (1981). Experimental therapeutic procedure for cancer. In: Recent Advances in Germfree Research Eds. Sasaki S, Ozawa A, Hashimoto K. Tokai University Press, Tokyo, Japan, p. 551.

Heneghan JB (1982). Inhibition of human colon tumor growth in nude mice by 5-flurouracil. Abstracts 20th Annual Meeting of the Association of Gnotobiotics, Lexington, Kentucky, p. 23.

Germfree Research: Microflora Control and Its Application
to the Biomedical Sciences, pages 319–321

RECENT STUDIES OF DISEASES DEMONSTRATING SIMILAR PATTERNS INCLUDING INFECTIOUS MONONUCLEOSIS (IM) AND KAPOSI'S SARCOMA (KS)

Audrey Fjelde, Ph.D., Biological Resources
Sureshkant Patel, M.D., Dermatology Department
Ole A. Holtermann, M.D., Ph.D. Dermatology Department
Roswell Park Memorial Institute
666 Elm Street, Buffalo, New York 14263

The course of events in IM and atypical KS indicates involvement of the immune system, a prominent feature of which is the acquired-immuno-deficiency syndrome (AIDS). Investigations by us and others revealed in IM patients, as in AIDS cases, an inversion of the Helper/Suppressor-Cytotoxic T cell ratio, corresponding to the appearance (polyclonal activation?) of the diagnostic Paul-Bunnell (PB) heterophile antibodies associated eventually with rheumatoid arthritis. In IM the H/S inversion, increased S-C lymphocytes, occurs after infection of B lymphocytes by the Epstein Barr virus (EBV), for which the B cells develop receptors during a stage of maturation. The host remains infected for life. We proposed that initially IM disease is controlled by the S-C cells but natural killer (NK) cells and large granular lymphocytes (LGL) control the life-long EBV infection. (Fjelde 1983) Downey, in studies related to IM originally described these cells as a separate line, arising direct from the RE System (Downey, McKinlay 1923). One atypical KS patient, with symptoms of immune deficiency did not show an inverted H/S ratio but had a lesion diagnosed as "early KS lesion suggestive of the patch stage of KS, or an unusual angioma". After excision of the lesion, the patient recovered. Peripheral blood studies showed EBV-VCA positive 1:40, Cytomegalo V. pos 1:160, normal IgM, IgA, high normal IgG values and a normal karyotype. The fast growing malignant KS lesions in young AIDS patients and association with EBV is different from the relatively benign classical KS. Normally IM patients recover from their disease and immuno deficiency, (Fjelde, Ozer, Kellam, Bennett 1982); immuno incompetent individuals do not recover from IM, but die from lymphoma (Klein, Purtillo 1981). It is not understood why AIDS individuals develop a malignant form of KS. We propose that the RE cell is the stem cell for NK-LGL's, and represents the most primitive line of immunological defense. The KS tumor is the result of the final defensive effort. Perhaps this line of cells is related to the diagnostic cell for Hodgkins' disease (HOD). We have included in our studies of unusual patients cases of HOD related IM; one patient showed many NK-LGL cells in peripheral blood. The (HTLV) human T cell leukemia virus has been suggested as the cause of AIDS (Gallo 1984). The H/S inversion giving a H/S ratio less than 1, is caused in IM by excessive growth of S cells, in AIDS by killing of H cells. HTLV and EBV are expressed by AIDS patients. Because AIDS patients develop IM, (insertion of viral information into B cell genes years earlier) and because pre KS lesions progress to rapidly growing malignancies, we have done histochemical studies of NK-LGL's and also adenosine deaminase activity (ADA) activity in human spleens (Fjelde, Kellan, Bennett, Stelluto, Tritsch, 1983). Using affinity of substrate (Km) and reaction velocity at high substrate concentration (Vmax) for ADA, we found for HOD similar Km and appreciable range of Vmax. At the present time, we are testing the theory that the azur granules in the LGL and NK population are part of the oxygen dependent killing system, through the production of lethal oxidants in these granules by the enzyme in the pathway (adenosine→uric acid) of purine metabolism. Preliminary studies suggested that the granules were toxic. We have long been interested in the effect of infection on chromosomes (Fjelde, Levan, Rask-Nielsen, 1962; Fjelde, Holtermann, 1962; Fjelde, Trenton, Bryan, 1964); and on the induction of malignancy (Fjelde, Sorkin, 1960; Fjelde, Turk, 1975). Recently, Mori, Sudo and Kano (1983) showed that the Forssman antigen is expressed in the genome of malignant human adeno 12 tumors. We have entered a new area of research in which virology, immunology and genetics are expressed in molecular terms; and a better understanding of the role of oncogenes and regulating factors is possible, and the significance of the myc gene, and reciprocal translocations between chromosomes 8-14 in Burkitt's Lymphoma (probably EBV related).

Acknowledgements: Technical, P. Bennett, J. Brzykcy, J. Kellam; Support, Onan Family Foundation, W. Walker Fund, Investors Diversified Services, K-M Wahlberg, W. Brock, USPHS Grant R01-CA-26945-03.

Downey H, McKinlay CA (1923). Acute lymphadenosis compared with acute lymphatic leukemia. Arch Inst Med 32:82.

Fjelde A, Sorkin E (1960). The occurrence of malignancies and other pathology in PRI mice treated with PRI skin tumor extracts. Acta Path et Microbiol Scand 50:757.

Fjelde A, Holtermann O (1962). Chromosome studies on the HEP 2 tissue culture cell line during infection with measles virus. The Life Sciences 12:683.

Fjelde A, Levan A, Rask-Nielson R (1962). The chromosomes of four transplantable murine plasma cell leukemias characterized by varying pathological serum protein changes and/or amyloid formation. Hereditas 48:630-644.

Fjelde A, Trentin J, Bryan E (1964). The chromosomes of tumors induced in mice by human adenovirus Type 12. Proc Soc Exp Biol & Med 116:1102.

Fjelde A, Turk J (1965). Induction of an immune response in local lymph nodes by chemical carcinogens. Nature 205:512.

Fjelde A (1975). Helper activity for the defective friend leukemia virus in human malignant cell cultures, and a review of instructive DNA polymerase and higher molecular weight virus-related RNA and human lymphoma. In Fogh J (ed): "Human Tumor Cells in Vitro", New York: Plenum, Chapter 16.

Fjelde A, Ozer H, Kellam J, Bennett P (1982). Characterization of peripheral lymphoid cells (Downey) during infectious mononucleosis (IM), using monoclonal antibodies. In Vitro 18:309.

Fjelde A, Kellam J, Bennett P, Stelluto E, Tritsch G (1983). Adenosine Deaminase in Isolated Human Spleen Cells. Proceedings 74th annual AACR Meeting, 24:48.

Fjelde A (1983). The Nature of Infectious Mononucleosis and the Role of Downey Cells. In Cohen E, Singal D (eds): "Non-HLA Antigens in Health, Aging and Malignancy", New York: Alan R. Liss, p 224.

Gallo RC (1984). See overview in Science 224:497.

Kano K, Fjelde A, Milgrom F (1977). Paul-Bunnell antigen in lymphoma and leukemia spleens. J Imm 19:146.

Klein G, Purtillo D (1981). Symposium on Epstein-Barr virus-induced Lymphoproliferative Diseases in Immunodeficient Patients. Cancer Res 41:4209.

Milgrom F, Kano K, Fjelde A (1973). Studies on heterophile antigen in lymphoma and leukemia spleens by means of absorption of infectious mononucleosis sera. Int Arch for Allergy and Applied Immunology 45:631.

Mori T, Sudo T, Kano T (1983). Expression of heterophile Forssman antigen on cultured malignant cell lines. J Natl Cancer Inst 70:42.

VII IMMUNOLOGY

Germfree Research: Microflora Control and Its Application to the Biomedical Sciences, pages 325–328

IMMUNORESPONSES AND NATURAL KILLER ACTIVITY OF NUDE AND NORMAL MICE UNDER CONVENTIONAL AND GERMFREE CONDITIONS

Tatsuo Suzuki, Yumiko Suzuki, Noriyuki Maruo, Takaya Yamada, Shizuo Katagiri, Eiji Kawamura, Tetsuro Okano*, Yoshiyuki Kubota*, Akira Ghoda*, Ryoko Yamashita**, Makoto Shimizu**, Kikuo Nomoto***

The Kitasato Institute Hospital, Tokyo, Japan. *Kitasato University, Tokyo, Japan. **Keio University, Tokyo, Japan. ***Kyushu University Fukuoka, Japan.

To understand the actual conditions of the in vivo defense mechanism including immunity, it is necessary to understand clearly the effects of microorganisms as an environmental factor and their relation with the host. However, the microorganisms present in the normal environment have complex antigenicity and diverse activities, and they are considered to have complicated effects on the differentiation and maturation of the cells responsible for immunity in the host and the appearance of their functions. Therefore, to elucidate the role of stimulation in natural environment, especially the normal flora, we used germfree animals in clarifying the in vivo defense mechanisms. The results are reported here.

Materials and Methods

(1) Animals : The animals used were ICR, NC, BALB/c, nude mice of BALB/c origin, nude mice of ICR origin and NOD mice of ICR origin which serve as a model for juvenile onset diabetes. They were reared under conventional and germfree conditions.

(2) Erythrocyte suspension : Hamster red blood cells (HRBC) were used as the antigen.

(3) Tumor cells : YAC cells were used as the target cells to measure NK activity.

(4) NK activity measurement (Habu S 1979) : The target cell cytotoxicity test by means of ^{51}Cr release was used.

Results

(1) Immunoresponse

The appearance of antibody production for HRBC in a germfree environment was investigated (Table 1). One x 10^9 HRBC was administered intraperitoneally to specific pathogen free (SPF) and germfree (GF) ICR mice, and the HA antibody titers on the 4th day were measured. GF mice showed the same or slightly higher antibody titers than SPF ICR mice against HRBC antigen when large doses of 1 x 10^9 of antigen were administered. However, there were individual differences in HA antibody titers among the SPF ICR mice. HRBC was repeatedly administered intraperitoneally to SPF and GF ICR and NC mice on days 0, 4, 8, 12 and 16, and the HA antibody titers were measured on the 4th, 8th, 12th, 16th and 22nd days. In both the SPF and GF animals, there was moderate HA antibody production in ICR mice due to the primary immunization but 2-ME resistant antibodies were not detected on the 4th day in both types of mice. There were increases in HA antibody titers on the 8th and 12th days and shift to 2-ME resistant antibodies due to the 2nd and 3rd immunizations. However, NC mice showed lower HA antibody titers than ICR mice on the 4th day, but on the 8th day the antibody titers were the same as those of ICR mice. The HA antibody titer reached a peak on the 16th day, and almost all of the antibodies were 2-ME resistant antibodies in both strains (Sell S 1965). Differences in the HA antibody titer in the ICR mice on the 4th day after the start of immunization were the same in both SPF and GF mice. This result is considered to be due to the lack of uniformity in the genetic background of the ICR mice as a closed colony. The differences in HA antibody titer seen in the SPF NC mice were almost completely absent in the GF NC mice, which is considered to indicate the effects of modification due to the presence of microorganisms. Therefore, it is considered necessary to use germfree animals with no presensitization to obtain more precise solutions with respect to immunogenetics and to observe nonspecific in vivo responses.

Table 1.
Antibody Response Repeated Injection of HRBC in GF, SPF mice

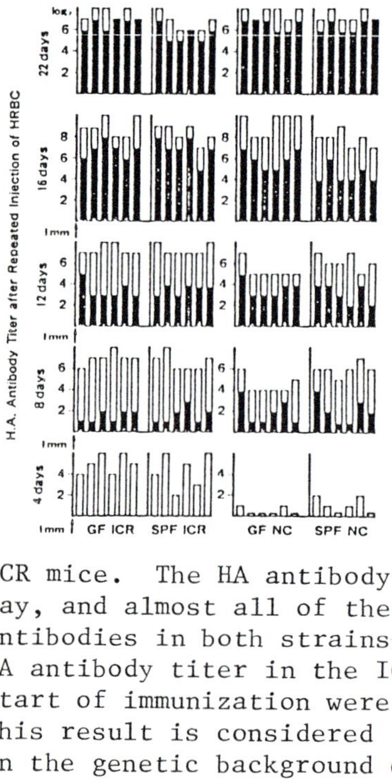

(2) Natural killer activity

Next we observed the effect of normal flora on what has been called natural killer activity (Table 2) (Lotzová E 1984). Changes in the NK activity with age were investigated in GF and SPF BALB/c mice, and SPF BALB/c nude mice. The times of appearance of NK activity in the spleen, peritoneal cavity, regional lymph nodes and thymus is considered to be somewhat earlier in nude mice, but in both GF and SPF BALB/c mice, NK activity appeared from after the 3rd week of age and reached a peak at around the 8th week of age. It then tended to decrease with age. In GF and SPF mice, the NK activity was higher in the SPF mice at 3 weeks, 6-8 weeks and 15 weeks of age, and the role of indigenous bacteria was strongly suggested. In GF mice, the NK activity had decreased to almost the detectable limit at 15 weeks of age, but there was no decrease in NK activity in SPF nude mice and it appears that the congenital lack of a thymus plays some role from the standpoint of in vivo defense.

Table 2.
Changes in the NK Activity with Age in Germfree and SPF BALB/c or Nude mice

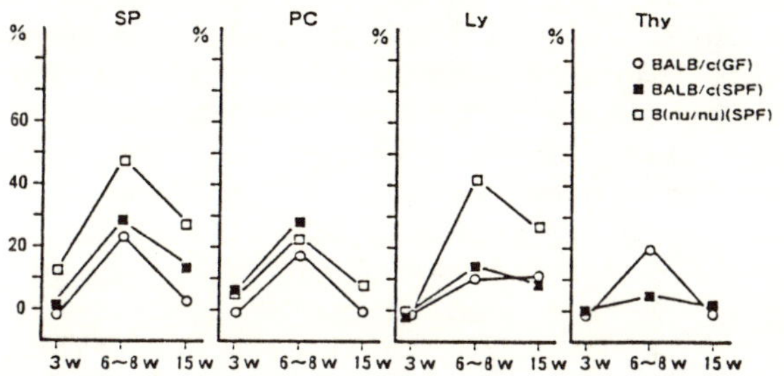

NK activities were compared between GF and SPF environments in mutant ICR mice including NOD mice which have been established genetically as disease models for juvenile-onset diabetes (Table 3). Compared with the NK activity in GF ICR mice, the NK activity in GF NOD mice was lower, and the NK activity was even lower in diabetes-positive NOD mice. After the onset of diabetes in GF NOD mice, the mice soon died. In NOD nude mice,

Table 3.
Compared NK Activities between Germfree and SPF Juvenile-Onset Diabetes(NOD) mice or NOD nude mice

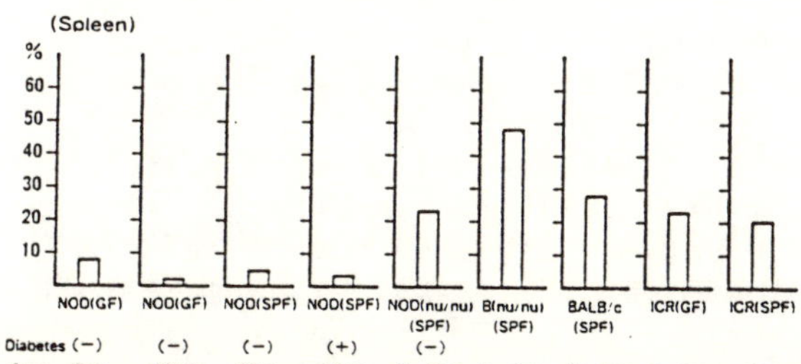

there was no onset of diabetes by 30 weeks of age, and NK activity was maintained to the same degree as in GF ICR mice. These results suggested that T cells are involved in the onset of diabetes in NOD mice and NK activity is related to progress of the symptoms. Next a comparison was made between GF and SPF mice concerning the potentiating effects on NK activity using various types of immunoactivators. The biologically active substance NSP extracted from the skin of rabbits which were inoculated subcutaneously with vaccine virus, a hemolytic streptococcus preparation OK-432 and formalinized Pseudomonas aeruginosa M-24 cell were used for immunoactivators. GF and SPF ICR mice were inoculated intraperitoneally with the immunoactivators 5 consecutive times, and the changes in NK activity were compared. The results showed that OK-432 increased intraperitoneal NK activity in 25 week old ICR mice with decreased NK activity, and NSP increased the NK activity in the regional lymph nodes. M-24 increased the NK activity in the peritoneal cavity and in the regional lymph nodes. The above results appear to indicate that NK activity can activate at specific localized sites according to the type of the stimulation.

Conclusion

1. In a GF environment, immunoresponse gene dominance is expressed more faithfully.
2. NK activity appears even when there is no stimulation of indigenous bacteria, etc.
3. Because of presensitization by indigenous bacteria, the NK activity showed strong secondary response when immunoactivators were administered. Therefore, these effects are included in animals reared under conventional conditions.
4. As seen in NOD mice, it is necessary to prepare disease models from GF animals for etiological studies to avoid effects of indigenous bacteria.

Habu S, Hayakawa K, Okumura K, Tada T (1979). Surface markers on natural killer cells of the mouse. Eur J Immunol 9:938.

Lotzová E, Savary CA (1984). Analysis of natural killer cell activity in random bred Rowett athymic rats. Immune-deficient animals. 4th Int. Workshop on Immune-Deficient Animals in Exp. Res. Chexbres, 1982. (Karger, Basel 1984) Exp Cell Biol 52:53.

Sell S (1965). Mercaptoethanol-sensitive antibody production in germ-free mice and guinea pigs. J Immunol 95:300.

Germfree Research: Microflora Control and Its Application to the Biomedical Sciences, pages 329–332

Effect of Recombinant Human Interferons on NK and K Cell Activities of Gnotobiotic Miniature Swine[1]

Yoon Berm Kim, M.D., Ph.D. and Tae June Chung, M.D.

Department of Microbiology and Immunology
University of Health Sciences/The Chicago Medical School, North Chicago, IL 60064

Interferons (IFN) may be one of the most important regulatory substances for natural killer (NK) cells. IFN might activate preexisting NK cells and/or induce pre-NK cells to become active NK effector cells (Targan and Dorey 1980; Djeu et al 1981). Whether IFN also augments killer (K) cells for antibody-dependent cellular cytotoxicity (ADCC) is controversial (Trinchieri et al 1978; Herberman et al 1979).

We have investigated the role of IFN in regulating NK and K cell activities in the gnotobiotic miniature swine. We found that the human leukocyte interferon (Cantell's HuIFN-α, specific activity of 10^6 unit/mg protein) is a good augmentor of porcine NK activity but not ADCC activity (Chung et al 1982). To rule out a possibility that impurities in HuIFN-α preparations may cause enhancement of porcine NK activity, we have examined the effect of highly purified recombinant human interferons (rIFN-αA and rIFN-αD, Hoffmann-LaRoche, Inc., Nutley, NJ) (Goeddel et al 1981) and recombinant human interferon (rIFN-γ, Genentech, Inc., South San Francisco, CA) on porcine NK and ADCC activities.

Animals, preparation of lymphoid cells, culture systems, and assay for NK and ADCC and calculation of cytotoxic activity were as described elsewhere (Kim et al 1980).

First we have titrated the optimal concentrations and time of incubation to study the effects of rIFN on NK and K

[1] This work was supported in part by US PHS Grant CA-38336 awarded by the National Cancer Institute.

Table 1

Effect of Recombinant Human Leukocyte Interferons on NK and ADCC Activities of Peripheral Blood Lymphoid Cells from SPF Adult Miniature Swine

		% Specific Lysis[a]			
Animal No.	IFN Treatment[b]	K562	TNP-SB	TNP-SB + Anti-TNP	ADCC[c]
SPF 3114-8	-	7.0	7.7	70.3	62.6
	rIFN-αA	20.8	12.3	70.1	57.8
	rIFN-αD	21.9	12.8	54.4	41.3
SPF 3129-5	-	34.9	4.4	40.6	36.2
	rIFN-αA	54.1	11.3	46.6	35.3
	rIFN-αD	64.0	10.9	50.0	39.1
SPF 3162-7	-	20.8	7.7	36.9	29.1
	rIFN-αA	38.3	18.6	42.2	23.6
	rIFN-αD	37.1	19.1	38.9	19.8
SPF 4029-3	-	18.0	6.2	36.3	30.1
	rIFN-αA	31.5	6.0	36.8	30.8
	rIFN-αD	32.6	6.5	39.8	33.3
SPF 4030-3	-	35.4	1.9	48.2	46.3
	rIFN-αA	50.0	9.2	60.4	51.2
	rIFN-αD	50.8	10.1	60.8	50.7
SPF 4105-1	-	34.1	11.1	36.1	25.0
	rIFN-αA	50.8	19.1	50.2	31.1
	rIFN-αD	54.1	17.1	41.7	24.6

[a]Effector-to-target cell ratios were 100:1.
[b]Effector cells (PBL) were incubated without (-) or with 1.5 ng of rIFN-αA or rIFN-αD/$5x10^6$ cells/ml for 3 hr at 37^oC and washed three times with cold BSS before assaying.
[c]Net ADCC were calculated by subtracting % specific lysis of TNP-SB from that of TNP-SB plus anti-TNP.

cell activities. It was found that as low as < 0.1 ng/$5x10^6$ cells/ml had augmented NK activity and had maximum augmentation at 1.5 ng/ml equivalent to specific activity of 300 units (NIH standard units)/ml, and 3 hr incubation at 37°C gave a maximum augmentation with rIFN-α and 18 hr at 37°C with rIFN-γ (data not shown).

Table 1 shows the results of the effects of rIFN-αA and -αD on NK and K cell activities by using pretitrated optimal dose and incubation time (300 units/ml, 3 hr at 37°C). The rIFN-α augmented NK activity of all PBL from SPF miniature swine but not ADCC activity. Note that the apparent enhancement of ADCC (TNP-SB + anti-TNP) was due to the enhancement of spontaneous cytotoxicity against TNP-SB in the TNP-SB in the absence of added anti-TNP antibody; therefore, it is necessary to calculate net ADCC by subtracting percent specific lysis of TNP-SB from that of TNP-SB plus anti-TNP to determine the effects on ADCC only.

Table 2. Summary of Effect of Recombinant Human Leukocyte Interferons on the Porcine NK and K cell Activities of Various States of Miniature Swine

States of Animals (No. Tested)	Preexisting NK Activity	Augmentation of NK Activity by rIFN	Preexisting ADCC Activity	Augmentation of ADCC Activity by rIFN
Germfree				
<3 wk old (7)	-	-	+	-
>3 wk old (7)	+	+	+	-
SPF				
<2 wk old (4)	-	-	+	-
>2 wk old (7)	+	+	+	-

Effects of rIFN on NK activity were further investigated to determine whether IFN might activate preexisting NK cells and/or induce pre-NK cells to become active NK effector cells. As summarized in Table 2, rIFN pretreatment of PBL from greater than 3-week old germfree piglets or greater than 2-week old SPF piglets augmented preexisting NK activity but failed to induce NK activity of PBL from less than 3-week old germfree piglets or less than 2-week old SPF piglets whose NK activity has not yet developed.

In conclusion, these data suggest that pure interferons are insufficient inducers of NK cell differentiation from pre-NK cells while they are a good activator of NK cells and/or enhancer of NK activity. Furthermore, interferons augment porcine NK cell activity but not porcine K cell activity.

Acknowledgements: The authors wish to acknowledge Dr. Patrick W. Trown, Hoffmann-LaRoche, Inc., Nutley, NJ for his gift of purified recombinant human interferons αA and αD and Dr. Chris Platsoucas for sharing purified recombinant human interferon-γ (Genentech, Inc., South San Francisco, CA).

REFERENCES

Chung TJ, Huh ND, Kim YB (1982). Differential effects of interferons on porcine NK and K cell activities. In Herberman RB (ed): "NK Cells and Other Natural Effector Cells". New York, Academic Press, p. 381.

Djeu JY, Stocks N, Varesio L, Holden HT, Herberman RB (1981). Metabolic requirements for the in vitro augmentation of mouse natural killer activity by interferon. Cell Immunol 58:49.

Goeddel DV, Loung DW, Dull TJ, Gross M, Lawa RM, McCandliss R, Seeburg PH, Ullrich A, Yelverton E, Gray PW (1981). The structure of eight distinct cloned human leukocyte interferon cDNAs. Nature 290:20.

Herberman RB, Ortaldo JR, Bonnard GD (1979). Augmentation by interferon of human natural and antibody-dependent cell-mediated cytotoxicity. Nature 277:221.

Kim YB, Huh ND, Koren HS, Amos DB (1980). Natural killing (NK) and antibody-dependent cellular cytotoxicity (ADCC) in specific pathogen-free (SPF) miniature swine and germ-free piglets. I. Comparison of NK and ADCC. J Immunol 125:755.

Targan S, Dorey F (1980). Interferon activation of "pre-spontaneous killer" (pre-SK) cells and alteration in kinetics of lysis of both "pre-SK" and active SK cells. J Immunol 124:2157.

Trinchieri G, Santoli D, Koprowski H (1978). Spontaneous cell-mediated cytotoxicity in humans: role of interferon and immunoglobulins. J Immunol 120:1849.

Germfree Research: Microflora Control and Its Application to the Biomedical Sciences, pages 333–336

MODULATION OF MURINE NATURAL KILLER CELLS BY CRYPTOCOCCUS NEOFORMANS

Kenneth F. Bartizal, Jr., Ph.D., Cindy A Salkowski, B.A., and Edward Balish, Ph.D.
Dept. of Microbiology/Immunology, Kirksville College of Osteopathic Medicine, Kirksville, MO [KFB], Depts. of Surgery & Medical Microbiology, Univ. of Wisconsin Medical School, Madison, WI [CAS; EB]

Large granular lymphocytes called natural killer (NK) cells are capable of killing tumor cells without prior exposure to them. Modulation of NK cell activity appears to be influenced by many factors. Splenic NK activity in mice is influenced by colonization of their alimentary tract with microorganisms, but this is mainly dependent on the composition and complexity of the gastrointestinal microflora. Our previous studies have shown that colonization of athymic and euthymic mice with a complex conventional (CV) or limited defined-flora (FD) augmented NK cell activity when compared to germfree (GF) counterparts. Conversely, colonization of the alimentary tract of GF athymic and euthymic mice with a pure culture of *Candida albicans*, or diassociation with *C. albicans* and a *Bacillus* sp., did not alter NK cell activity when compared to GF counterparts. In addition, colonies of inbred CV mice were found to have significantly different NK cell activity depending on their diet and the microbial status of the particular mouse colony (Bartizal, Salkowski, Balish 1983; 1984). Other investigators have demonstrated an enhancement of NK cell activity followed by suppression or return to baseline natural cytotoxicity levels after CV mice were colonized or infected with microbes (Kamiyama, Hagiwara 1982; Wolfe, Tracy, Henney 1976; Djeu, Huang, Herberman 1980). These induced NK cells were considered to play an effector role as a natural defense mechanism against infectious agents.

This study was designed to investigate the modulation of splenic NK cell activity by *Cryptococcus neoformans*, an opportunistic fungal pathogen. *C. neoformans* is an encapsulated yeast that can cause fatal disseminated disease in man and experimental animals. It is not understood how the immune response prevents disseminated disease, but NK cells have been implicated as a host defense mechanism in cryptococcosis and possibly other fungal diseases. Athymic nude mice having high NK cell activity were able to eliminate C. neoformans from tissues more readily than normal heterozygous mice within the first 7 days after infection (Cauley, Murphy 1979). NK levels of high and low reactive mouse strains correlated with the ability of NK effector cells to limit the in vitro growth of Cryptococcus (Murphy, McDaniel 1982). In addition, cryptococcal capsular polysaccharide antigens have been implicated as causing immunological unresponsiveness, involving both antibody and cell mediated immunity (CMI). Injection of cryptococcal antigen into mice induces populations of lymph node T suppressor cells that suppress delayed-type hypersensitivity and CMI to cryptococcal antigens which also may be responsible for modulation of NK cell reactivity in these animals. To investigate the ability of C. neoformans to modulate NK cell activity we determined the effects of host microbial status, host immunocompetence (athymic, euthymic or NK deficient beige mice), route (oral association, IV inoculation), and form (live, dead or capsular polysaccharide) of cryptococcal challenge on NK activity in GF, FD and CV mice.

Inbred athymic (nu/nu) and euthymic heterozygous (+/nu) GF and FD Balb/c mice as well as GF and CV beige (bg/bg) C57Bl and heterozygous (+/bg) littermate mice were used in this study. Gnotobiotic mice were housed in flexible film isolators following established gnotobiotic procedures and ranged between 6-10 weeks of age.

A colony of C. albicans monoassociated (MA) Balb/c mice was obtained by orally inoculating GF mice with C. albicans strain B311. GF Balb/c and C57Bl mice were also MA with C. neoformans (serotype A). Candida MA mice were intravenously (IV) inoculated in the tail vein with 2×10^4 live Candida cells. GF Balb/c mice were IV inoculated with either 1×10^5 live Cryptococcus or 250 µg of cryptococcal capsular polysaccharide. Capsular polysaccharide was prepared by an ethanol precipitation method (Goren, Middlebrook 1967), and the concentration of the polysaccharide was determined by the anthrone reaction (Gadebusch 1961). FD Balb/c mice were IV inoculated with either 1×10^5 live or dead cryptococcal cells. To determine the number of viable cells, each yeast inoculum was diluted, aliquots plated on Sabouraud's dextrose agar, and colonies were counted. Mice were sacrificed by cervical dislocation at various time points after challenge with yeast organisms. NK cell activity was determined by a direct 4-hr ^{51}Cr-release assay as described previously (Bartizal, Salkowski and Balish 1983; 1984).

Candida MA nu/nu Balb/c mice, when IV challenged with *C. albicans* had enhanced splenic NK cell activity compared to uninoculated *Candida* MA mice while their heterozygous littermates had unaltered NK cell activity following IV challenge. However, both GF and FD nude and heterozygous mice exhibited significantly elevated NK cell activity after oral or IV challenge with *C. neoformans*. Augmented NK cell activity occurred by 24 hrs after challenge and persisted for 14 days. Live or heat killed *Cryptococcus* or its capsular polysaccharide enhanced NK cell activity in Balb/c mice. *Cryptococcus* augmented NK cell activity to a greater extent in GF nude mice than in their heterozygous littermates, while the opposite effect occurred in FD nude and heterozygous mice. Augmentation of splenic NK cell activity, above GF baseline NK cell levels, was highest in mice challenged with viable *C. neoformans*. Increased NK cell activity was also observed in orally associated mice and, to a lesser extent, in capsular polysaccharide injected mice (Table 1). NK cell activity of spleen cells

Table 1. Augmentation of Splenic NK Cell Activity After *C. neoformans* Challenge in Balb/c Mice.

	Increased cytotoxicity over unchallenged controls							
	Day 0		Day 3		Day 7		Day 14	
Germfree	Nude	Het	Nude	Het	Nude	Het	Nude	Het
IV Capsular Polysaccharide	ND*	ND	6.9	4.2	11.0	2.8	7.9	0
Oral Association	ND	ND	12.5	8.6	16.3	5.9	34.6	5.4
IV (live) Injection	ND	ND	26.7	16.3	39.6	23.2	37.1	6.6
Flora Defined								
IV (live) Injection	2.7	1.1	1.5	0	15.8	28.8	dead	41.6
IV (heat killed) Injection	8.6	3.0	9.0	13.1	4.2	6.3	ND	ND

*ND = not done

from GF or CV beige (bg/bg) C57Bl mice was not enhanced by IV challenge with *C. neoformans*. However, heterozygous (+/bg) mice had elevated NK cell activity after IV cryptococcal challenge. It was surprising to find that *C. neoformans*-enhanced

NK cell activity did not correlate with the ability of systemically challenged mice to clear or limit cryptococcal infection since all (nu/nu, +/nu) and C57Bl (bg/bg, +/bg) mice eventually succumbed to overwhelming disseminated cryptococcosis.

These studies demonstrate that the composition (Cryptococcus, Candida), route (oral association, IV inoculation), and form (live, dead, capsular polysaccharide) are important in modulation of splenic NK cell activity in mice. Our results are not consistent with the postulated natural defense function of NK cells against fungal pathogens in vivo, since C. neoformans-augmented NK cell activity did not prevent dissemination or death after systemic fungal challenge. This suggests that other immune mechanisms may be more important in controlling systemic cryptococcal infections. Thus, enhancement of splenic NK cell activity may occur after mice are exposed (oral or IV) to C. neoformans but the level of enhancement appears also dependent on the immune status (athymic, euthymic or beige) of mice after systemic challenge.

Bartizal KF, Salkowski C, Balish E (1983). The influence of a gastrointestinal microflora on natural killer cell activity. J Reticuloendothel Soc 33:381.

Bartizal KF, Salkowski C, Balish E (1984). The effect of microbial flora, diet, and age on the tumoricidal activity of natural killer cells. J Leukocyte Biol (In press).

Cauley LK, Murphy JW (1979). Response of congenitally athymic (nude) and phenotypically normal mice to Cryptococcus neoformans infection. Infect Immun 23:644.

Djeu JY, Huang KY, Herberman RB (1980). Augmentation of mouse natural killer activity and induction of interferon by tumor cells in vivo. J Exp Med 151:781.

Gadebusch HH (1961). Carbohydrate estimation. In Kabat EA, Mayer MM (eds): "Experimental Immunochemistry", Springfield, IL p. 526.

Goren MB, Middlebrook GM (1967). Protein conjugates of polysaccharide from C. neoformans. J Immunol 98:901.

Kamiyama T, Hagiwara T (1981). Augmented followed by suppressed levels of natural cell-mediated cytotoxicity in mice infected with Toxoplasma gondii. Infect Immun 36:628.

Murphy JW, McDaniel DO (1981). In vitro reactivity of natural killer (NK) cells against Cryptococcus neoformans. J Immunol 128:1577.

Wolfe SA, Tracy DE, Henney CS (1976). Induction of "natural killer" cells by BCG. Nature 262:584.

Germfree Research: Microflora Control and Its Application to the Biomedical Sciences, pages 337–341

EFFECT OF BIOLOGICAL RESPONSE MODIFIER-STREPTOCOCCAL PREPARATION OK-432 ON NK/K CELL SYSTEM OF GNOTOBIOTIC MINIATURE SWINE[1]

Yoon Berm Kim, M.D., Ph.D.
Department of Microbiology and Immunology
University of Health Sciences/The Chicago Medical School
North Chicago, IL 60064

There has been increasing evidence that the natural killer (NK) and killer (K) cell system can play an important role in the first line of host defense mechanisms against neoplastic growth and tumor metastasis as well as resistance against microbial and parasitic infections. We have been investigating the ontogenic development, cellular nature and lineages, and regulatory mechanisms for NK activity in our gnotobiotic miniature swine by taking advantage of the fact that immunologically "virgin" piglets would not be contaminated with "natural" antibodies *in vivo* and they could be maintained in a controlled environment (Kim 1975). We have found that K cells for antibody-dependent cellular cytotoxicity (ADCC) develop early in ontogeny and before the development of NK cells, that NK and K cells have different tissue distributions, that NK cells are not K cells armed with "natural" antibodies, and that heterologous "anti-NK" sera plus C' eliminate NK cells without affecting ADCC, suggesting that porcine NK cells and K cells are distinct subpopulations (Huh et al 1981b; Kim et al 1980, 1981, 1982). This is further supported by the differential effects of interferons on NK and ADCC and the differential effects of recently developed monoclonal antibodies against NK cells on NK and K cells activities (Chung et al 1982; Kim et al 1984).

Also, we found that NK cells develop two weeks and three to four weeks after birth in specific pathogen-free

[1]This work was supported in part by US PHS Grant HD-19420 and CA-38336 awarded by the National Institutes of Health.

(SPF) and germfree (GF) piglets, respectively, and when GF piglets were associated with anaerobic flora (*Lactobacillus sp.* and *Streptococcus sp.* from SPF swine), they developed significant NK activity within two weeks of age, indicating that microbial flora is responsible for acceleration of maturation or differentiation of NK cells during ontogeny (Huh *et al* 1981a; Kim *et al* 1982). The fact that OK-432, a streptococcal preparation (Picibanil, Chugai Pharmaceutical Co., Ltd., Tokyo, Japan), has been reported to be a potent augmentor of human NK cell activity by both activating NK cells and reducing suppressor cells (Uchida and Hoshino, 1980; Uchida and Micksche, 1981, 1983), prompted our attention to examine the effects of OK-432 on our porcine NK and K cell system *in vitro* as well as *in vivo* to investigate the regulation of NK and K cell activities.

Animals, preparation of lymphoid cells, culture systems, and assays for NK and ADCC and calculations of cytotoxic activity were described elsewhere (Kim et al 1980).

First we have titrated the optimal dose and time of incubation to study the effects of OK-432 on NK and K cell activities. It was found that preincubation of effectors for 18 hrs at 37°C with 5-50 μg/ml of OK-432 gave the optimal augmentation of NK activity from the peripheral blood lymphoid cells (PBL) from SPF piglets who have preexisting NK cells. However, there was little effect on ADCC at any dose levels. Monocytes were not required for augmentation of NK activity by OK-432 as demonstrated by equal enhancement of NK activity of unseparated peripheral blood mononuclear cells and of monocytes-depleted PBL which were incubated with OK-432 50 μg/ml for 18 hrs at 37°C (data not shown).

Table 1 shows the results of the effects of OK-432 on NK and K cell activities by using pretitrated optimal dose and incubation time (50 μg/ml, 18 hrs at 37°C). The OK-432 augmented NK activity in PBL of all 10 SPF miniature swine at least twofold; in contrast, only 2 of 10 tested (SPF 4117-6, SPF 4135-5) had some augmentation of ADCC and 8 of 10 animals had no significant change or slightly depressed ADCC indicating that OK-432 specifically augments NK cell activity but not K cell activity.

In addition, the OK-432 induces NK activity from

Table 1

EFFECT OF OK-432 ON PORCINE NK AND K CELL ACTIVITIES _IN VITRO_

Animal No.	% Specific Lysis[a]			
	NK (K562)		ADCC (TNP-SB + anti-TNP)	
	Control	OK-432	Control	OK-432
SPF 4108-7	7.8	14.5	73.2	64.3
SPF 4117-3	12.6	23.0	57.2	52.2
SPF 4117-6	14.8	32.9	60.8	72.5
SPF 4119-5	31.1	63.8	58.9	66.7
SPF 4127-2	24.5	42.1	41.4	45.3
SPF 4135-1	12.6	30.8	53.7	60.6
SPF 4135-5	15.5	39.2	46.2	62.7
SPF 4135-6	21.1	46.7	72.5	63.4
SPF 4166-3	17.2	45.9	81.5	78.1
SPF 4172-2	23.1	64.2	78.0	71.2

[a]PBL from SPF miniature swine were incubated with medium (control) or OK-432 50 μg/ml for 18 hr at 37°C and washed three times with BSS before assay. Effector-to-target ratios were 100:1.

"background"-free newborn SPF and germfree piglets when PBL were incubated with OK-432 50 μg/ml for 3 days at 37°C, and also a significant acceleration of maturation or induction of NK activity within 3-13 days post injection of OK-432 0.5-1 mg intraperitoneally. It is clear that OK-432 does more than simply induce interferon (IFN) because the purified recombinant IFN augmented only preexisting NK cell activity but failed to induce NK activity of PBL in less than 3-week old GF piglets or less than 2-week old SPF piglets whose NK activity had not yet developed (Chung et al 1982; Kim and Chung 1984). In contrast, OK-432 induced NK activity in vitro as well as in vivo from "background"-free newborn piglets (Table 2). Perhaps, the streptococcal preparation, OK-432 (Picibanil, Chugai Pharmaceutical Co., Ltd., Tokyo, Japan) prepared by incubating the culture of low-virulent Su Strain of type 3, group A _Streptococcus pyogenes_ of human origin, treated with penicillin G potassium and heating at 45°C for 30 min, followed by lyophilization may act very similarly to living microbial association of GF piglets (Huh et al 1981a; Kim et al 1982). Recently, we are

Table 2. SUMMARY OF EFFECT OF RECOMBINANT HUMAN INTERFERONS (rIFN) AND OK-432 ON THE PORCINE NK ACTIVITY OF VARIOUS STATES OF MINIATURE SWINE

States of Animals	Preexisting NK Activity	Augmentation of NK Activity by	
		rIFN	OK-432
Germfree piglets			
< 3 weeks old	-	-	+
> 3 weeks old	+	+	+
SPF piglets			
< 2 weeks old	-	-	+
> 2 weeks old	+	+	+

investigating the mechanisms of activation and/or induction of differentiation of NK cells by OK-432. Preliminary results show that OK-432 induces NK activating factor (NKAF) as reported by Ichimura et al (1984), and NKAF may in turn stimulate NK cells and/or activate or induce pre-NK cells to active NK cells. Thus, NKAF may be one of the mechanisms for the observed induction of NK activity from "background"-free newborn gnotobiotic piglets.

In conclusion, the streptococcal preparation OK-432 augmented porcine NK activity in vitro and as well as in vivo, but had little effect on ADCC. The OK-432 is a strong inducer of neonatal NK cells. Thus, the OK-432 is a good biological response modifier particularly in enhancement of porcine NK cell activity.

Acknowledgement: The author wishes to acknowledge Mr. Gene A. Monson for his excellent technical assistance for this study.

Chung TJ, Huh ND, Kim YB (1982). Differential effects of interferons on porcine NK and K cell activities. In Herberman RD (ed): "NK Cells and Other Natural Effector Cells". New York:Academic Press, p 381.

Huh ND, Kim YB, Koren HS, Amos DB (1981a). Natural killing and antibody-dependent cellular cytotoxicity in specific pathogen-free miniature swine and germfree piglets. II. Ontogenic development of NK and ADCC. Int J of Cancer 28:175.

Huh ND, Kim YB, Amos DB (1981b). Natural killing (NK) and antibody-dependent cellular cytotoxicity (ADCC) in specific pathogen-free (SPF) miniature swine and germfree piglets. III. Two distinct effector cells for NK and ADCC. J Immunol 127:2190.

Ichimura O, Suzuki S, Sugawara Y, Osawa T (1984). Characterization of mouse natural killer cell activating factor (NKAF) induced by streptococcal preparation OK-432. In Hoshino T, Koren HS, Uchida A (eds): "Natural Killer Activity and Its Regulation", Amsterdam-Princeton-Geneva-Tokyo: Excerpta Medica, p 208.

Kim YB (1975). Developmental immunity of the piglet. In Bergsma D, Good RA, Finnstad J (eds): "Immunodeficiency Diseases in Man and Animals". Sunderland, MA: Sinauer Associates, Inc. Birth Defects: Original Articles Series, Vol XI, No 1, p 549.

Kim YB, Huh ND (1981). Natural killer and killer (NK/K) cell systems in gnotobiotic miniature swine. In Sasaki S, Ozawa A, Hashimoto K (eds): "Recent Advances in Germfree Research". Tokyo, Japan: Tokai University Press, p 585.

Kim YB, Chung TJ (1984). Effect of recombinant human interferons on NK and K cell activities of gnotobiotic miniature swine. In the Proceedings of the Eighth International Symposium on Germfree Research (Abstract p. 80). New York: Alan R. Liss, Inc., p 000.

Kim YB, Huh ND, Koren HS, Amos DB (1980). Natural killing (NK) and antibody-dependent cellular cytotoxicity (ADCC) in specific pathogen-free (SPF) miniature swine and germfree piglets. I. Comparison of NK and ADCC. J Immunol 125:755.

Kim YB, Huh ND, Koren HS and Amos DB (1982). Ontogenic development of porcine NK and K cells. In Herberman RD (ed): "NK Cells and Other Natural Effector Cells". New York: Academic Press, p 341.

Kim YB, Czajkowski M, Monson GA (1984). Distinct subpopulations of NK and K cells in porcine natural cell-mediated immunity. In Hoshino T, Koren HS, Uchida A (eds): "Natural Killer Activity and Its Regulation". Amsterdam-Princeton-Geneva-Tokyo: Excerpta Medica, p 250.

Uchida A, Hoshino T (1980). Reduction of suppressor cells in cancer patients treated with OK-432 immunotherapy. Int J Cancer 26:401.

Uchida A, Micksche M (1981). In vitro augmentation of natural killing activity by OK-432. Int J Immunopharmacology 3:365.

Uchida A, Micksche M (1983). Intrapleural administration of OK-432 in cancer patients: activation of NK cells and reduction of suppressor cells. Int J Cancer 31:1.

Germfree Research: Microflora Control and Its Application to the Biomedical Sciences, pages 343–345

NATURAL KILLER CELL ACTIVITY IN GERMFREE AND CONVENTIONAL RATS

S. Tazume, A. Wade, and M. Pollard

Lobund Laboratory, University of Notre Dame

Notre Dame, IN 46556 U.S.A.

In recent years it has become increasingly evident that natural killer (NK) activity plays a central role in antiviral (Welsh 1981) and anti-tumor (Hanna 1981) immunity. NK activity has been demonstrated in many species, including mice (Kiessling 1978), guinea pigs (Herberman 1978), rats (Herberman 1978), pigs (Kim 1981), and dogs (Krakowka 1983). Germfree (GF) animals have been reported to be deficient in NK activity (Kim 1981). This study analyzes NK activity in GF and conventional (CV) Sprague-Dawley rats, with respect to variation with age and influence of intestinal flora.

Animals: Lobund germfree and conventional Sprague-Dawley rats of both sexes from 5 to 20 weeks of age were used in this study.

NK Cell Assay: Partially-purified NK cell suspensions were prepared by pressing spleens obtained from ether-anesthetized rats through stainless-steel mesh, and then passing the suspension through nylon wool columns. Nonadherent cells were collected and washed in RPMI-1640 before being used as effector cells in NK cell assays. Mouse YAC-1 lymphoma cells maintained in suspension culture in vitro were used as targets in NK cell assays. Target cells were ^{51}Cr-labeled and suspended at 10^5 cells/ml in RPMI-1640 supplemented with 10% fetal calf serum. Effector cells were suspended in the same medium. Effector:target cell ratios of 200:1, 100:1, 50:1, 25:1 were prepared at a volume of 200 μl per well in a 96-well microtiter plate and incubated for 4 hrs. at 37°C and 5% CO_2. After the incubation period, the plate was centrifuged and a 100 μl amount of supernatant was removed. Counts per minute were then determined on a gamma counter. The percent of NK cytotoxicity was calculated as: (experimental release-spontaneous release)/(maximum release-spontaneous release)x100.

NK activity of spleen cells from GF and CV rats was compared. As shown in Table 1, data from this study indicate that spleen cells from GF rats had significantly more cytotoxicity for YAC-1 cells than those from CV rats. It is also evident from these data (Table 1) that NK activity of spleen cells from GF and CV rats was highest between 5 and 10 weeks of age. In previous studies, few studies have shown differences in NK activity of spleen cells from GF and CV animals. Kim and Huh (1981) reported no significant difference in NK or antibody-dependent cellular cytotoxicity activity between GF and specific pathogen-free animals, but they suggested that the intestinal flora may enhance maturation of NK activity.

Table I. Natural Killer Activity in Sprague - Dawley Germfree and Conventional Rat Spleen

Microbial Status	Age (Weeks)	No. of Rats	% Cytotoxicity (Effector : Target Ratio)			
			200:1	100:1	50:1	25:1
Germfree	5~10	4	50.1±2.2 ※	30.8±2.2 ※	17.3±2.5 ※※	10.3±1.0 ※
Conventional	5~10	6	36.3±5.3	21.4±5.6	12.5±2.5	7.4±1.1
Germfree	11~15	10	37.4±11.4 ※	25.3±8.6 ※	15.7=5.1 ※	11.5±4.0 ※
Conventional	11~15	8	20.2±3.7	11.7±2.0	6.5±1.7	3.7± 1.9
Germfree	16~20	6	36.7±4.6 ※	23.8±5.8 ※※	14.7±1.9 ※	10.2±1.2 ※
Conventional	16~20	6	25.3±6.0	14.6±5.4	8.3±2.1	5.5±2.4

Data are mean ± S.D.
※ $p < 0.01$
※※ $p < 0.05$

It was clearly shown here that differences in the NK cytotoxicity of spleen cells from GF and CV rats exist. These findings suggest that the intestinal flora does exert some influence. In particular, it appears that the intestinal flora suppresses NK activity. However, as shown in Table 2, spleen cells from GF rats which had been conventionalized for 2-4 weeks exhibited the same NK activity as GF rat spleen cells. These data suggest that a longer time of exposure than four weeks is necessary for the stimulus of intestinal flora to exert its effect.

Table 2. NK Cytotoxicity of Sprague-Dawley Germfree, Conventional, and Conventionalized Germfree Rats.

Microbial Status	Age Weeks	No. of Rats	% Cytotoxicity (Effector: Target Ratio) 200:1	100:1	50:1	25:1
Germfree	14	2	24.7	17.0	11.2	8.4
Conventional	14	2	17.7	11.6	6.3	3.0
Conventionalized[a]	14	2	27.3	16.8	11.2	5.1
Germfree	16	2	29.2	19.8	13.3	9.0
Conventional	16	2	24.0	16.4	9.2	6.8
Conventionalized[b]	16	2	32.2	19.3	13.6	8.0

Data are mean.

a.b. Rats were conventionalized by exposing of 2 weeks and 4 weeks in a conventional animal room.

REFERENCES

Hanna N, Burton RC (1981). Definitive evidence that natural killer (NK) cells inhibit experimental tumor metastasis in vivo. J Immunol 127:1754.

Herberman RB, Holden HT (1978). Natural cell-mediated immunity. Adv Cancer Res 27:305.

Kiessling R, Klein E, Wigzell H (1975). "Natural" killer cells in the mouse 1. Cytotoxic cells with specificity for mouse Moloney leukemia cells. Specificity and distribution according to genotype. Eur J Immunol 5:112.

Kim YB, Huh ND (1981). Natural killer and killer (NK/K) cell system in gnotobiotic miniature swine. In Sasaki S (ed): "Recent Advances in Germfree Research", Tokyo: Tokai University Press, p. 585.

Krakowka S (1983). Natural killer cell activity in adult gnotobiotic dogs. Am J Vet Res 44:635.

Welsh RM (1981). Natural cell-mediated immunity during viral infections. In Haller O (ed): "Natural Resistance to Tumors and Viruses", Berlin: Springer-Verlag, p. 83.

Germfree Research: Microflora Control and Its Application to the Biomedical Sciences, pages 347–349

INFLUENCE OF ANTIGEN-FREE DIET, MICROBIAL FLORA AND HOST STRAIN ON NATURAL KILLER CYTOTOXIC ACTIVITY IN MICE

S. Tazume, K. Umehara, K. Hashimoto, and S. Sasaki

Department of Microbiology
Tokai University School of Medicine
Isehara, Japan

Natural cytotoxicity is an important host defense mechanism against neoplastic cells and viruses and may also be important in immunoregulation. Various factors influence the expression of NK activity in mice, including age and genetic background (Kiessling 1975). NK activity can be enhanced by interferon (Gidlund 1978) or conversely, NK activity can be suppressed by carrageenan (Cudkowicz 1980). However, few investigators have analyzed the effect of intestinal flora (Kim 1981) and diet (Pleasants 1981) on NK activity. The objective of this study is to determine the effect of antigenic stimuli derived from the intestinal flora and diet on NK activity. NK activity of germfree mice reared on an "antigen-free" diet (AgF-GF) was compared with that of germfree (GF), conventionalized-germfree (EX-GF), and conventional (CV) mice reared on a pellet diet at ages 5 to 15 weeks.

Animals: Mice of the outbred ICR strain and the inbred C3H/He strain were purchased from Clea Japan, Inc. (Tokyo, Japan) at 4 weeks of age. The "antigen-free" diet has been described by Hashimoto et al (1978).

NK Cell Assay: ^{51}Cr labeled YAC-1 (strain A Moloney virus-induced T-cell lymphoma) cells were used as targets. Spleens were obtained from mice sacrificed by cervical dislocation, and cell suspensions were prepared by pressing the spleens through stainless-steel mesh. The spleen cell suspensions were treated with 1 ml of red cell lysing buffer on ice for 5 min. to lyse red blood cells. Effector-target cell ratios of 100:1 were set up in a volume of 200 µl in 96-well microtiter plates and incubated for 4 hrs. at 37°C and 5% CO_2. The plates were then centrifuged and a 100 μl amount of supernatant was removed. Counts per minute were measured on a gamma counter. The percent of cytotoxicity was calculated as: (experimental release-spontaneous release)/(maximum release-spontaneous release)x100.

NK activity was analyzed in spleen cells from AgF-GF, GF, EX-GF, CV, ICR and C3H/He mice from 5 to 15 weeks old. Results are summarized in Figure 1. In the two strains tested, in all groups (AgF-GF, GF, EX-GF, and CV) NK activity of C3H/H3 mice tended to be higher than that of ICR mice. Other studies (Lanza 1982) have previously demonstrated that C3H/HeN mice have high splenic NK activity. In the ICR mice, NK activity of CV mice was much higher than that of EX-GF, GF, and AgF-GF mice. Data from the C3H/He mice were inconclusive. Furthermore, in both the ICR and C3H/He mice, NK activity of CV mice was already at high levels by 5 weeks; however, that of other groups including the EX-GF did not rise during the test period. Freimark and Bealmear (1981) reported that specific pathogen-free (SPF) and CV mice exhibited NK activity at an earlier age than GF mice, and, with regard to microflora, NK activity was not shown to be significantly different among GF, SPF, and CV mice. In a separate study using SPF and GF piglets, Kim and Huh (1981) suggested that the microbial flora enhances maturation of NK activity.

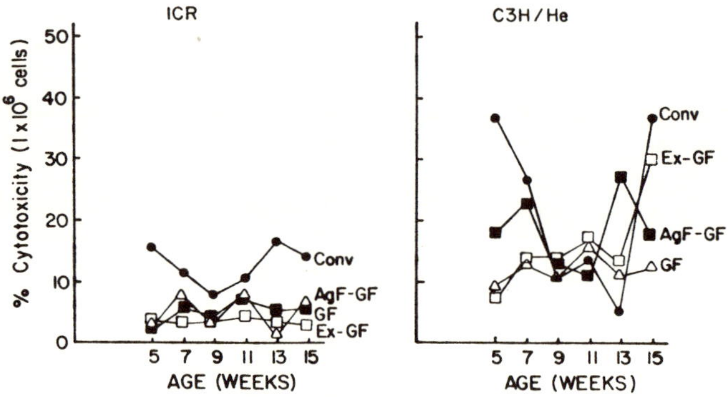

Figure 1 NK cytotoxicity of ICR and C3H/He mice

We observed differences in the cytotoxicity of spleen cells from GF and CV mice. These findings suggest that the antigenic stimulus derived from the diet does not affect NK activity, but that from the intestinal flora does exert some effect. In this study, it appears that the intestinal flora enhances NK activity. However, the data from EX-GF mice indicate that a longer time of exposure than 5 to 15 weeks is necessary for the stimulus of intestinal flora to exert its effect.

REFERENCES

Cudkowicz G, Hochman PS (1980). Carrageenan-induced decline of natural killer activity. I. In vitro activation of adherent non-T-suppressor cells. Cell Immunol 53:395.

Freimark B, Bealmear PM (1981). Influence of strain, microflora, and the thymus on natural killer (NK) cell levels in mice. In Sasaki S (ed): "Recent Advances in Germfree Research", Tokyo: Tokai University Press, p. 607.

Gidlund M, Orn A, Wigzell H, Senik A, Gresser I (1978). Enhanced NK cell activity in mice infected with interferon and interferon inducers. Nature 273:759.

Hashimoto K, Handa H, Umehara K, Sasaki S (1978). Germfree mice reared on an "antigen-free" diet. Lab Animal Sci 28:38.

Kiessling R, Klein E, Wigzell H (1975). "Natural" killer cells in the mouse 1. Cytotoxic cells with specificity for mouse Moloney leukemia cells. Specificity and distribution according to genotype. Eur J Immunol 5:112.

Kim YB, Huh ND (1981). Natural killer and killer (NK/K) cell system in gnotobiotic miniature swine. In Sasaki S (ed): "Recent Advances in Germfree Research", Tokyo: Tokai University Press, p. 585.

Lanza E, Djeu JY (1982). Age-independent natural killer cell activity in murine peripheral blood. In Herberman RB (ed): "NK Cells and Other Natural Effector Cells", New York: Academic Press, p. 335.

Pleasants JR, Schmitz HE, Wostmann BS (1981). Immune responses of germfree C3H mice fed chemically defined ultrafiltered "Antigen-Free" diet. In Sasaki S (ed): "Recent Advances in Germfree Research", Tokyo: Tokai University Press, p. 535.

Germfree Research: Microflora Control and Its Application to the Biomedical Sciences, pages 351–353

THE EFFECT OF THE ANTIGENIC STIMULI DERIVED FROM THE DIET AND MICROBIAL FLORA ON IMMUNOGLOBULIN LEVELS

Umehara, K., S. Tazume, K. Hashimoto and S. Sasaki
Department of Microbiology, Tokai University School of Medicine, Isehara, Japan

Animals under natural circumstances are constantly exposed to various antigenic stimuli, therefore one may assume that such stimuli may have certain modifying effects on the immune response of the animals. For the purpose of studying the effect of antigenic stimuli derived from the microbial flora and diet on the production of immunoglobulin, we examined the level of IgG, IgM, IgA (Hashimoto et al. 1971) and IgG subclasses, in outbred ICR and inbred C3H/He mice at 5 or 11 weeks. Four experimental groups were included. These are 1. antigen-free-germfree (AgF-GF) mice: second generation of germfree mice switched to an AgF diet (Hashimoto et al. 1978, Pleasants, et al. 1970), 2. germfree (GF) mice reared on a pellet diet, 3. exposure-germfree (Ex-GF) mice: second generation of GF mice exposed to conventional environment reared on a pellet diet and 4. conventional (CV) mice reared on a pellet diet.

Table 1 and 2 indicate the levels of classes of immunoglobulin in the serum, the intestinal wall and its contents of ICR and C3H/He mice respectively. The level of IgG in the serum of AgF-GF mice was the lowest among the four experimental groups. That is to say the effect on immunoglobulin level by antigenic stimuli derived from the microbial flora and diet was especially shown with the level of IgG in the serum. With Ex-GF ICR mice, the level of IgG at 5 weeks was the same as that of GF mice but at 11 weeks had risen to the level of that of CV mice. On the contrary, the level of IgG in the serum of Ex-GF C3H/He mice at 11 weeks did not rise. On examining the concentrations of subclasses of IgG in ICR and C3H/He mice (Fig. 1), the levels of IgG1

Table 1 Titer of classes of immunoglobulin in the serum, the intestinal wall and its contents of ICR mice

	5 weeks			11 weeks		
	IgG	IgM	IgA	IgG	IgM	IgA
Serum						
CV	5.2±0.5[a]	1.2±0.8	1.8±0.8	5.4±0.6	1.6±0.6	2.8±0.8
Ex-GF	2.4±0.6	1.0±0.7	0±0	5.6±0.9	2.0±0.7	1.8±0.5
GF	2.2±0.5	1.4±0.6	-1.0±0	4.0±1.0	2.8±0.5	-1.0±0
AgF-GF	0.5±0.6	0.3±1.0	-1.0±0	2.8±0.5	0.5±0.6	-1.0±0
Intestinal wall						
CV	-0.2±1.1	-1.0±0	2.2±0.5	0.2±1.1	-1.0±0	4.0±0
Ex-GF	-1.0±0	-1.0±0	2.8±0.5	1.0±0.7	-1.0±0	3.4±0.6
GF	-1.0±0	-1.0±0	-1.0±0	-1.0±0	-1.0±0	-0.4±1.3
AgF-GF	-1.0±0	-1.0±0	-1.0±0	-1.0±0	-1.0±0	-1.0±0
Intestinal contents						
CV	-1.0±0	-1.0±0	4.8±0.5	1.2±2.3	1.6±2.5	4.4±0.6
Ex-GF	-0.6±0.9	-1.0±0	4.6±0.9	-1.0±0	-1.0±0	3.6±0.6
GF	-1.0±0	-1.0±0	-0.4±0.9	-1.0±0	-0.4±1.3	-0.4±1.3
AgF-GF	-1.0±0	-1.0±0	-1.0±0	-0.3±1.5	0.5±1.7	-1.0±0

a : Mean of $\log_2$titer ± SD, titer is reciprocal of dilution titer lower than $1:2^0$ regarded as $1:2^{-1}$

Table 2 Titer of classes of immunoglobulin in the serum, the intestinal wall and its contents of C3H/He mice

	5 weeks			11 weeks		
	IgG	IgM	IgA	IgG	IgM	IaA
Serum						
CV	4.4±0.6	1.8±0.5	0.4±0.6	6.2±0.5	3.0±0	2.2±0.5
Ex-GF	2.0±0.7	2.0±0	-1.0±0	3.6±0.9	2.8±0.5	0.8±0.5
GF	0.8±0.5	1.8±0.5	-1.0±0	1.0±0	2.8±2.1	-1.0±0
AgF-GF	0±0.7	1.6±0.6	-1.0±0	2.3±0.7	1.5±0.5	-0.8±0.5
Intestinal wall						
CV	0.6±0.9	-1.0±0	3.2±1.1	1.4±0.6	-1.0±0	3.8±0.5
Ex-GF	-1.0±0	-1.0±0	2.8±0.5	-1.0±0	-1.0±0	4.8±0.5
GF	-1.0±0	-1.0±0	-1.0±0	-1.0±0	-1.0±0	-0.8±0.5
AgF-GF	-1.0±0	-1.0±0	-1.0±0	-1.0±0	-1.0±0	0.8±1.9
Intestinal contents						
CV	-1.0±0	-0.4±1.3	4.4±0.9	-0.6±0.9	0.6±2.3	4.2±0.8
Ex-GF	-1.0±0	-1.0±0	3.4±0.9	-1.0±0	-1.0±0	4.8±0.8
GF	-1.0±0	-1.0±0	-1.0±0	-1.0±0	-1.0±0	-1.0±0
AgF-GF	-1.0±0	-1.0±0	-1.0±0	-0.1±1.6	1.0±1.9	0±1.9

and IgG2a of ICR Ex-GF and GF mice at 11 weeks were high, while those of C3H/He mice were lower. The IgG2b and IgG3 concentrations of all groups were low and did not differ significantly. Therefore it seems that the difference in IgG level resulting from antigenic stimuli was mainly due to difference of IgG1 and IgG2a level.

Our data agree with previous report (Nielsen et al. 1980) that the level of IgM in the serum of GF mice was a little higher or the same as that of CV mice, however it is

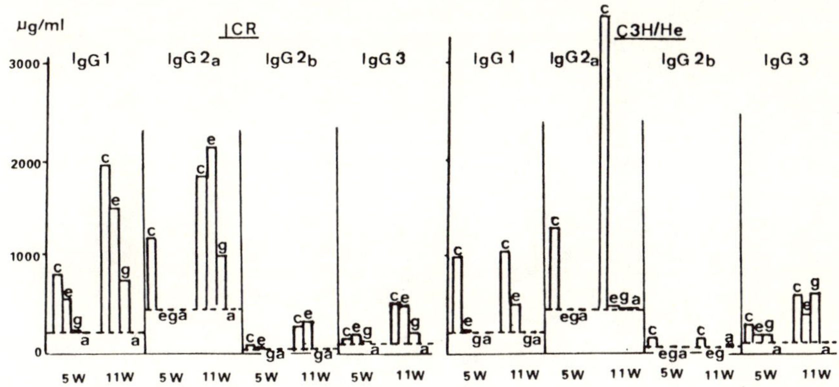

Fig. 1. The concentration of subclasses of IgG in ICR and C3H/He mice (serum).
Dotted line represents the lowest detectable concentration
c : CV mice, e : Ex-GF mice, g : GF mice, a : AgF-GF mice

interesting that that of AgF-GF mice remained at lower level.

While the levels of IgA in the intestinal wall and contents of GF mice were barely detectable, those of Ex-GF mice were already the same as that of CV mice at 5 weeks. This indicates that the antigenic stimuli derived from resident microbial flora as a result of exposure to conventional environment affect the production of secretory IgA.

We are indebted to Dr. T. Shimamura for supplying the antisera.

Hashimoto K, Yoshikawa M, Sugihara Y, Sasaki K (1971). Intestinal resistance in the experimental enteric infection of mice with a mouse adenovirus. II. Determination of the neutralizing substance in the intestinal tract as an IgA antibody. Jpn J Microbiol 15:499.

Hashimoto K, Handa H, Umehara K, Sasaki K (1978). Germfree mice reared on an "antigen-free" diet Lab Ani Sci 28:38.

Nielsen E, Friis C W (1980). Influence of an intestinal microflora on the developmemt of the immunoglobulins. Acta Path Microbiol Scand Sect C 88:121.

Pleasants J R, Reddy B S, Wostmann B S (1970). Qualitative adequacy of a chemically defined diet for reproducing germfree mice. J Nutr 100:498.

Germfree Research: Microflora Control and Its Application to the Biomedical Sciences, pages 355–358

ISOTYPE DISTRIBUTION AND SPECIFICITY REPERTOIRE OF 'SPONTANEOUSLY' OCCURRING ('BACKGROUND') IMMUNOGLOBULIN-SECRETING CELLS IN GERMFREE MICE FED CHEMICALLY DEFINED ULTRAFILTERED 'ANTIGEN-FREE' DIET

H. Hooijkaas, N. Bos, R. Benner, J.R. Pleasants[1], B.S. Wostmann[1]

Dept. Cell Biology and Genetics, Erasmus University, Rotterdam, The Netherlands, and [1]Lobund Laboratory, University of Notre Dame, Notre Dame, IN, USA

The possible essential role that exogenous antigens and/or mitogens might play in the development of the B cell antibody specificity repertoire as expressed by 'spontaneously' occurring ('background') immunoglobulin (Ig)-secreting cells in nonimmunized mice (Benner et al., 1982), has neither been properly established nor excluded. With the successful breeding of germfree (GF) C3H/HeCr mice fed an ultrafiltered solution of chemically defined (CD) low molecular weight nutrients, these exogenous stimuli are reduced to the presently achievable minimum (Wostmann et al., 1982). Previous studies showed that these mice normally produced IgM, while the IgG and IgA production was drastically decreased or even undetectable (Benner et al., 1981a; Hashimoto et al., 1978; Wostmann et al., 1971).

Since data on the immunological activity of mice deprived from exogenous antigenic and/or mitogenic stimuli can be of value for the contemporary thinking about internal regulation in the immune system we decided to study the background IgM-, IgG- and IgA-secretion in spleen and bone marrow (BM) of GF-CD and conventional (CV) C3H/HeCr mice fed natural ingredient (NI) (Kellogg, Wostmann, 1969). For this purpose the protein A plaque assay was used (Van Oudenaren et al., 1981). Furthermore, a part of the antibody specificity repertoire of the IgM-secreting cells in the spleen was analyzed using plaque assays specific for sheep red blood cells (SRBC) haptenated with either 4 mg nitroiodophenyl (NIP), 2 mg 4-hydroxy-3,5-dinitrophenyl (NNP) or 30 mg 2,4,6-trinitrophenyl (TNP) per ml washed packed SRBC (Hooijkaas et al., 1983).

Spleen and BM cell suspensions were prepared as described (Benner et al., 1981b).

The numbers of background Ig-secreting cells in the spleen and BM of CV-NI and GF-CD C3H/HeCr mice are given in Table 1. In the spleen of GF-CD mice the numbers of IgM- and IgA-secreting cells were significantly reduced while the number of IgG secreting cells was not significantly different from CV-NI mice. In the BM of GF-CD mice the numbers of IgM-, IgG- and IgA-secreting cells were 3.5, 5 and 27 times lower than in CV-NI controls.

Table 1. Numbers of background Ig-secreting cells in the spleen and BM of conventional C3H/HeCr mice fed natural ingredient (CV-NI) and germfree C3H/HeCr mice fed chemically defined diet (GF-CD)[a]

Organ	Ig-isotype	Ig-secreting cells/organ[b]/10^{-3}	
		CV-NI	GF-CD
Spleen	IgM	899(865-934)[c]	382(337-456)[d]
	IgG	86(72-103)	115(107-124)[e]
	IgA	54(51-58)	8.8(8-10)[d]
BM	IgM	162(153-173)	47(39-57)[d]
	IgG	132(127-137)	26(24-28)[d]
	IgA	68(56-83)	2.5(2-3)[d]

a. All mice were 8 to 12-wk-old.
b. Background Ig-secreting cells were calculated using the number of Ig-secreting cells in the protein A plaque assay and the total cell yield per organ. Ig-secreting cells in the BM were determined in the femoral BM. We have adapted a conversion factor of 7.9 to calculate the number of Ig-secreting cells in the total BM from the number of Ig-secreting cells in two femurs (Benner et al., 1981b).
c. Figures represent the geometric mean $\pm$ 1 SE of 10 individually tested mice.
d. $p < 0.001$. Statistical evaluation was performed with the Student *t* test.
e. Not significantly different.

Table 2. Relative frequencies of background IgM-secreting cells specific for NIP0.4-SRBC, NNP0.2-SRBC and TNP30-SRBC in the spleen of conventional C3H/HeCr mice fed natural ingredient (CV-NI) and germfree C3H/HeCr mice fed chemically defined diet (GF-CD)[a]

Antigen	CV-NI	GF-CD
NIP0.4-SRBC	1 in 109 (101-118)[b]	1 in 114 (93-104)[c]
NNP0.2-SRBC	1 in 369 (315-433)	1 in 320 (215-477)[c]
TNP30-SRBC	1 in 107 (94-120)	1 in 79 (65-96)[c]

a. All mice were 8 to 12-wk-old.
b. Figures represent the mean ratio of specific IgM-antibody secreting cells to the total number of IgM-secreting cells as determined in the protein A plaque assay, The geometric mean $\pm$ 1 SE has been calculated. In the spleen of 8 to 12-wk-old CV-NI and GF-CD mice (n = 10) 3113 (3040-3187) and 4482 (4121-4874) IgM-secreting cells per 10^6 nucleated cells were found, respectively.
c. Not significantly different.

The results given in Table 2 show that in GF-CD mice the frequencies of IgM-secreting spleen cells specific for NIP0.4-SRBC, NNP0.2-SRBC and TNP30-SRBC, respectively, were not significantly different from those found in CV-NI age-matched controls.

From our findings it can be concluded that as far as the pool of background IgM-secreting cells is concerned, and with the limitation of having assessed only a minor fraction of the specificity repertoire, exogenous stimuli do not affect the antibody specificity repertoire that these cells express. This suggests that available antibody repertoires are established independently of exogenous antigenic and/or mitogenic stimulation. Therefore, endogenous stimulation must account for the activation of background Ig-secreting cells. This places the driving force for the intrinsic activity of the immune system within the individual itself, which is in direct support of Jerne's network ideas, postulating that the immune system is autonomous and complete in recognizing antigens (Jerne, 1974; Coutinho, 1980; Coutinho et al., 1983).

Benner R, Van Oudenaren A, Björklund M, Ivars F, Holmberg D (1982). 'Background' immunoglobulin production: measurement, biological significance and regulation. Immunol. Today 3:243.
Benner R, Van Oudenaren A, Haaijman JJ, Slingerland-Teunissen J, Wostmann BS, Hijmans W (1981a). Regulation of the 'spontaneous' (background) immunoglobulin synthesis. Int Archs Allergy appl Immun 66:404.
Benner R, Van Oudenaren A, Koch G (1981b). Induction of antibody formation in mouse bone marrow. In Lefkovits I, Pernis B (eds): 'Immunological Methods', Vol. 2, New York: Academic Press, p 247.
Coutinho A (1980). The self-nonself discrimination and the nature and acquisition of the antibody repertoire. Ann Immunol (Inst Pasteur) 131D:235.
Coutinho A, Forni L, Holmberg D, Ivars F (1983). Is the network theory tautologic? In Möller E, Möller G (eds): 'Genetics of the Immune Response', New York: Plenum Press, p 273.
Hashimoto K, Handa H, Umehara K, Sasaki S (1978). Germfree mice reared on an 'antigen-free' diet. Lab Animal Sci 28:38.
Hooijkaas H, Preesman AA, Van Oudenaren A, Benner R, Haaijman JJ (1983). Frequency analysis of functional C and V gene expression in murine B cells at various ages. J Immunol 131:1629.
Jerne N (1974). Towards a network theory of the immune system. Ann Immunol (Inst Pasteur) 125C:373.
Kellogg TF, Wostmann BS (1969). Stock diet for colony production of germfree rats and mice. Lab Animal Care 19:812.
Van Oudenaren A, Hooijkaas H, Benner R (1981). Improvement of the protein A plaque assay for immunoglobulin secreting cells by using immunoglobulin-depleted guinea pig serum as a source of complement. J Immunol Methods 43:219.
Wostmann BS, Bruckner-Kardoss E, Pleasants JR (1982). Oxygen consumption and thyroid hormones in germfree mice fed glucose-aminoacid liquid diet. J Nutr 112:552.
Wostmann BS, Pleasants JR, Bealmear P (1971). Dietary stimulation of immune mechanisms. Fed Proc 30:1779.

Germfree Research: Microflora Control and Its Application to the Biomedical Sciences, pages 359–362

ISOTYPE EXPRESSION AND SPECIFICITY REPERTOIRE OF LIPOPOLYSACCHARIDE-REACTIVE B CELLS IN GERMFREE MICE FED CHEMICALLY DEFINED ULTRAFILTERED 'ANTIGEN-FREE' DIET

H. Hooijkaas, A.A. van der Linde-Preesman, W.M. Bitter, R. Benner, J.R. Pleasants[1] and B.S. Wostmann[1]
Dept. Cell Biology and Genetics, Erasmus University, Rotterdam, The Netherlands, and, [1]Lobund Laboratory, University of Notre Dame, Notre Dame, IN, USA

It is not known whether exogenous antigenic or mitogenic stimuli play a role in the generation of immunocompetent B cells and their specificity repertoire. Germfree (GF) mice, raised on natural diets, have been used to study this question, but these mice cannot be regarded as truly devoid of exogenous antigenic/mitogenic stimuli, since substantial mitogenic and/or antigenic activity can be exerted by the diet (Wostmann et al., 1971). With the successful breeding of GF C3H/HeCr mice fed an ultrafiltered solution of chemically defined (CD) low molecular weight nutrients, exogenous stimuli are reduced to the presently achievable minimum (Wostmann et al., 1982). We used these mice to determine their ability to produce immunocompetent B cells. Therefore, we employed an *in vitro* lipopolysaccharide (LPS) culture system that activates only newly formed short-lived B cells (Freitas et al., 1982) to analyze their expression of immunoglobulin (Ig) isotypes and their IgM antibody specificity repertoire.

Spleen and bone marrow (BM) cells, frozen and thawed according to standard protocols, were cultured under conditions that allow the clonal proliferation and maturation of every LPS-reactive B cell into a clone of IgM-secreting cells, that are capable of switching to the secretion of other Ig isotypes (Hooijkaas et al., 1983). The secretion of IgM, IgG and IgA was assessed in the protein A plaque assay (van Oudenaren et al., 1981), while specific IgM antibody secreting cells were detected in plaque assays specific for sheep red blood cells (SRBC) and differently haptenated SRBC (Hooijkaas et al., 1983). For the hapten-specific plaque assays

Table 1. Ig-secreting cells in LPS-activated cultures of spleen and BM cells from conventional C3H/HeCr mice fed natural ingredient (CV-NI) and germfree C3H/HeCr mice fed chemically defined diet (GF-CD)

Organ	Ig isotype	8- to 12-wk-old		52-wk-old	
		CV-NI	GF-CD	CV-NI	GF-CD
Spleen	IgM	6085 ± 2548[a)]	10,952 ± 1550	8335 ± 3020	10,147 ± 2336
	IgG	2921 ± 1279	5228 ± 941	1963 ± 443	2373 ± 507
	IgA	11 ± 2.1	14 ± 2.1	14 ± 2.5	12 ± 1.5
BM	IgM	865 ± 243	2629 ± 568	1508 ± 200	3945 ± 745
	IgG	530 ± 282	1955 ± 514	701 ± 245	1863 ± 448
	IgA	7 ± 3.2	7 ± 2.8	14 ± 3.6	13 ± 6.6

a. The numbers of Ig-secreting cells were determined in cultures of LPS-stimulated spleen cells (1200 viable nucleated cells per well) and BM cells (6000 viable nucleated cells per well). At these and at lower cell inputs the numbers of Ig-secreting cells generated are linearly related to the number of B cells cultured. IgM-secreting cells were determined on day 5 and IgG- and IgA-secreting cells on day 6 of culture. The data represent the arithmetic mean ± SE of 4 different experiments using the pooled spleen cells and pooled BM cells of 2 mice.

4-hydroxy-3,5-dinitrophenyl (NNP) and nitroiodophenyl (NIP) were coupled to SRBC by adding 2 and 0.4 mg of the hapten per ml of washed packed SRBC, respectively.

Table 1 shows the results obtained with LPS-activated cultures of spleen and BM cells from conventional (CV) C3H/HeCr mice fed natural ingredient (NI) (Kellogg, Wostmann, 1969) and GF C3H/HeCr mice fed CD diet. In general, the numbers of IgM-, IgG- and IgA-secreting cells tended to be higher in spleen and BM of 8 to 12-wk-old and 52-wk-old GF-CD mice as compared with their age-matched CV controls, although the differences were not significant in all cases. As it was shown that IgG- and IgA-secreting cells develop as subclones within clones that initially secrete IgM (Coutinho et al., 1982) we conclude that young and old GF-CD mice generate B

Table 2. Frequencies of LPS-reactive splenic B cells specific for NNP2-SRBC, NIP0.4-SRBC and SRBC in conventional C3H/HeCr mice fed natural ingredient (CV-NI) and germfree C3H/HeCr mice fed chemically defined diet (GF-CD)[a)]

Plaque assay	CV-NI	GF-CD
Protein A-SRBC + rabbit-anti-mouse-IgM	1 in 13[b)] (4;11;25)	1 in 6 (4;4;11)
NNP2-SRBC	1 in 19[c)] (10;12;35)	1 in 19 (17;20)
NIP0.4-SRBC	1 in 120 (94;122;143)	1 in 142 (99;152;174)
SRBC	1 in 1470 (864;977;2567)	1 in 1640 (1122;1540;2258)

a. All mice were 8 to 12-wk-old.
b. Cells were cultured under limiting dilution conditions and the frequency of reactive cells was determined by confirming the fraction of negative cultures to the zero-term of the Poisson distribution. Frequencies of LPS-reactive B cells were expressed as fraction of all viable nucleated cells. Figures represent the arithmetic mean of the data from different experiments. The figures from the individual experiments are given in parentheses.
c. Frequencies of antigen-specific B cells were calculated as fraction of all LPS-reactive B cells.

cells capable of switching from IgM-secretion to the secretion of other isotypes.

These experiments do not reveal the frequency of LPS-reactive B cells among the spleen cells of 8 to 12-wk-old CV-NI and GF-CD C3H/HeCr mice. Therefore, we also determined the frequency of LPS-reactive B cells and a part of their antibody specificity repertoire under limiting dilution culture conditions. The results are shown in Table 2. Again, the GF-CD mice tended to have a higher frequency of LPS-reactive B cells than CV-NI mice. The frequencies of B cells specific for NNP2-SRBC, NIP0.4-SRBC and SRBC among these LPS-reactive B cells were, however, identical in GF-CD and CV-NI mice.

In conclusion, within the limitations of having determined the switch capacity of IgM to IgG and IgA in bulk cultures only, and having assessed only a minor fraction of the total B cell specificity repertoire, the data indicate that GF-CD mice, although devoid of exogenous stimulation by antigens and/or mitogens, generate B cells with a similar switching capacity and a similar IgM antibody specificity repertoire as CV mice.

Coutinho A, Benner R, Björklund M, Forni L, Holmberg D, Ivars F, Martinez-A C, Pettersson S (1982). A 'trans' perspective on the control of immunoglobulin C gene expression. Immunol Rev 67:87.

Freitas AA, Rocha B, Forni L, Coutinho A (1982). Population dynamics of B lymphocytes and their precursors: demonstration of high turnover in the central and peripheral lymphoid organs. J Immunol 128:54.

Hooijkaas H, Preesman AA, van Oudenaren A, Benner R, Haaijman JJ (1983). Frequency analysis of functional C and V gene expression in murine B cells at various ages. J Immunol 131:1629.

Kellogg TF, Wostmann BS (1969). Stock diet for colony production of germfree rats and mice. Lab Animal Care 19:812.

Van Oudenaren A, Hooijkaas H, Benner R (1981). Improvement of the protein A plaque assay for immunoglobulin secreting cells by using immunoglobulin-depleted guinea pig serum as a source of complement. J Immunol Methods 43:219.

Wostmann BS, Bruckner-Kardoss E, Pleasants JR (1982). Oxygen consumption and thyroid hormones in germfree mice fed glucose-aminoacid liquid diet. J Nutr 112:552.

Wostmann BS, Pleasants JR, Bealmear P (1971). Dietary stimulation of immune mechanisms. Fed Proc 30:1779.

Germfree Research: Microflora Control and Its Application to the Biomedical Sciences, pages 363–366

CINEMATOGRAPHIC STUDIES ON THE PHAGOCYTIC ACTIVITY OF GRANULOCYTES DERIVED FROM GNOTOBIOTIC AND CONVENTIONAL MICE

Ohnishi, N.*, Ozawa, A.*, Hashimoto, K.*, Watanabe, K.**, Nakane, P.K.*** and Sasaki, S.
Departments of Microbiology*, Pathology** and Cell Biology***. Tokai University School of Medicine, Bohseidai, Isehara, Kanagawa, Japan

The resistance to bacterial infection by animals maintained in the germfree environment is much less than that of animals maintained in the conventional animal quarters. For an example, the mortality of germfree mice following the subcutaneous injection of 8×10^4 Salmonella typhimurium is 6 out of 10 within seven to ten days, whereas the mortality of conventional mice when injected with a same dose of bacteria is 0 out of 10 for the ten day duration (Ozawa et al, 1982).

Among various defense mechanisms available to the host against the bacterial infection, circulating specific antibodies against the bacteria play a major role for the protection of the host. Since the germfree mice lack the specific antibodies, and only 1/10 of total IgG as compared to that of the conventional mice is found to the sera of germfree mice, although the concentration of total IgM is similar, the high mortality of germfree animal is thought to result (Hashimoto et al, 1978). However, in what manner or manners these specific antibodies participate in vivo against the bacterial infection have not been clearly elucidated.

As the preliminary experiments to define the role that the specific antibody plays against bacterial infection, the histopathological events taking place in the tissues of germfree mice and conventional mice following Salmonella typhimurium were examined. When the Salmonella-inoculated conventional mice were examined, the liver was characterized by acute inflammation within four to five days. The acute inflammation was then replaced occasionally by granulomas within one to two weeks. The granulomas were composed of mainly macrophages when examined at the ultrastructural level. When germfree mice were infected with _Salmonella typhimurium_, there was a massive accumulation of polymorphonuclear leucocytes in the area of the hepatic portal triad within two to three days that persisted for two to three weeks; at the same time multiple granulomas composed of mainly peroxidase-positive pro-monocytes were developed throughout the liver (Ozawa et al, 1982). This hyperreaction by the germfree mice has suggested to us that in the absence of specific antibody either the macrophages or neutrophils must behave differently.

The first set of experiments were carried out using macrophages isolated from the peritoneal cavities of conventional mice and germfree mice. When the macrophages were exposed to Salmonella _in vitro_, the cells from conventional mice and from germfree mice phagocytized the bacteria with an equal efficiency (Aiba et al, 1983). The result suggested to us that the macrophages are functionally similar with or without the specific antibodies.

Whether or not the neutrophils from conventional mice and those from germfree mice behave similarly when the cells encounter bacteria was investigaged _in vitro_. In these studies, granulocytes were isolated from the peripheral blood from the conventional, the germfree mice mono-contaminated with _Salmonella typhimurium_ 8 weeks prior to sacrifice and the germfree mice. The isolated cells were washed extensively. The granulocytes were then added to plastic culture dishes containing either _E. coli_ as an example of the extracellular bacteria or _Salmonella typhimurium_ as an example of the facultative intracellular bacteria, and the activity of the granulocytes was recorded cinematographically. There were considerable differences in the behaviors of granulocytes from the conventional mice and those from the germfree mice. The granulocytes from the conventional mice were attracted to the bacteria and phagocytized the bacteria efficiently. On the other hand, the granulocytes from germfree mice either skirted the bacteria or may be attracted to the bacteria, but rarely phagocytized the bacteria. Those bacteria which came in contact with the granulocytes from germfree mice stuck on to the surface of the cells and were dragged by the cells as the cells moved about. The mucoid material which cemented the bacteria on the cell surface was visible as the bacteria were dragged by the cell. The phagocytic activity of the granulocytes isolated from the germfree mice mono-contaminated with _Salmonella typhimurium_ was slightly less vigorous than that of the conventional mice. The differences in the behavior of granulocytes were noted regardless of the type of bacteria used as the target. In order to define the role specific antibody against bacteria plays in this experimental system, Salmonella typhimurium was first reacted with antibodies against Salmonella typhimurium. As a control to this, some of the bacteria were incubated with normal serum. When the granulocytes from germfree mice were exposed to the antibody-coated Salmonella typhimurium, the cells phagocytized the bacteria efficiently, although the phagocytic activity was not as vigorous as that of the granulocytes from conventional mice. The granulocytes exposed to the bacteria which had been incubated with the normal serum did phagocytize the bacteria occasionally, but not as much as the antibody-coated bacteria.

TABLE

BEHAVIOR OF GRANULOCYTES WITH SALMONELLA TYPHIMURIUM

Granulocytes from	Target	Behavior with Salmonella	
		Attracted	Phagocytize
conventional mice	Salmonella	+++	+++
germfree mice	Salmonella	+	-/+
germfree mice	Salmonella coated with anti-Salmonella	++	++
germfree mice	Salmonella coated with N.IgG	+	+
germfree mice mono-contaminated with Salmonella (8W)	Salmonella	++	++

Salmonella=Salmonella typhimurium. N.IgG-VenoI (Midori-Juji)

These results are summarized in the TABLE. The activities of granulocytes are scored in two aspects, the degrees of attraction to Salmonella and the degrees of phagocytosis. The granulocytes from conventional mice were attracted to the bacteria and phagocytized the bacteria vigorously. On the other hand, the granulocytes from germfree mice poorly recognized the presence of bacteria and even when the cells recognized the bacteria, they did not phagocytize the bacteria efficiently. Since the bacteria were free from any serum factors, these results suggest that the ability of granulocytes to recognize the presence of bacteria and phagocytize the bacteria depends upon the conditioning of the granulocytes by the prior exposure of the mice to the bacteria. The failure of the granulocytes from germfree mice to phagocytize bacteria were overcome partly when the bacteria were coated with antibodies against the bacteria. This observation suggested that the presence of antibody on the bacteria is one of the necessary factors for the granulocyte to phagocytize the bacteria efficiently and that the nature of the target also plays a role for the recognition and phagocytosis of the target by the granulocytes. Although the chemical and biological natures of the recognition mechanisms are not yet defined, in order for bacteria to be effectively phagocytized by granulocytes, both the granulocytes as well as the bacteria must be properly conditioned. The germfree mice lack the conditioning of both the granulocytes and the bacteria, and they are more sensitive to bacterial infection.

Aiba Y, Ohnishi N, Ozawa A, Watanabe K, Izumi S (1983). Modification of host responses by intestinal bacterial flora. Japanese J of Bac 38:211.

Hashimoto K, Handa H, Umehara K, Sasaki S (1978). Germfree mice reared on an "antigen-free" diet. Laboratory Animal Science 28:38.

Ozawa A, Ohnishi N, Ikigai H, Seiki T, Wawamur S, Watanabe K, Izumi S (1982). The role of intestinal bacteria in host defense mechanisms against infection. Proc IInd IPCR Symposium on Intestinal Flora. pp. 191-211.

Germfree Research: Microflora Control and Its Application to the Biomedical Sciences, pages 367–370

COMPARISON OF BRONCHOALVEOLAR LAVAGE CELLS FROM GERMFREE AND CONVENTIONAL LAMBS

Darrel D. Joel, D.V.M.,Ph.D., Arjun D. Chanana, M.D., Rachelle A. Weiss, Ph.D.

Medical Research Center
Brookhaven National Laboratory
Upton, L.I., New York 11973

It is well known that newborn infants and animals have a diminished resistance to a variety of infections including respiratory infections. We have employed newborn lambs as experimental subjects in studies designed to establish the basis for this increased susceptibility (Weiss, Chanana, Joel 1984). Of interest was the observation that 1 day after birth the majority of cells in bronchoalveolar lavage (BAL) fluid were polymorphonuclear leukocytes (PMN) and not alveolar macrophages (AM). During the first week of life a marked decline in the number and proportion of PMNs was seen concurrently with a rapid rise in the concentration of AM. These changes in cell populations were accompanied by an improvement in phagocytic and bactericidal capacities.

The question arose as to whether the rapid postnatal influx of PMNs into the lung was a consequence solely of environmental exposure, particularly the inhalation of microorganisms. This transient postnatal elevation in BAL PMN may have important clinical consequences since PMNs in the presence of O_2 concentrations frequently employed in the therapy of the preterm infants, have been implicated as mediators of lung tissue injury (Fox, Hoidal, Brown, Repine 1981). The current study was thus undertaken to determine the influence of a germfree environment on the concentration and function of BAL cells.

Germfree (GF), colostrum-deprived lambs were obtained by aseptic Caesarean section of dams in their 141st to 143rd day of gestation, maintained in flexible plastic isolators, and fed sterile evaporated milk. Verification of their germfree status was done by culturing feces and milk from feeding bottles for aerobic and anaerobic bacteria. At 1, 4, 8-10 or 21 days of age, lambs were removed from the isolator and within 10 minutes lung free cells were obtained aseptically by BAL (15 aliquots of 10 ml sterile saline) using a pediatric fiberoptic bronchoscope.

In comparison to conventional (CV) lambs, cell counts in BAL samples from 1-day old GF lambs were substantially lower (2.2 ± 1.4 vs $1.5 \pm 0.05 \times 10^5$/ml) due exclusively to a marked reduction in PMNs (neutrophils) (Figure 1a). PMNs accounted for ~75% of BAL cells in CV lambs as opposed to ~14% in GF lambs. In 8-10 day old GF lambs the proportion of PMN was less than 1% and by day 21 they were totally absent. In contrast, the early appearance and increase in concentration of AM was essentially the same in both groups of animals (Figure 1b). On day 8-10, GF lambs had comparatively lower numbers of AM; however, on day 21 this difference was not apparent.

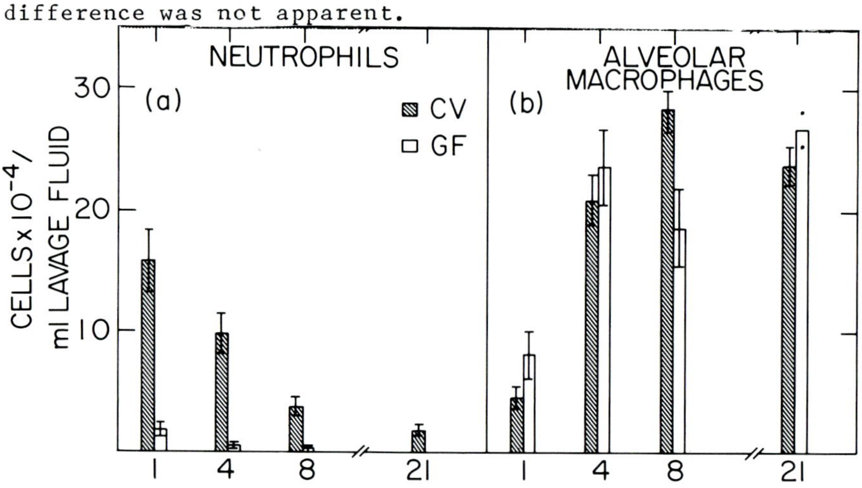

Figure 1. The mean concentration (± SEM) of neutrophils (a) and alveolar macrophages (b) in bronchoalveolar lavage fluid from germfree (GF) and conventional (CV) lambs.

The capacity of BAL cells to bind and/or phagocytize preopsonized bacteria (S. aureus) was assessed in vitro as previously described (Weiss, Chanana, Joel, 1984). As shown in Table 1 no difference between GF and CV BAL cells was apparent in any of the age groups tested. In both groups of animals substantial improvement in this function was evident between days 1 and 8-10. The assay was initiated using a bacteria: cell ratio of 2:1. Since BAL cells from GF lambs contained predominately AM, the relatively poor performance found on days 1 and 4 cannot be ascribed to PMN contamination, as might be the case with BAL cells from CV lambs.

Table 1. The percent (± SEM) of preopsonized bacteria (*S. aureus*) bound and/or phagocytized by BAL cells following 90 minutes incubation (39°C with gentle mixing) at a target: effector cell ratio of 2:1. Number of conventional (CV) or germfree (GF) lambs sampled is in parenthesis.

	Lamb Age (Days)		
	1	4	8-10
CV	62.6 ± 5.6 (5)	81.2 ± 2.9 (8)	94.4 ± 0.8 (6)
GF	63.3 ± 4.8 (3)	83.3 ± 3.3 (4)	93.2 ± 5.8 (4)

AM from neonatal lambs were actively proliferating as evidenced by the fraction of cells incorporating tritiated thymidine *in vitro*. In CV lambs a peak labeling index (LI) of about 24% was seen on day 4. It was anticipated that in the absence of bacteria and/or other particulate contamination the LI of AM from GF lambs might be reduced. As shown in Table 2 this was not the case.

Table 2. The percent (± SEM) of alveolar macrophages incorporating tritiated thymidine *in vitro*. Number of lambs sampled is in parenthesis.

	Lamb Age (Days)		
	1	4	8-10
CV	18.7 ± 2.7 (9)	23.9 ± 5.1 (9)	17.1 ± 3.4 (9)
GF	15.5 ± 8.6 (4)	23.0 ± 5.8 (4)	18.8 ± 2.6 (4)

The results of this study clearly show that the influx of PMNs into lung air spaces is markedly reduced in GF lambs. CV lambs are exposed to a variety of inert particulate pollutants in addition to microrganisms. Nonviable particles could theoretically stimulate AM to elaborate chemotactic factors, however, we have shown that BAL AM from 8-day old lambs are poor elaborators of chemotaxins for PMNs (Weiss, Chanana, Joel 1984). Therefore it is likely that the influx of PMNs into the lung is largely in response to chemotactic factors of bacterial origin.

Of interest was the observation that the influx, phagocytic function and proliferative capacity of AM were essentially the same in GF and CV lambs. These findings are, in part, contradictory to those of Rothlein and Kim 1981 who reported a significant reduction in both PMN and AM in GF piglets. Reasons for this discrepancy are unknown. The GF lamb may represent an ideal model with which to study the role of PMNs as mediators of lung damage following exposure to hyperoxia. It has been suggested (Harada, Vatter, Repine 1984) that hyperoxia stimulates AM to elaborate factors chemotactic for PMNs, which, when exposed to high O_2 levels in the lung, initiate tissue damage. GF lambs have very few lung PMNs. If, in addition, their AM are not capable of elaborating chemotaxins, it may be possible to expose such lambs to hyperoxia in the absence of a PMN influx. It would be of importance to determine if under such conditions there is reduced lung tissue damage. Such findings would permit further investigations into the relative roles of hyperoxia, mechanical ventilation, and bacterial colonization in the genesis of bronchopulmonary dysplasia of preterm infants.

Fox RB, Hoidal JR, Brown DM, Repine JE (1981). Pulmonary inflammation due to oxygen toxicity: Involvement of chemotactic factors and polymorphonuclear leukocytes. Am Rev Resp Dis 123:521.

Harada RN, Vatter AE, Repine JE (1984). Macrophage effector function in pulmonary oxygen toxicity: Hyperoxia damages and stimulates alveolar macrophage to make and release chemotaxins for polymorphonuclear leukocytes. J Leuk Biol (formerly RES) 35:373.

Rothlein R, Gollily R, Kim YB (1981). Development of alveolar macrophages in specific pathogen-free and germ-free Minnesota minature swine. J Reticuloendothel Soc 30:483.

Weiss RA, Chanana, AD, Joel, DD (1984). The status of pulmonary host defense in the neonatal sheep: Cellular and humoral aspects. NY Acad Sci (in press).

Germfree Research: Microflora Control and Its Application to the Biomedical Sciences, pages 371–375

FC RECEPTOR-DEPENDENT ACTIVATION OF EFFECTOR FUNCTION OF GNOTOBIOTIC PORCINE PULMONARY ALVEOLAR MACROPHAGES[1]

Robert Rothlein, Ph.D. & Yoon Berm Kim, M.D., Ph.D.

Sloan-Kettering Institute for Cancer Research, Rye, NY 10580 and University of Health Sciences/The Chicago Medical School, North Chicago, IL 60064

Macrophages are extremely sensitive to the microbial environment of their host in terms of their expression in tissue (Blusse van Oud Alblas et al., 1981; Rothlein et al., 1981) and their activation for nonspecific and specific cytolytic functions (Hibbs et al., 1972; Pace et al., 1981). In fact, it has been clearly shown that the magnitude of macrophage-mediated antibody-dependent cellular cytotoxicity (ADCC) is very much related to the "activated" state of the macrophage (Ralph & Nakoinz, 1977; Nathan et al., 1980a; Koren et al., 1981). To reduce the variability that microbial stimulation may play on the results of experiments designed to study the effects of Fc Receptor (FcR) modulation on various functions of macrophages, we have used pulmonary alveolar macrophages (PAM) derived from the lungs of germfree and specific-pathogen free (SPF) miniature swine maintained in a barrier-sustained facility with sterile feed, water and air.

Our initial observation seen in Table 1 was that PAM cultured on immobilized immune complexes (IIC), chicken red blood cells (CRBC)/anti-CRBC complexes in this case, became nonspecifically cytotoxic to ^{51}Cr-labeled autologous pig red blood cells (PRBC) and to ^{51}Cr-labeled human cell lines K562 and SB in an 18 hr. ^{51}Cr-release assay (Rothlein & Kim, 1982). Control wells in which PAM are exposed to no immune complexes, anti-CRBC antiserum alone or CRBC/anti-CRBC immune complexes in suspension showed little nonspecific cytotoxicity towards "bystander" targets.

[1] This work was supported in part by US PHS Grant CA-38354 awarded by the National Cancer Institute, DHHS.

Table 1. Lysis of RBC and Tumor Targets by Macrophages Exposed to Immobilized Immune Complexes

	% lysis of ^{51}Cr-labeled targets by PAM exposed to:			
target	untreated	anti-CRBC	CRBC/anti-CRBC in suspension	immobilized CRBC/anti-CRBC
PRBC[a]	-0.3 ± 0.5	0.1 ± 1.2	10.4 ± 5.3	60.7 ± 10.4
K562[b]	1.7 ± 3.5	2.3 ± 3.0	3.9 ± 2.8	46.1 ± 9.2
SB[b]	-0.7 ± 3.5	-2.2 ± 2.2	-2.3 ± 2.1	21.7 ± 5.5

[a] Effector to target cell ratio of 1:1
[b] Effector to target cell ratio of 10:1

We next tried to determine the lytic mechanism by which the "bystander" targets were lysed. Since previous work has implicated peroxide as a major component of some types of macrophage-mediated cytolysis (Nathan et al., 1979; Nathan & Cohn, 1980b) and it has been shown that macrophages exposed to IIC secrete elevated levels of superoxide (Johnston et al., 1976), it seemed reasonable to test whether peroxide played a role in the nonspecific cytotoxicity generated by exposing PAM to IIC. Table 2 shows that when the peroxide scavengers catalase or thioglycollate were added to PAM exposed to IIC there was considerable inhibition of cytolysis (100% for catalase and 50% for thioglycollate) while conventional ADCC towards the CRBC in the CRBC/anti-CRBC complexes in suspension remained uninhibited. These data in conjunction with previously published data (Rothlein & Kim, 1983) suggest that ADCC and the nonspecific cytotoxicity generated by exposing PAM to IIC differ at least in part in their lytic mechanisms in that ADCC by PAM is primarily peroxide-independent while "bystander" killing is primarily peroxide-dependent under the conditions of this assay.

We also examined the effects of IIC on other PAM functions. Figure 1 shows that not only is "bystander" killing augmented but also PGE secretion is enhanced when PAM are cultured with immobilized CRBC/anti-CRBC complexes in suspension.

Further studies which are not presented show that the nonspecific cytotoxicity and enhanced PGE secretion by PAM exposed to IIC may share a common trigger (the IIC) but are independent functions of the PAM. These findings may prove to be extremely important when one considers the potential roles of prostaglandins in disease states where antibodies are directed against immobilized self antigens or antigens on immobilized tissue.

Table 2. Effect of Peroxide Scavengers on ADCC of CRBC and "Bystander" Lysis of PRBC by PAM

animal no.	% inhibition of ADCC (CRBC)[a] in the presence of: catalase[c]	thioglycollate[d]	% inhibition "bystander" lysis (PRBC)[b] in the presence of: catalase[c]	thioglycollate[d]
SPF4112-1	-11.0	-12.3	99.9	56.8
SPF4161-3	25.2	- 7.7	99.8	71.3
SPF4161-4	- 6.5	- 8.3	100.0	21.4
average % inhibition ($\bar{x}$ ± SD)	2.6 ± 19.7	- 9.4 ± 2.5	99.9 ± 0.1	49.8 ± 25.7

[a] Lysis of CRBC in the absence of peroxide scavengers was 62.3 ± 3.6%.
[b] Lysis of PRBC in the absence of peroxide scavengers was 53.2 ± 5.9%.
[c] Catalase was used at a final concentration of 250 µg/ml.
[d] Thioglycollate broth was used at a final concentration of 3.0 mg/ml.

Finally, we found that by exposing PAM to CRBC/anti-CRBC complexes in suspension in conjunction with the phagocytic inhibitor cytochalasin B, we were able to generate nonspecific cytotoxicity towards PRBC (Rothlein & Kim, 1982). Since the PAM cannot internalize IIC or immune complexes in suspension in conjunction with cytochalasin B, this suggested that it was the inability to internalize the ligand-bound FcR on the PAM which was responsible for "bystander" killing and enhanced PGE secretion. In fact, we postulate that the macrophage becomes lytic and secretes PGE when exposed to immune complexes and that internalization of these immune complexes signals the PAM to exit from its "activated" mode. By preventing the internalization of the ligand-bound FcR, the macrophage remains "turned on" and thus has enhanced lytic and secretory capacities.

In conclusion, we described a unique mechanism by which

macrophages become activated for enhanced effector functions. By exposing PAM to IIC, they become nonspecifically cytotoxic to tumor and RBC targets via a peroxide-dependent lytic mechanism. Furthermore, PAM exposed to IIC secrete more PGE than do PAM exposed to immune complexes in suspension. This model may have in vivo relevance in disease states where antibodies are directed against immobilized self-antigens or against viral antigens expressed on immobilized tissue. In instances such as these it is possible that macrophages enter the site of inflammation and destroy healthy tissue via a nonspecific cytotoxic mechanism while secreting important immunological mediators which will contribute to the pathogenesis of the disease.

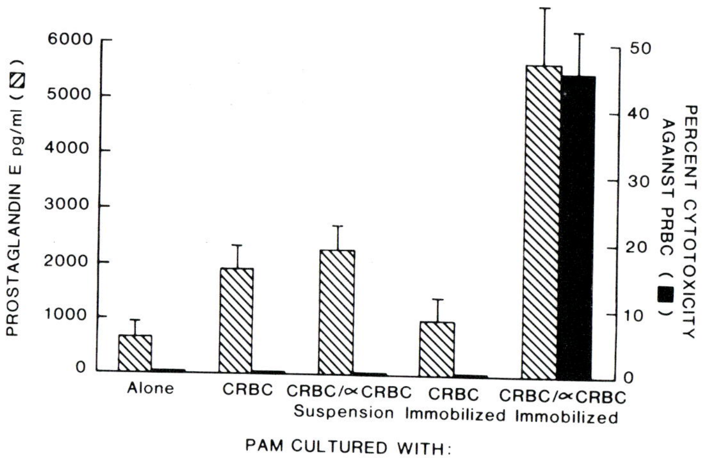

Fig. 1. PGE secretion and nonspecific cytotoxicity of PAM cultured with IIC or immune complexes in suspension. In parallel experiments, PAM were cultured under appropriate conditons and either ^{51}Cr-PRBC were added to the cultures for cytotoxicity assays or complete medium was added to assays for PGE measurement.

Blusse van Oud Alblas A, van der Linden-Schruer B, van Furth R (1981). Origin and kinetics of pulmonary macrophages during an inflammatory reaction induced by intravenous

administration of heat-killed bacillus Calmette-Guerin. J Exp Med 154:235.
Hibbs JB, Lambert LH, Remington JS (1972). Possible role of macrophage-mediated nonspecific cytotoxicity in tumor resistance. Nature 235:48.
Johnston RB, Lehmeyer JE, Guthrie LA (1976). Generation of superoxide and chemiluminescence by human monocytes during phagocytosis and on contact with surface-bound immunoglobulin G. J Exp Med 143:1551.
Koren HS, Anderson SJ, Adams DO (1981). Studies on the antibody-dependent cell-mediated cytotoxicity (ADCC) of thioglycollate-stimulated and BCG-activated peritoneal macrophages. Cell Immunol 57:51.
Nathan C, Brukner L, Kaplan G, Unkeless J, Cohn Z (1980a). Role of activated macrophages in antibody-dependent lysis of tumor targets. J Exp Med 152:183.
Nathan C, Cohn Z (1980b). Role of oxygen-dependent mechanisms in antibody-induced lysis of tumor cells by activated macrophages. J Exp Med 152:198.
Nathan C, Silverstein SC, Brukner LH, Cohn ZA (1979). Extracellular cytolysis by activated macrophages and granulocytes II. Hydrogen peroxide as a mediator of cytotoxicity. J Exp Med 149:100.
Pace JL, Taffet SM, Russell SW (1981). The effect of endotoxin in eliciting agents on the activation of mouse macrophages for tumor cell killing. J Reticuloendothel Soc 30:15.
Ralph P, Nakoinz I (1977). Antibody-dependent killing of erythrocyte and tumor targets by macrophage-related cell lines: Enhancement by PPD and LPS. J Immunol 119:1977.
Rothlein R, Gallily R, Kim YB (1981). Development of alveolar macrophages in specific pathogen-free and germfree Minnesota miniature swine. J Reticuloendothel Soc 30:483.
Rothlein R, Kim YB (1982). Role of Fc receptor modulation by immobilized immune complexes in generation of nonspecific (bystander) cytotoxicity for autologous and xenogeneic targets by porcine alveolar macrophages. J Immunol 129:1859.
Rothlein R, Kim YB (1983). Two distinct mechanisms of cytotoxicity by porcine alveolar macrophages in antibody-dependent and immobilized immune complex-dependent cellular cytotoxicity. J Immunol 131:1438.

Germfree Research: Microflora Control and Its Application to the Biomedical Sciences, pages 377–379

CELLULAR AND STRUCTURAL CHANGES IN THE SPLEEN OF GERMFREE MICE IN THE PRIMARY IMMUNE RESPONSE

Junpei Asai, Hiroshi Nagura and Takashi Koshikawa

Laboratory of Germfree Life Research Institute for Disease Mechanism & Control, Nagoya University School of Medicine, Showa-ku, Nagoya 466, Japan

Understanding of immune responses in the spleen has advanced rapidly in recent years (Dresser 1972). However, much of the confusion regarding the histological changes that develop after immunization has been due to having to use conventional animals which have already been exposed to multiple external immunological stimuli. In order to clarify the mechanism of primary immune response, we studied immunohistochemically the population and localization of immunocompetent cells in the spleen of germfree mice which had minimum immunological noise.

Both germfree and conventional mice, ICR strain, were used at the age of 6 weeks. One group of each experiment consisted of 5 male mice. Sheep red blood cells (sRBC)(Nippon Biotest Lab. Inc.) were use as the antigenic stimulus. Mice were immunized by intraperitoneal injection of $2x10^8$ sRBCs. The spleens were examined histologically and immunohistochemically on the 2nd, 4th and 7th day after the injection of sRBC by the direct or indirect immunoenzyme technique of staining with the enzyme-labeled antibodies of Ia, Thy 1.2, Lyt 1, Lyt 2, IgA, IgG, IgM and J chain(Nakane 1975). Agglutinins to sRBC were determined by a microtitre procedure. The 2-mercaptoethanol resistant hemagglutinin titration was carried out after a 2hr incubation of serum with an equal volume of 0.2M 2-mercaptoethanol to destroy IgM. The numbers of plaque-forming cells(PFC) in the spleen were estimated by the liquid monolayer technique of Cunningham.

In both germfree and conventional mice, weights of the spleens increased temporarily on the 2nd day after the injection and decreased gradually down to the control weight. Mean values of the hemagglutinin titres of IgM increased sufficiently on the 4th day after the injection. However, any significant differences were not recognized between germfree and conventional cases. On the contrary, germfree mice contained lower numbers of direct PFCs(58x1000 cells per spleen) anti-sRBC IgM in the spleens than conventional ones(113x1000 cells per spleen) on the 4th day after the injection. The white pulp with certain numbers of germinal centers in the spleens of germfree mice developed remarkably after the injection, although germfree mice had hypoplastic white pulp before the injection. In the immunohistochemical studies, germfree mice increased more drastically in number of the cells which were stained positively with anti-Ia, anti-IgA, anti-IgM, anti-IgG or anti-J chain antibody after the injection than conventional ones. As contrasted, there were not observed any significant changes in number of the cells which were stained with anti-Thy 1.2, anti-Lyt 1 or anti-Lyt 2 antibody after the injection in both germfree and conventional mice It was a clear-cut finding that non-lymphoid cells stained positively with anti-Ia antibody increased obviously in number in the T cell zone of white pulp in germfree mice on the 2nd day after the injection (Fig. 1A), while there were few Ia-positive non-lymphoid cells in the T cell zone in control germfree mice. In comparison with this cell, the acid phosphatase-positive cells in the T cell zone did not change in number before and after the injection in both germfree and conventional mice. The IgA-positive cells without J chain increased significantly in number in the marginal area of the B cell zone of white pulp in germfree mice after the injection (Fig. 1B), although control germfree mice had few IgA-positive cells in the B cell zone. The IgM-positive cells with J chain increased in number in the germinal center of the B cell zone in germfree mice after the injection Moreover, the IgG-positive cells increased in number in both the T cell and B cell zone on the 4th day of the experiment.

Little information has been available on the cellular responses induced by immunological stimuli in the spleens of germfree animals, nevertheless the fundamental morphological studies on the lymphoid tissues of germfree animals were reported by some investigators (Thorbecke 1959). We could succeed in demonstrating the localization and distribution of the immunocompetent cells in the

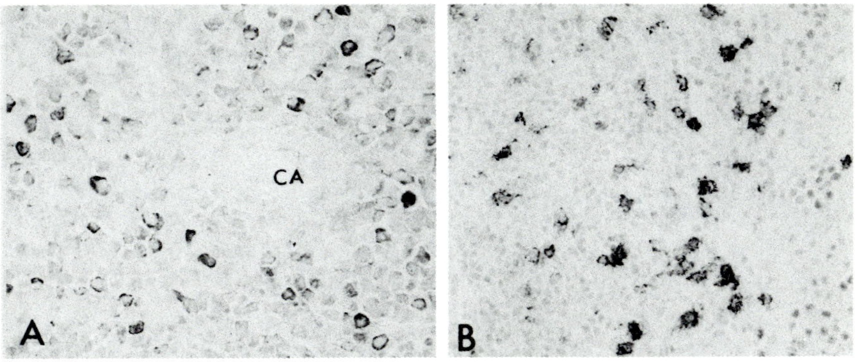

Fig. 1 Immunohistochemical localization and distribution of Ia antigen and IgA in the spleens of germfree mice after the injection of sRBC. A:Ia-positive non-lymphoid cells in the T cell zone. CA;central artery. B:IgA-positive lymphocytes in the B cell zone.

spleen in primary immune response by using germfree mice. Steinman had pointed out that in the spleen of germfree mouse all the dendritic cells were Ia-positive but 90 to 95 percent of macrophages were Ia-negative, whereas the spleen adherence cells were one of the best sources of Ia-bearing macrophages in conventional animals(Steinman 1982). Considering the space where the cellular recognition was performed, it might be a noteworthy evidence that Ia-positive non-lymphoid cells in the T cell zone of white pulp increased remarkably in number at the early stage of primary immune response. Based on the appearance of numbers of IgA-positive cells without J chain in the B cell zone of the white pulp after an intraperitoneal immunological stimulation, it is strongly suggested that the spleen is one of the major organs to produce serum monomeric IgA.

Dresser DW(1972) The role of T cells and adjuvant in the immune response of mice to foreign erythrocytes. European J Immunol 2:50

Nakane PK(1975) Recent progress in peroxidase-labeled antibody Method. Ann NY Acad Sci 254:203

Stainman RM(1982) In Mizuno D, Cohn ZA, Takeya K, Ishida(eds): "Self-defense Mechanisma: Role of Macrophages", Univ Tokyo Press, Elsvier Biomedical, p 178

Thorbecke GJ(1959) Some histological and functional aspects of lymphoid tissue in germfree animals: 1. Morphological studies. Ann NY Acad Sci 78:237

Germfree Research: Microflora Control and Its Application to the Biomedical Sciences, pages 381–385

ISOLATION AND FUNCTION OF MUCOSAL LYMPHOCYTES IN THE MOUSE

Susan B. Dillon & Thomas T. MacDonald Ph.D.

Department of Microbiology
Jefferson Medical College
Philadelphia, PA 19107

The major stimulus for the appearance of lymphocytes in the gut is the normal bacterial flora, since mucosal lymphocytes are dramatically reduced in germfree animals (Crabbe, et. al., 1970; Ferguson and Parrott, 1972; Ropke and Everett, 1976). However, when germfree mice are reconventionalized IEL return to normal levels (Crabbe, et. al., 1970: Glaister, 1973). In order to investigate T and B lymphocyte function in the gut mucosa we isolated intra-epithelial lymphocytes (IEL) from murine small intestine and characterized these cells *in vitro*.

METHODS

Male (C57BL/6 X DBA/2) F_1 (BDF_1) and C3H/HeJ mice were purchased from the Jackson Labs, Bar Harbor, Me. Male C3H/HEN mice were from Charles River Breeding Labs, N. Wilmington, Ma. All mice were used between 6-12 weeks of age. The small intestines of 3 mice were pooled for isolation of IEL exactly as described previously (Dillon and MacDonald, 1984). Lamina propria lymphocytes (LPL) were isolated as described by Davies and Parrott (1981). Surface phenotypes of IEL and LPL (by immunofluorescence) and -proliferation assays were carried out exactly as previously described (Dillon and MacDonald, 1984). To quantitate cytoplasmic immunoglobulin positive (cIg^+) cells, cytocentrifuged slides were fixed in absolute methanol before staining with fluorescein labeled $F(ab')_2$ goat anti-mouse IgG.

RESULTS AND DISCUSSION

Surface phenotypes of IEL and LPL. Freshly isolated IEL contained 57% Lyt 2^+ cells, 17% Lyt 1^+ cells, 30% Thy 1.2^+ cells, 3% Mac-1^+ cells and 6% sIg^+ cells. About 40% of Giemsa stained IEL contained fine cytoplasmic granules. In contrast to IEL, LPL contained 38% sIg^+ and 23% cIg^+ lymphocytes, and only 1.3% LPL had cytoplasmic granules. Since the populations isolated were clearly distinct, it is unlikely that IEL were significantly contaminated with LPL.

Response to Mitogens. IEL respond to the T cell mitogen Concanavalin A (Con A), and to allogeneic stimulator cells in an *in vitro* mixed lymphocyte reaction. (Dillon and MacDonald, 1984). Supernatants from Con A activated IEL contained low levels of Interleukin 2 (IL-2) (10 U/ml after 48 hours incubation compared to 70 U/ml in supernatants from equivalent numbers of spleen cells stimulated with Con A). Lyt 2^+ IEL (isolated by panning) contained 50% granulated cells; and also responded to Con A (stimulation index = 10-20 on day 4 of culture). IEL also proliferate in response to either phenol-water extracted LPS (PW-LPS) (*E. coli* 0111:B4; Difco) or whole heat-killed lyophilized mouse indigenous *E. coli* (a gift from Dr. Rodney Berg) (Table 1). As evidence that B cells from IEL were responding to LPS we studied the development of cells containing cytoplasmic immunoglobulin in LPS-stimulated IEL. Freshly isolated IEL contained only 0-0.3% cIg^+ cells; however by day 4 of culture, 26% of IEL were cIg^+ cells.

Specificity of the response of IEL to LPS.

In order to determine if the IEL B cell response to LPS was a mitogenic response or a specific B-cell response to enteric LPS, we tested LPS-stimulated cultures of IEL for the presence of B-cells secreting LPS specific antibody with a plaque-forming cell assay; however none were detected (data not shown). In order to approach the question in another way, we compared the ability of IEL isolated from either C3H/HeN or C3H/HeJ mice for responsiveness to a phenol water extracted LPS preparation or to killed whole *E. coli* organisms. We reasoned that since C3H/HeJ lymphocytes do not proliferate in response to lipid A (Goodman and Sultzer, 1979), a positive response by C3H/HeJ IEL cultured with PW-LPS would indicate specific stimulation via recognition of the antigenic O side chain sugar portion of the LPS molecule. However, the data in

Fig. 1 show that IEL isolated from C3H/HeJ mice were LPS non-responsive. Thus it would appear that IEL B cells are responding to LPS in a mitogenic fashion rather than in an antigen-specific fashion. Further work is therefore necessary to determine if IEL T or B cells have specificities for gut bacterial antigens.

TABLE 1. Proliferative Responses of BDF_1 IEL Cultured with Con A, LPS (Phenol-Water Extract); or Whole, Heat-Killed Bacteria.

2.5×10^5 IEL Cultured With:[a]	Day 4 CPM (Stimulation Index)[b]	
	Exp. 1	Exp. 2
Medium	487	134
Con A	9,652 (20)	2,306 (17)
PW-LPS	61,968 (127)	46,147 (377)
Whole *E. coli*	46,798 (96)	35,325 (237)
Whole *S. pneumoniae* (R36a)	351 (<1)	N.D.

[a] Final concentration Con A = 5 ug/ml; PW-LPS and heat-killed bacteria = 10 ug/ml.

[b] Results are shown as the mean cpm of triplicate wells; standard errors were between 10-20% of cpm values. Stimulation indices were calculated by dividing the mean cpm in cultures with mitogens or heat killed bacteria by the mean cpm in cultures with medium alone.

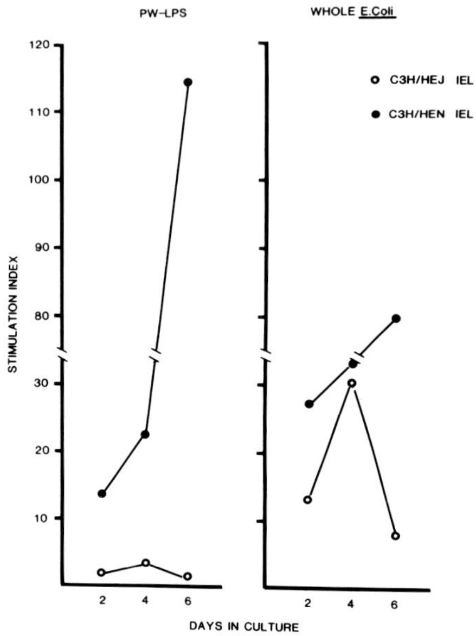

Fig. 1 Time course of the *in vitro* proliferative response of IEL (2.5 x 10^5 per well) from C3H/HeJ or C3H/HeN mice cultured with 10 ug/ml PW-LPS or 10 ug/ml whole *E. coli*. Stimulation indices were calculated as described in the legend of Table 1; standard errors were between 10-20% of cpm values.

Acknowledgements

This work was supported by Public Health Service Award AI 16628 from the National Institutes of Health. S.B.D. is a recipient of a predoctoral Percival E. and Ethel Brown Foerderer Foundation fellowship.

REFERENCES

Crabbe, P.A., Nash, D.R., Bazin, H., Eyssen, H. and Heremans, J.F. (1970). Immunohistochemical observations on lymphoid tissues from conventional and germ free mice. Lab. Invest. 22:448.

Davies, M.D.J. and Parrott, D.M.V. (1981). Preparation and purification of lymphocytes from the epithelium and lamina propria of murine small intestine. Gut 22:481.

Dillon, S.B. and MacDonald, T.T. (1984). Functional properties of lymphocytes isolated from murine small intestinal. epithelium. Immunology 52: in press.

Ferguson, A. and Parrott, D.M.V. (1972). The effect of antigen deprivation on thymus-dependent and thymus-independent lymphocytes in the small intestine of the mouse. Clin. and Exp. Immunol. 12:477.

Glaister, J.R. (1973). Factors affecting the lymphoid cells in the small intestinal epithelium of the mouse. Int. Arch. Allergy 45:719.

Goodman, G.W. and Sultzer, B.M. (1979). Further studies on the activation of lymphocytes by endotoxin protein. J. Immunol. 122:1329.

Ropke, C. and Everett, N.B. (1976). Kinetics of intra-epithelial lymphocytes in the small intestine of thymus-deprived mice and antigen deprived mice. Anat. Rec. 185:101.

Germfree Research: Microflora Control and Its Application to the Biomedical Sciences, pages 387–390

MODULATION OF IgE ISOTYPE EXPRESSION ON B LYMPHOCYTES IN PEYER'S PATCHES OF GERMFREE RATS

Helen G. Durkin, Ph.D. and Seto M. Chice, M.S.

Department of Pathology, State University of New York, Downstate Medical Center, Brooklyn, N.Y. 11203

Previous studies in this laboratory (1) have established that Peyer's patches (PP) of germfree (GF) rats contained high numbers of cells with surface IgE (sIgE), one-half of which were simultaneously positive for sIgA, whereas sIgE+ cells were absent from PP of conventionally raised (C) rats. GF rat PP also differed from those of C rats in that they contained elevated numbers of sIgA+, but reduced numbers of sIgM+ cells. We have now succeeded in modulating IgE isotype expression on PP cells of GF rats so that they resemble those of C rats. The data suggest that exposure to gut flora influenced these changes, either directly or through the mediation of sThy 1+ regulatory T cells (see 2,3).

Germfree Sprague-Dawley rats (males, 8-10 weeks old) were anaesthetized, 2-3 PP removed, and their cell surface markers analyzed using indirect immunofluorescence (1). The rats were then either thymectomized or fed unautoclaved chow for one day, at which time they were killed and the remaining PP removed for analysis. Indirect immunofluorescence assay of cells for membrane bound immunoglobulin was carried out with anti-IgM, A and E antisera prepared by Dr. Hervé Bazin, University of Louvain, Brussels, Belgium (4,5), as previously described (1). T cell surface markers were studied using anti-RT 7.1 (formerly ART 1a) and anti-Thy 1 antisera, prepared as previously described (6-8). Fluorescein and rhodamine conjugates (goat anti-rabbit or rabbit anti-goat Ig) were purchased from N.L. Cappel Laboratories, Inc. Data are expressed as per cent total cells in suspension.

TABLE 1

CHANGES IN CELLULAR COMPOSITION OF GERMFREE RAT PEYER'S PATCHES AFTER THYMECTOMY (Tx) OR FEEDING UNAUTOCLAVED CHOW 1)

Day	Treatment	Surface Markers RT+ * Thy⁻	RT+ Thy+	RT⁻ Thy+	Ig-Bearing Cells M	G	A	E	EA
		(% Total Cells) Effect of Thymectomy							
	GF Rat No. 1								
0	3 PP removed + Tx	12	18	4	6	5	45	20	11
1	Remaining PP removed	26	2	0	20	8	20	1	1
	GF Rat No. 2								
0	3 PP removed + sham Tx	18	18	4	5	4	29	14	7
1	Remaining PP removed	16	17	4	4	5	26	14	8
		Effect of Feeding Unautoclaved Chow							
	GF Rat No.3								
0	3 PP removed + unautoclaved chow	23	20	5	10	4	38	29	13
1	Remaining PP removed	30	1	1	21	7	20	1	1
	GF Rat No.4								
0	3 PP removed + autoclaved chow	19	11	4	7	5	26	21	8
1	Remaining PP removed	20	15	3	8	5	31	25	11
	Conventional Rat								
	None	35	1	1	14	5	17	1	1

1) Representative experiments, 1 of 3, each group, with comparable data. Indirect immunofluorescence, cell suspensions. >1000 cells counted. *RT=RT 7.1

Supported by USPHS grants AI15603 and R0549122 and American Cancer Society Grant IM-145.

RESULTS

In agreement with our earlier studies (1), GF rat PP contained high numbers of cells with sIgE (~20%), whereas sIgE+ cells were virtually absent from PP of C rats (Table 1). The numbers of sIgA+ cells were also much higher in PP of GF (26-45%) when compared with C (17%) rats. About one-half of sIgE+ cells were also sIgA+. PP of GF rats also contained high numbers of T cells bearing sThy 1 (suppressor phenotype) (2,3) (~20%); either sThy 1+ (RT 7.1+) or sThy 1+ (RT 7.1^-). sThy 1+ cells were virtually absent from C rat PP (1%). Further, the numbers of RT 7.1+ (Thy 1^-) (helper phenotype) (7, see also 2,3) T cells were reduced in GF (~15%) when compared with C rat PP (35%).

The cellular composition of GF rat PP could be converted to that resembling a C rat within 1 day after either thymectomy or feeding unautoclaved chow (Table 1), with each procedure resulting in (1) loss of sIgE+ B cells and sThy 1+ T cells, (2) increase in sIgM+ B cells and RT 7.1+ (Thy 1^-) T cells, and (3) decrease in sIgA+ B cells from the PP. Neither treatment increased the numbers of either sIgE+ or sThy 1+ cells detected in bone marrow, other lymphoid organs or blood (data not shown), excluding the possibility that these cells emigrated from PP.

Thus, these findings are consistent with the idea that exposure to gut flora influenced IgE isotype switching in PP, either directly or through the mediation of sThy 1+ regulatory cells. It is, therefore, possible that elucidation of a mechanism for specific suppression of IgE precursor cells will provide a new means by which IgE antibody production can be controlled.

1. Durkin, H.G., H. Bazin and B.H. Waksman. 1981. Origin and fate of IgE-bearing lymphocytes. I. Peyer's patches as differentiation site of cells simultaneously bearing IgA and IgE. J. Exp. Medicine 154:640.
2. Durkin, H.G., S.M. Chice and I. Goldschneider. 1984. Origin and fate of antigen-specific suppressor T cells. Ann. New York Acad. Sciences, in press.
3. Durkin, H.G., S.M. Chice and I. Goldschneider. 1984. Migration pathways of antigen specific thymic suppressor cells. J. Immunol., in press.

4. Bazin, H., A. Beckers and P. Querinjean. 1974. Three classes and 4 subclasses of rat immunoglobulins, IgM, IgA, IgE and IgG1, IgG2a, IgG2b and IgG2c. Eur. J. Immunol. 4:44.
5. Bazin, H. and A. Beckers. 1976. In Molecular and Biological Aspects of the Acute Allergic Reaction. S.G.O. Johannson, K. Strandberg and B. Uvnas, editors. Plenum Publishing Corporation, New York. 125-152.
6. Lubaroff, D.M. 1973. An alloantigenic marker on rat thymus and thymus-derived cells. Transpl. Proc. 25:1625.
7. Lubaroff, D.M., H.D. Hunt and G.T. Rasmussen. 1983. Serologic evidence that the rat antigens ART-1 and L-C are separate and distinct. Transpl. Proc. 25:1625.
8. Crawford, J.M. and I. Goldschneider. 1980. Thy 1 antigen and B lymphocyte differentiation in the rat. J. Imm. 124:969.

Germfree Research: Microflora Control and Its Application to the Biomedical Sciences, pages 391-394

EFFECT OF AN *ESCHERICHIA COLI* STRAIN ON INTESTINAL IgA PLASMOCYTE STIMULATION AND SERUM ANTIBODY RESPONSE IN GNOTOBIOTIC MICE.

Moreau M.-Ch., R. Ducluzeau, M.-Cl. Muller, P. Raibaud
Laboratoire d'Ecologie Microbienne
INRA - CNRZ
78350 Jouy-en-Josas -France-

Oral vaccination as a practical method for protection against enteritic infections has been widely investigated (Collins 1970, Pierce 1978, Porter 1983). However, no efficient vaccination schedule has yet been developed. There are still discrepancies about the route of immunization (Keren 1983), antigen nature (Collins 1972) or doses. The purpose of this work was to determine the effect of oral antigen stimulation length on the immune response of gnotobiotic mice. Living or heat-killed *Escherichia coli* were used in these studies. Immunity state was evaluated by intestinal IgA plasmocyte number and serum antibody synthesis. Axenic mice were chosen for this work. In such animals, the immune system is poorly developed. Intestinal microflora induce the full development of the intestinal IgA immune system (Crabbe 1968, Moreau 1978) after 3 or 4 weeks of stimulation.

Live *E.coli* in the gut of monoxenic mice became established at 5.10^8 bacteria/g of feces. In experiments carried out with dead *E.coli*, bacteria were given in drinking water at the dose of 10^9 cells/ml. In order to stop the stimulation by live *E.coli*, the gut was decontaminated with neomycin (2 mg/ml of drinking water) at different periods of time. In all experiments, mice were killed 4 weeks after the first day of peroral stimulation.

Results show that the numbers of intestinal IgA plasmocytes and serum anti *E.coli* levels were similar in mice harboring either live or dead *E.coli* (Tables 1 and 2).

MICE	IGA PLASMOCYTES PER VILLUS
AXENIC + LIVING E.COLI	21.3+1.5 } NS
AXENIC + DEAD E.COLI	24.5+0.8
CONTROLS: AXENIC	5.7+1.4
HOLOXENIC	41.0+1.1

TABLE 1: INTESTINAL PLASMOCYTES RESPONSE IN GNOTOBIOTIC MICE HARBORING LIVING OR DEAD E.COLI IN THE DIGESTIVE TRACT FOR 4 WEEKS

MICE	ANTI E.COLI ANTIBODIES
AXENIC + LIVING E.COLI	2.6+0.1 } NS
AXENIC + DEAD E.COLI	2.5+0.1

TABLE 2: SERIC ANTIBODIES RESPONSE IN GNOTOBIOTIC MICE HARBORING LIVING OR DEAD E.COLI IN THE DIGESTIVE TRACT FOR 4 WEEKS

When E.coli was present less than 4 days, no IgA plasmocyte response was observed. A maximum of stimulation was recorded after 14 days (Fig. 1).

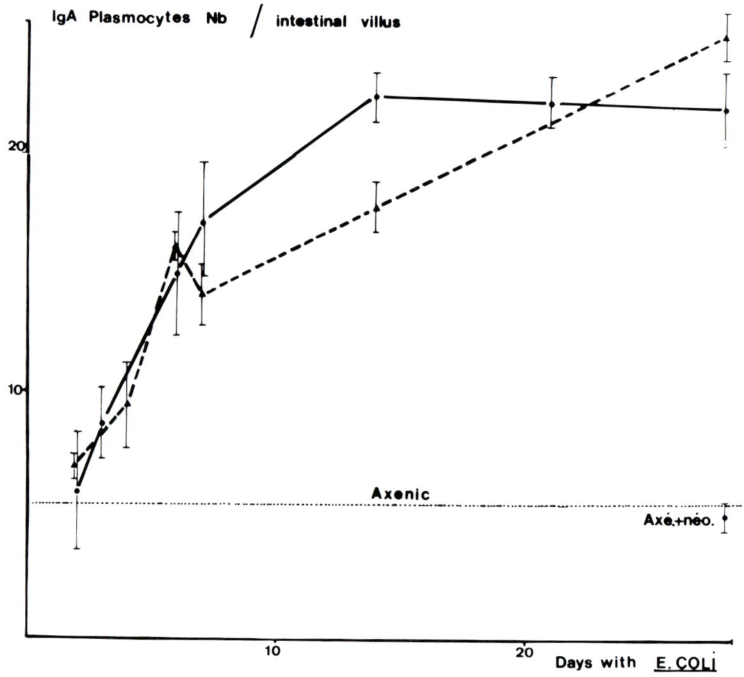

Fig. 1 - Relationship between the number of IgA plasmocytes and the presence of live (●——●) or dead (▲- - -▲) E.coli in the digestive tract.

In another experiment with dead E.coli, the effect of 8 consecutive days of stimulation was compared to short stimulations of 2 days each once a week for 4 weeks. The amounts of IgA plasmocytes were similar in the two groups (8 days or 4x2 days) whereas no stimulation occurred in the control group (only one 2-day stimulation) (Table 3).

	ORAL ANTIGENIC STIMULATION DURATION (DAYS)			
	2	8	4X2	28
IGA PLASMOCYTES	9.4 +0.8	13.3 +1.0	13.5 +1.1	24.5 +0.8
	P<0.02	NS	P<0.01	
ANTI E.COLI ANTIBODIES	1.6 +0.1	1.9 +0.1	2.2 +0.1	2.5 +0.1
	NS	P<0.05	NS	
		P<0.001		

TABLE 3 : IMMUNE RESPONSES AFTER SINGLE OR REPEATED ANTIGENIC STIMULATIONS IN THE DIGESTIVE TRACT OF AXENIC MICE

Serum antibody levels against E.coli were measured by the E.L.I.S.A. technique. Levels of E.coli antibodies were low in groups receiving dead E.coli for 2 or 8 consecutive days compared with the control group receiving the strain for 28 consecutive days (Table 3). The level was larger in the group receiving E.coli 4x2 days than in the group receiving it 8 consecutive days ($P < 0.05$).

In conclusion, live or dead E.coli in mouse digestive tract induce a similar effect on intestinal IgA plasmocyte stimulation and similar levels of serum anti E.coli antibodies. The same IgA plasmocyte response can be obtained with only one period of stimulation or with very short periods totalling the same length of time. This result suggests an additive stimulatory effect located in the gut. Concerning serum antibodies, several short stimulation periods (4x2 days) induced a response larger than that obtained after a single stimulation of the same duration. This result suggests systemic memory stimulation effect after several oral antigenic stimulations.

Collins FM (1970). Immunity to enteric infection in mice. Infect Immun 1:243.

Collins FM, Carter PB (1972). Comparative immunogenicity of heat-killed and living oral Salmonella vaccines. Infect Immun 6:451.

Crabbe PA, Bazin H, Eyssen H, Heremans JF (1968). The normal microbial flora as a major stimulus for proliferation of plasma cells synthesizing IgA in the gut. Int Arch Allergy Apply Immunol 34:362.

Keren DF, Scott PJ, Mc Donald RA, Wiatrak M (1983). Effect of parenteral immunization on the local immunoglobulin A response in the intestine to Shigella flexneri antigens. Infect Immun 42:202.

Moreau MC, Ducluzeau R, Guy-Grand D, Muller MC (1978). Increase in the population of duodenal immunoglobulin A plasmocytes in axenic mice associated with different living or dead bacterial strains of intestinal origin. Infect Immun 21:532.

Pierce NF (1978). The role of antigen form and function in the primary and secondary intestinal immune responses to cholera toxin and toxoid in rats. J Exp Med 148:195.

Porter P, Linggood MA (1983). Development of oral vaccines for preventing diarrhea caused by enteropathogenic Escherichia coli. J Infect 6:111.

Germfree Research: Microflora Control and Its Application to the Biomedical Sciences, pages 395–398

EFFECTS OF MONOASSOCIATION WITH CARIOGENIC STREPTOCOCCI ON LYMPHOID TISSUE IN THE INTESTINE AND SPLEEN OF HARVARD RATS

Barbara A. Peri, Ph.D.

The University of Chicago
Chicago, IL 60637

In contrast to conventional rodents, germfree mice have smaller lymph nodes, Peyer's patches (P.p.) and spleens, germinal centers and plasma cells are infrequent, and lamina propria is thin with few mononuclear cells (Pollard, Sharon, 1973). Following monoassociation of germfree mice with salmonella, germinal centers and plasma cells appear coincident with circulating antibody, but no response is detected when Streptococcus faecalis ND547 is administered (Pollard, Sharon 1970). However, rats monoassociated with S. faecalis frequently produce high titers of agglutinating antibodies in serum and moderate salivary antibody levels (Peri, Wagner 1977). Lymphoid development as related to antibody response to cariogenic streptococci in the monoassociated rat is presented here.

Harvard caries resistant (HCR) or caries susceptible (HCS) rats were monoassociated at 21 days of age with either cariogenic Streptococcus faecalis ND 547 (rat oral flora) or Streptococcus mutans, 6715 (human oral flora). Bacterial colonization 60 days later was measured in various parts of the intestinal tract as colony-forming units/gram of organ and cross sections of spleen and ileal P.p. were compared for lymphoid maturity. In spleens, periarteriolar sheaths and marginal zones were scored while number, size and thymus dependent areas were compared in P.p. Follicles, germinal centers, and plasma cells were scored in both organs. The most highly developed structure was used as a reference and scored as 3+.

In both rat strains S. faecalis was more numerous than

S. mutans in the ileum (for HCR, $\log_{10}$ 9.1 vs<3) and duodenum (HCR $\log_{10}$ 5.9 v<3). Bacterial numbers in caecum and colon were approximately equal ($\log_{10}$ 9-10).

Germfree rats showed marked changes in lymphoid tissue when monoassociated with either organism or when colonized by normal flora (conventionalized). Peyer's patches in the HCR rat monoassociated with either organism were numerous, large, and contained conspicuous germinal centers. The intestinal wall was thick and adjacent villi contained plasma cells (Fig. 1a). Similar maturation was seen in conventionalized HCR rats. Considerably less response to monoassociation was seen in HCS animals, particularly in germinal centers and plasma cells (Fig. 2). Lymphoid development did not always correlate with serum or salivary agglutinin titers. However, in S. faecalis monoassociated HCR rats, those with serum antibody had greater lymphoid development than similar animals with no antibody response (Fig. 2). As expected, in germfree animals P.p. were few, without germinal centers, and the intestinal wall was thin. Fig. 1b shows the most highly developed area seen in such sections.

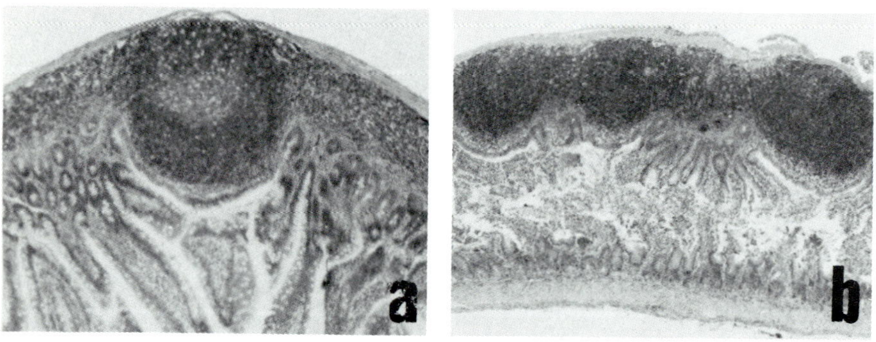

Fig. 1 Maturation of ileal Peyer's patches with a microbial stimulation. (a) HCR S. faecalis - associated rat with large numerous Peyer's patches, germinal centers well developed, thick intestinal wall, and plasma cells in adjacent villi. (b) HCR germfree rat with small Peyer's patches no germinal centers or plasma cells in adjacent villi and thin intestinal wall. (a and b x 32).

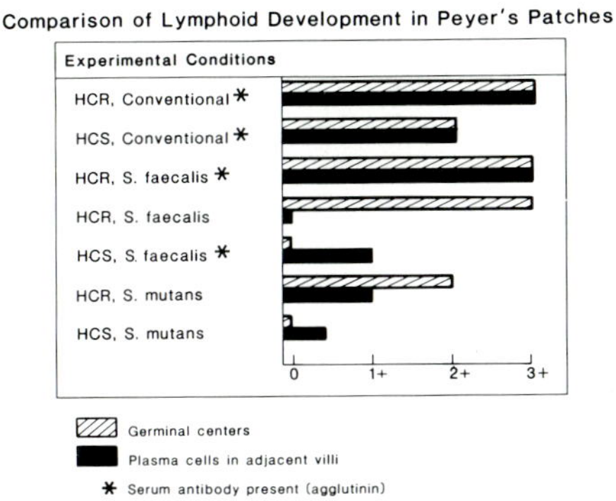

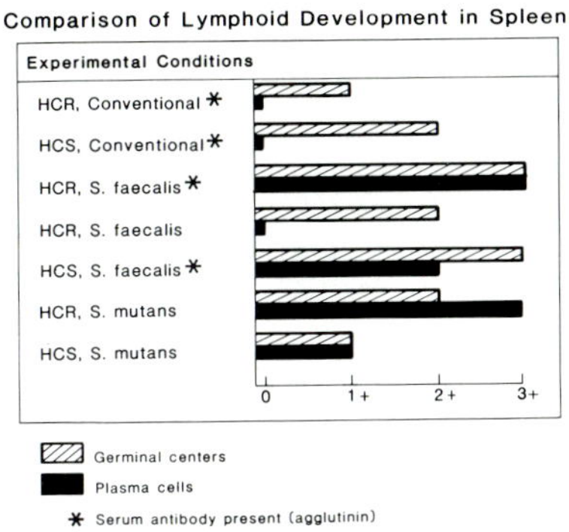

Fig. 2 Lymphoid development in Peyer's patches and spleen.

When monoassociated with S. mutans, a similar pattern was observed in P.p. but development was less pronounced (Fig. 2). The HCS animals resembled the germfree rat in the lack of germinal centers and the thymus dependent area appeared to be even less developed than under germfree conditions. Agglutinating antibody was not detected in serum of any rat exposed to S. mutans and salivary antibody titers were low.

As in P.p., the spleens from the same animals exposed to various microbial stimulation showed marked changes from the germfree state. Monoassociated HCR rats developed at least as many germinal centers as their conventional counterparts and plasma cells were more numerous (Fig 2). Development of the marginal zone in both strains was minimal with S. mutans but extensive with S. faecalis. Again responses of HCR rats exceeded those of the HCS animals.

Antibody response to S. mutans appears to be highly dose dependent (Michalek, McGhee, Babb, 1978) and the considerable difference between S. faecalis and S. mutans populations in the P.p. areas may explain the differing lymphoid responses to the two organisms. However, antigenic stimulation clearly does occur in HCR rats, and the absence of serum antibody to S. mutans could be due to suppression. Antigens in the intestinal tract are often suppressive (Silverman, Peri, Rothberg 1982; Kagnoff 1982), sometimes acting selectively (Michalek, McGhee, Mestecky, Arnold, Bozzo, 1976), and T suppressor cells are induced by S. mutans (Lamb, Kontianen, Lehner 1979). The lack of maturation in HCS rats seems to be a strain difference

(Peri, Wagner 1977) in immune responses to streptococci since bacterial populations were equal to those in HCR rats. Although agglutinating antibody levels did not correlate with caries resistance or lymphoid maturation, the consistent differences between the HCR and HCS rat strain suggest that the immune system may be involved in resistance to dental caries.

Kagnoff MF (1982). Immunological unresponsiveness after enteric antigen administration. In Strober W, Hanson LA, Sell KW (eds): "Recent Advances in Mucosal Immunity," Raven Press: p 99.

Lamb JR, Kontianen S, Lehner T (1979). Generation of specific T-cell suppressor function induced by *Streptococcus mutans* in monkeys and mice. Infect Immun 26:903.

Michalek SM, McGhee JR, Babb JL (1978). Effective immunity to dental caries: dose dependent studies of secretory immunity by oral administration of *Streptococcus mutans* to rats. Infect Immun 19:217.

Michalek SM, McGhee JR, Mestecky J, Arnold RR, Bozzo L (1976). Ingestion of *Streptococcus mutans* induces secretory IgA and caries immunity. Science 192:1238.

Peri BA, Wagner M (1977). Immune response and dental caries incidence in gnotobiotic Harvard caries - resistant and caries susceptible rats monoassociated with *Streptococcus faecalis*. Infect Immun 16:805.

Pollard M, Sharon N (1970). Responses of the Peyer's patches in germfree mice to antigenic stimulation. Infect Immun 2:96.

Silverman GA, Peri BA, Rothberg RM (1982). Systemic antibody responses of different species following ingestion of soluble protein antigens. Devel Comp Immunol 6:737.

Germfree Research: Microflora Control and Its Application to the Biomedical Sciences, pages 399–401

IMMUNE RESPONSES TO MONO-ASSOCIATED BIFIDOBACTERIUM LONGUM IN MICE

Yamazaki,S.,Machii,K.,Ueda,K.,
*Momose,H.,*Tsuyuki,S.,&*Kawashima,T.

The Institute of Public Health. Tokyo,Japan
*Morinaga Milk Ind. Co., B.M.R.L. Tokyo,Japan

Bacterial translocation of Bifidobacterium longum into internal organs in mono-associated mice is interesting because the translocation ceases 4 weeks after administration of the organisms. Thus, investigation was made on immune responses of B.longum-mono-associated mice to the organisms in order to explore whether immunological events are involved in the phenomenon: prevention of translocation.

Both sexes of GF nude (nu/nu) and heterozygous (nu/+) mice of BALB/c background at 8-12 wks of age were administered intragastrically B.longum grown in GAM broth. Recovery of live B.longum from the liver, kidney and mesenteric lymph node, anti-B.longum antibody measured by ELISA using a cell wall fraction of B.longum as antigen as well as total IgA in tissue fluids, and cell-mediated immunity, were examined after the mono-association.

DETECTION OF VIABLE B.LONGUM FROM ORGANS

A low number of B.longum were recovered from the internal organs of most nu/+ mice killed at wks 1 and 2 after p.o. administration with no recovery at wk 4 and later. The organisms consistently recovered in the organs of nu/nu mice. Both mice harboured more than 10^9 per gram cecal content.

ANTIBODY AND IgA RESPONSE

Serum Antibody: In nu/+ mice, a low level of anti-B.

longum antibody (using antiIgG) was detected at 1 week after administration. No IgA antibody was detected.

Anti-B.longum IgA Antibody: IgA antibody was detected in bile at wk 1 and a moderate titer was maintained throughout. IgA antibody was also detected at a lower titer in cecal content and ileal wall extract at wk 2 and wk 6, respectively. (Fig. 1)

Anti B.longum IgA antibody after monoassociation with B..longum

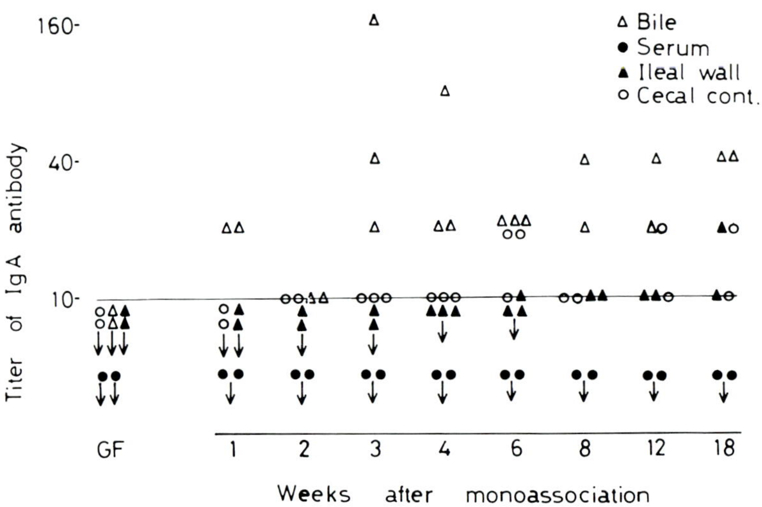

IgA Level: IgA was detected in every sample of pretreated GF mice. After B.longum-association a gradually increased IgA level was obtained in all samples, among them the bile and ileal wall extract was rather conspicuous. At the end of experiment, wk 6, increased titers; 5-fold in the serum, 10-fold in the bile and ileal wall extract and 25-fold in the cecal content compared to the titers of pre-association was seen.

CELL MEDIATED IMMUNITY

Delayed footpad reaction was seen at wk 4 after association but not at wk 2 in nu/+ mice. Nu/nu mice did not show delayed swelling at wks 2 and 4, while a low level of immediate swelling was seen at wk 4. Migration inhibition of peritoneal exudate cells in the presence of B.longum antigen was positive nu/+ mice at wk 4. No migration inhibition was detected at wk 2 in nu/+ mice and wks 2 and 4 in nu/nu mice. (Table 1)

Table 1
Cell mediated immune responses in B.longum-monoassociated nude mice

Mice	Weeks after administr.	Footpad swelling* (0.1 mm ± S.D.) 3 hr	24 hr	Macrophage migration inhibition test** (% migration)
nu/+	2	1.3 ± 0.6	0.8 ± 0.4	98.8
	4	3.8 ± 1.2	4.6 ± 0.8	59.1
nu/nu	2	1.3 ± 0.5	0.6 ± 0.4	102.4
	4	3.4 ± 0.9	0.8 ± 0.7	106.0

* B.longum protein fraction, 10 μg/0.04 ml/mouse.
** B.longum protein fraction, 25 μg/ml

Recovery of B.longum from organs and delayed footpad reaction to B.longum was examined in nu/nu mice iv transferred 4 wks previously with 5×10^6 spleen T cells. 4 wks after association of B.longum, reconstituted mice were devoid of viable B.longum (Table 2) in liver and mesenteric nodes.

Table 2 Recovery of B.longum from organs and footpad reaction in lymphocyte transferred B.longum-monoassociated nude mice

Lymphocyte transfer	Recovery of B.longum from*			Footpad swelling***
	Liv.	M.LN	Cecal cont.	24 hr
+	0/4** (2.0)	0/4 (2.0)	4/4 (9.6 ± 0.2)	3.2 ± 1.5****
-	4/4 (3.0 ±0.3)	4/4 (3.5 ± 0.6)	4/4 (9.7 ± 0.1)	0.4 ± 0.7

* 4 wks after B.longum-monoassociation
** positive/examined, (viable counts/organ or g. of cecal cont.)
*** Bifidus-protein fraction, 10 μg/0.04 ml/mouse
**** 0.1 mm ± S.D.

CONCLUSION

The time when translocation ceases (4 weeks) seems correlated well with the appearance of CMI to the monoassociated bacteria rather than with the IgA response.

Germfree Research: Microflora Control and Its Application to the Biomedical Sciences, pages 403–406

RESISTANCE TO INVASION BY *ESCHERICHIA COLI*

Gerd Pluschke[+] and Mark Achtman

Max-Planck-Institut für Molekulare Genetik
D-1000 Berlin 33, Germany [+]present address:
Basel Institute for Immunology, Basel, Switzerland

Escherichia coli is a common cause of neonatal septicemia and meningitis. More than 80% of the *E.coli* strains isolated from cases of meningitis possess the K1 (sialic acid) capsular antigen (Siegel and McCracken, 1981). K1 *E.coli* bacteria are also frequently isolated from feces of healthy babies and adults. Most of the isolates belong to a limited spectrum of O-lipopolysaccharide (LPS) antigen types. We have observed distinct differences in virulence between K1 strains of the common LPS types O1, O7 and O18 (Pluschke *et al.*, 1983). Ninety-five strains, obtained from various sources including both diseased and healthy individuals, were tested in a rat infection model. After oral inoculation, strains of all three serotypes were able to multiply extensively in the gut of newborn rats and to translocate to the mesenteric lymph nodes. With only a few exceptions, O7:K1 and O18:K1 strains were able to cause bacteremia, while O1:K1 strains could not. Mixed infection experiments suggested that the virulent O7:K1 and O18:K1 strains were able to multiply directly in the bloodstream of the newborn rats, while the O1:K1 bacteria were incapable of such multiplication. No correlation between virulence and the source of the different strains was observed. An analysis of published epidemiological data revealed that in human populations as in the rat model, O7:K1 and O18:K1 bacteria were common and O1:K1 bacteria were rare among isolates from cases of meningitis. All three serotypes were common among fecal isolates.

Under stringent assay conditions, serum resistance or sensitivity is correlated with virulence or lack of virulence of the various isolates (Pluschke and Achtman, 1984). K1 bacteria of all three O serotypes were able to multiply for several hours in 90% serum deficient in classical pathway activity (MgEGTA-chelated sera or genetically C2- or C4- deficient sera). In contrast, K1-negative mutant derivatives were killed after a delay. Sialic acid seems to specifically impede alternative pathway activation by changing the affinity between certain complement components (B or H to C3b) (Pangburn, 1983). Therefore, cells that carry the K1 (sialic acid) capsule do not activate the alternative complement pathway. This may well be the major reason for the clear predominance of K1 strains in meningitis in newborns.

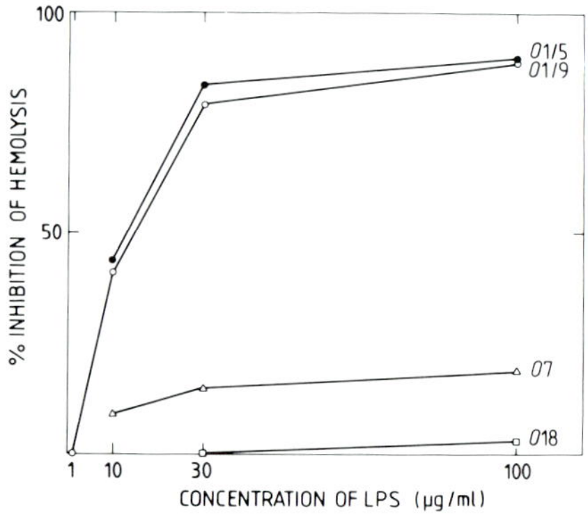

Fig. 1. Fixation of complement by isolated LPS. A 90 μl portion of GF-RS was mixed with 10 μl of isolated LPS. The figure shows the percentage of hemolytic activity consumed upon incubation of the mixture for 1 h at 37°C relative to control samples incubated without LPS. Both O1/5 LPS and O1/9 LPS, which are structurally somewhat different (Kusecek _et al._, 1984), activate complement.

Virulent O7:K1 and O18:K1 strains were able to multiply for several hours in 90% serum obtained from non-immunized rats or guinea pigs, while O1:K1 bacteria were efficiently killed after a delay. There was no difference in the killing capacity of serum obtained from specific-pathogen-free or from germ-free rats (GF-RS). The ability of different LPS types to activate complement directly, was tested by incubating isolated LPS in 90% GF-RS and measuring the residual hemolytic complement activity. Isolated O1 LPS had a far higher anticomplementary activity than O7 or O18 LPS (Fig. 1). GF-RS lacks significant amounts of antibodies specific for O1:K1 bacteria (Pluschke and Achtman, 1984). Thus, antibody-independent activation of the classical complement pathway by the O1 antigen seems to be responsible for serum sensitivity of O1:K1 cells, and differing capacities to activate the classical pathway seem to account for O antigen related differences in virulence of K1 strains.

O7:K1 and O18:K1 cells were not intrinsically resistant to the bactericidal activity of the terminal complex of complement but were efficiently killed by 90% adult rat serum in the presence of O-antigen specific antibodies. In human newborns the gastrointestinal tract seems to be the source of E.coli that invades the bloodstream with subsequent involvement of the central nervous system. Many babies carry K1 bacteria, but only a small percentage of babies colonized with potentially virulent strains develop bacteremia and meningitis (Siegel and McCracken, 1981). Antibodies against the common O antigen types of E.coli are commonly found in the serum of healthy adults. Therefore, we tested the protective capacity of anti-LPS antibodies in the newborn rat infection model (Table 1).

Table 1. Protective effect of monoclonal anti-O18 IgM. 5-day-old rats were challenged orally with 4 x 10^6 O18:K1 or O7:K1 cells and the number of bacteremic animals was determined three days later. 100 µl of saline or saline containing 25 µg anti-O18 IgM were injected i.p. before feeding the bacterial suspensions.

Challenge strain	Injected agent	Percentage of rats colonized	Percentage of rats bacteremic
O18:K1	saline	100	51 (38/74)
O18:K1	IgM	100	2 (1/45)
O7:K1	saline	100	43 (22/51)
O7:K1	IgM	100	40 (16/40)

The control experiments (injection of saline) yielded results similar to those described previously (Pluschke et al., 1983); that is about half of the animals challenged orally developed bacteremia. In contrast, only one of 45 rats that had received antibodies specific for the O18 antigen prior to oral challenge with O18:K1 bacteria yielded a positive blood culture. The protection was specific, since no significant effect was seen when the animals injected with anti-O18 IgM were challenged with O7:K1 bacteria. O18:K1 cells were not killed by serum obtained from 5-day old rats that had been injected with anti-O18 IgM. Thus opsonisation rather than antibody-mediated direct bactericidal activity of complement seems to be responsible for the protective effect of anti LPS antibodies in newborns.

The above results indicate that the chemical composition of both the capsular polysaccharide and the O antigen determine the degree of complement activation and hence the virulence of E.coli strains in meningitis in newborns. Antibody-independent complement activation mechanisms seem to provide the pre-immune host with efficient defence against invasion by most types of E.coli and - in addition - most newborns colonized with potentially invasive K1 bacteria are probably protected by maternal antibodies from sepsis and meningitis. Adult rats are highly resistant to K1 E.coli and this bacterium is only rarely associated with menigitis in older children or adults. The fully developed phagocytic systems of the noncompromised adult host thus seem to be able to clear efficiently any E.coli cell, irrespective of the serotype, from the circulation.

Kuscek B, Wloch H, Mercer A, Vaisänen V, Pluschke G, Korhonen T, Achtman M (1984). Lipopolysaccharide, capsule and fimbriae as virulence factors among O1, O7, O18 or O75 and K1, K5 or K100 Escherichia coli. Infect Immun 43:368.

Pangburn MK (1983). Activation of complement via the alternative pathway. Federation Proc 42:139.

Pluschke G, Mercer A, Kusecek B, Pohl A, Achtman M (1983). Induction of bacteremia in newborn rats by Escherichia coli K1 is correlated with only certain O (lipopolysaccharide) antigen types. Infect Immun 39:599.

Pluschke G, Achtman M (1984). Degree of antibody-independent activation of the classical complement pathway by K1 Escherichia coli differs with O antigen type and correlates with virulence of meningitis in newborns. Infect Immun 43:684.

Siegel JD, McCracken GH (1981). Sepsis neonatorum. N Engl J Med 304:642.

Germfree Research: Microflora Control and Its Application to the Biomedical Sciences, pages 407–410

SUSCEPTIBILITY OF CONGENITALLY IMMUNODEFICIENT MICE TO CRYPTOCOCCUS NEOFORMANS

Cindy A. Salkowski, B.S., Kenneth F. Bartizal, Jr., Ph.D., and Edward Balish, Ph.D.
Departments of Surgery and Medical Microbiology
University of Wisconsin Medical School
Madison, Wisconsin 53706

Cryptococcus neoformans is an encapsulated yeast that is capable of causing pulmonary, systemic, or meningeal disease in man. The infectious agent is an environmental fungus that is believed to enter the lung by inhalation. While the primary infection may be pulmonary, this mycosis commonly disseminates to the central nervous system. The true incidence of cryptococcosis is not known; however, clinical evidence indicates it may be one of the major mycoses in the world (Kaufman, Blumer 1977).

While the primary portal of entry is believed to be the respiratory tract, evidence indicates that oral ingestion of *C. neoformans* can also result in disseminated cryptococcosis. Disseminated cryptococcosis was reported in monkeys fed *C. neoformans* indicating that the alimentary tract may be a portal of entry for this organism (Takos 1956). When conventional (CV) mice were intragastrically challenged with cryptococci, disseminated cryptococcosis with mortality was observed (Green, Bulmer 1979). In another study, large numbers of cryptococci were isolated from the gastrointestinal (GI) tract of CV mice following exposure to soil aerosols contaminated with *C. neoformans* (Neilson, Bulmer 1975). Therefore, the GI tract may serve as a reservoir as well as a site for dissemination of *C. neoformans*. To date, no one has utilized germfree (GF) animals to study the pathogenesis of *C. neoformans*.

In this study, we orally challenged GF mice with *C. neoformans* to assess whether this pathogen can colonize the alimentary tract and cause systemic cryptococcosis. Also

GF, flora-defined (FD), and CV mice were intravenously challenged with C. neoformans to determine the effect of a bacterial flora on resistance to systemic cryptococcosis.

Athymic nude (nu/nu) and heterozygous (+/nu) GF and FD BALB/c mice as well as GF and CV beige (bg/bg) C57Bl mice and their heterozygous (+/bg) littermates were used in this study. All mice were housed in flexible-film isolators following established gnotobiotic procedures and ranged between 6-10 weeks of age.

Cryptococcus neoformans (serotype A) was grown in Sabouraud's dextrose broth supplemented with 10% glucose. Prior to animal challenge, the yeast cells were washed and resuspended in phosphate buffered saline (PBS). Germfree BALB/c and C57Bl mice were orally associated by swabbing their mouth with a suspension containing 1 X 10^7 C. neoformans/ml. Both BALB/c and C57Bl mice were intravenously inoculated in the tail vein with C. neoformans. Cryptococci were counted on a hemocytometer to standardize the i.v. inoculum to either 1 X 10^5 or 6.4 X 10^3 in 0.2 ml PBS. To verify the number of viable cells, the inoculum was diluted and plated on Sabouraud's dextrose agar (SDA).

Mice were killed by cervical dislocation at various intervals following challenge with C. neoformans. The kidneys, liver, lungs, spleen, brain, stomach, and ceca were aseptically removed and individual organs from three mice were pooled, homogenized in 5 ml of PBS, and plated in duplicate on SDA. Colonies were counted after incubation for 48 hrs at 30°C. Data are expressed as the number of colony forming units (cfu's) per gram dry weight of tissue homogenate.

The alimentary tracts of GF nu/nu and +/nu BALB/c mice were colonized within 24 hrs after oral challenge with C. neoformans. For 2 weeks after oral challenge, C. neoformans was recovered from the stomach and ceca of nu/nu and +/nu mice in low numbers ($<10^4$/gm dry wt). By day 19 after oral challenge the number of cryptococci in the stomach and ceca of nu/nu mice had dramatically increased to 10^6-10^8/gm dry wt, which coincided with recovery of C. neoformans from the internal organs. Nude (nu/nu) mice succumbed to disseminated cryptococcosis, with meningitis, 19-28 days after oral challenge. Despite isolation of C. neoformans from the lungs and brain of +/nu mice as early as the first 3 days after oral challenge and the persistent presence of low levels of cryptococci (<10^3/gm dry wt tissue) in the kidney, liver, and spleen throughout the 28 day study, no mortality was observed in the C. neoformans-colonized +/nu mice during a 12 week observation period. Surprisingly, GF +/bg mice were more susceptible to oral challenge with C. neoformans than immunodeficient bg/bg mice. Thirty days after oral challenge with C. neoformans, all +/bg mice had died, whereas mortality among bg/bg mice was only 33%.

Nude (nu/nu) mice, regardless of microbial status, were more susceptible to i.v. challenge with C. neoformans than their +/nu littermates. Germfree +/nu mice were more resistant to i.v. challenge than their FD counterparts (Table 1). However, all GF and FD BALB/c mice were incapable of controlling the infection and eventually succumbed to disseminated cryptococcosis with meningitis.

Table 1. Survival of Gnotobiotic BALB/c Mice Following Intravenous Challenge with 1 X 10^5 C. neoformans

	NU/NU	+/NU
Germfree	9-15 days*	38-41 days
Flora-Defined	12 days	15 days

*Survival time after challenge.

In contrast to oral challenge with C. neoformans, bg/bg mice that were intravenously challenged with cryptococci died earlier than their +/bg littermates (Table 2). Both bg/bg and +/bg mice developed disseminated cryptococcosis with meningeal involvement within approximately the same time period.

Table 2. Survival of C57Bl Mice Following Intravenous Challenge with 1 X 10^5 C. neoformans

	BG/BG	+/BG
Germfree	15-19 days	20+ days*
Conventional	16-19 days	19+ days*

*Moribund animals sacrificed Day 21.

The results presented in this paper demonstrate that *C. neoformans* can colonize the alimentary tract of GF mice. In nu/nu mice, heavy gastrointestinal colonization coincided with dissemination, and death occurred 19-28 days after oral challenge. The susceptibility of nu/nu mice and the relative resistance of +/nu mice to oral challenge is strong evidence that functional T-cells are required for the hosts' resistance to *C. neoformans*. Our results with bg/bg mice indicates that decreased leukocyte chemotaxis, lysosomal enzyme dysfunction, defective degranulation, and low natural killer cell activity (Brandt, Swank, Novak 1981) do not enhance their susceptibility to cryptococcosis.

Our gnotobiotic mice data are consistent with CV animal studies that demonstrated low levels of *C. neoformans* in the fecal pellets of CV mice for several weeks following oral inoculation. Disseminated cryptococcosis occurred in CV mice with the highest levels of viable *C. neoformans* in their feces (Green, Bulmer 1979). Thus, the GI tract may be a portal of entry for *C. neoformans* in both GF and CV mice. In CV rodents, the GI tract could also be a reservoir for dissemination of *C. neoformans* into the environment.

Brandt EJ, Swank RT, Novak EK (1981). The murine Chediak-Higashi mutation and other murine pigmentation mutations. In Gershwin ME, Merchant B (eds): "Immunologic Defects in Laboratory Animals," Vol 1. New York: Plenum Press, p 99.

Green JR, Bulmer GS (1979). Gastrointestinal inoculation of *Cryptococcus neoformans* in mice. Sabouraudia 17:233.

Kaufman L, Blumer S (1977). Cryptococcosis: The awakening giant. In "Pan American Health Organization, Proceedings of the IV Int'l. Conf. on the Mycoses." PAHO Scientific Publication No. 356, p 176.

Neilson JB, Bulmer GS (1975). The infectious particles of *C. neoformans*. In "Abstracts of the American Society for Microbiology," No F45, p 93.

Takos MJ (1956). Experimental cryptococcosis produced by ingestion of virulent organisms. NEJM 254:598.

Germfree Research: Microflora Control and Its Application to the Biomedical Sciences, pages 411–414

RESPONSE OF GNOTOBIOTIC PIGS TO IMMUNIZATION AND CHALLENGE WITH HEMOPHILUS PLEUROPNEUMONIAE

O.P. Miniats, D.V.M., M.Sc. and
S. Rosendal, D.V.M., Ph.D.
Ontario Veterinary College
University of Guelph
Guelph, Ontario, Canada N1G 2W1

INTRODUCTION

Colostrum-deprived gnotobiotic pigs are used for the establishment and repopulation of Minimal disease (MD) or Specific-pathogen-free (SPF) herds. After placement on farms at approximately 6 weeks of age these animals frequently succumb to infectious diseases because they possess no maternal antibodies and lack previous exposure to antigens of pathogenic micro-organisms. Attempts are therefore being made to find methods whereby primary SPF pigs can be immunized at an early age, while they are being raised in isolators, against some of the pathogens that they are likely to encounter in the new farm environment.

Hemophilus pleuropneumoniae, a disease of particular concern in recent years, has world wide distribution (Nicolet and Scholl, 1981; Sanford and Josephson, 1981). The etiological agent Hemophilus pleuropneumoniae is transmitted from pig to pig via the respiratory passages and causes explosive outbreaks of acute fibrinous pleuropneumonia with high morbidity and mortality or chronic debilitating infections. Intensive production methods and close confinement contribute to the problem. No adequate preventative measures have yet been found.

In this study the response of gnotobiotic pigs to infection with H. pleuropneumoniae and the effectiveness of its capsular antigen as an immunizing agent were investigated.

MATERIALS AND METHODS

Twenty Yorkshire pigs belonging to two litters were derived by open hysterotomy and reared in isolators (Miniats and Jol, 1978). At 3 and 5 weeks of age 10 piglets were immunized subcutaneously with 8 ml of a capsular extract of H. pleuropneumoniae Serotype 1, strain CM5 method (Kodama et al, 1981) mixed with aluminum hydroxide gel adjuvant (6 ml extract + 2 ml adjuvant). The immunized piglets and another 10 non-immunized controls were challenged at 7 weeks of age with the homologous strain suspended in saline to contain 5 x 10^6 C.F.U. per ml. One ml of the suspension was instilled into each nostril of the pigs. Observations were made on clinical response and pre- and post-vaccination complement fixing (C.F.) antibody levels (Lombin et al, 1982). The pigs were examined for pathological lesions and distribution of H. pleuropneumoniae in internal organs either after natural death or following euthanasia 7 days post-inoculation (p.i.).

RESULTS

Of the 10 non-immunized pigs, 6 died suddenly in less than 24 hours p.i., two showed clinical signs of the disease and died on day 3 and two survived until day 7 when they were euthanized and necropsied. Four of the vaccinated animals had clinical signs and died on days 2,3,5 and 6 p.i. and six survived until day 7. Seven of the non-immunized and 2 of the immunized pigs had a septicemia. All the non-immunized pigs had negative C.F. antibody titers while 8 of the vaccinated group had post vaccination titers ranging from 16 to 128. Three of the latter had positive titers 36, 128 and 128 respectively prior to immunization.

The clinical signs were fever, inappetance, lethargy and shallow breathing. Most of those with a septicemia had cyanosis of ventral parts of the body and a terminal epistaxis. At necropsy the main lesion was a severe necrotizing fibrinous pneumonia either bilateral involving the entire lung mass, or more frequently affecting mainly the cardiac lobe of the left lung. In some of the immunized pigs the lung lesions were of focal distribution. Most of the pigs had an increased volume of a sero-sanguinous pleural fluid, pleurisy, pericarditis and edematous swollen bronchial and mediastinal lymph nodes. The lesions in most cases were confined to the thoracic cavity. Some animals had swollen spleens and severely congested small or large intestines. In general the lesions were more severe in the non-immunized animals. Two of the immunized and one of the non-immunized pigs had no grossly visible lesions. A summary of the results is presented in Table 1.

Table 1

RESPONSE OF NON-IMMUNIZED AND IMMUNIZED PIGS TO CHALLENGE WITH HEMOPHILUS PLEUROPNEUMONIAE

Pig#	Clin. Signs	Surv. Days	Organ Lesions Resp.	Organ Lesions Others	Organs Isolated Resp.	Organs Isolated Others	Pre-Challenge C.F. Titer
			10 Non-Immunized Pigs				
1	F.D.*	1	+	-	+	+	0
2	F.D.	1	+	+	+	-	0
3	+	3	+	-	+	+	0
4	F.D.	1	+	+	+	+	0
5	F.D.	1	+	+	+	+	0
6	-	7K**	-	-	-	-	0
7	+	3	+	-	+	+	0
8	F.D.	1	+	+	+	+	0
9	-	7K	+	-	+	-	0
10	F.D.	1	+	+	+	+	0
			10 Immunized Pigs				
11	-	7K	+	+	+	-	64
12	+	5	+	-	+	+	0
13	+	3	+	-	+	+	16
14	-	7K	-	-	-	-	0
15	-	7K	+	-	+	-	16
16	-	7K	+	-	+	-	64
17	+	7K	+	-	+	-	128
18	+	2	+	-	+	-	128
19	-	6	+	-	+	-	128
20	-	7K	-	-	-	-	128

*FD - Found dead **K - Killed

DISCUSSION AND CONCLUSIONS

At the levels administered, infection with H. pleuropneumoniae in gnotobiotic pigs resulted in a syndrome and lesions similar, but more severe, than those seen in conventional pigs in cases of peracute outbreaks (Nicolet and Sholl, 1981; Sanford and Josephson, 1981). Immunization with the capsular antigen reduced the incidence of septicemia and mortalities and in some cases the severity of the lesions. Similar observations have been made in pigs vaccinated with commercial vaccines (Nicolet and Sholl, 1981). A capsular antigen was chosen for this trial, after immunization attempts with other vaccines because previous work (Nielsen 1976) indicated that young cultures of H. pleuropneumoniae which were more mucoid and probably had more capsular material than older cultures were more immunogenic than the latter.

Immunization with the capsular antigen stimulated C.F. antibody production but some apparently false positives were encountered in the pre-vaccination sera making the interpretation difficult. Thc level of C.F. antibody titers did not appear to be related to the severity of the disease. The presence of an innate immunity in some of the pigs against the organism suggests that in addition to search for more effective vaccines than those available today genetic selection for resistant pigs may be one approach for attempting to control Hemophilus pleuropneumonia in swine.

ACKNOWLEDGEMENTS

The authors thank the Ontario Department of Agriculture and Food and the Ontario Pork Producers Marketing Board for their financial support.

REFERENCES

Kodama H, Matsumoto M, Snow LM (1981). Immunogenicity of capsular antigens of Pasteurella multocida in turkeys. Am J Vet Res 42: 1838-1841.

Lombin LH, Rosendal S, Mitchell WR (1982). Evaluation of the complement fixation test for the diagnosis of pleuropneumonia in swine caused by Haemophilus pleuropneumoniae. Can J Comp Med 46: 109-114.

Miniats OP, Jol D (1978). Gnotobiotic pigs - derivation and rearing. Can J Comp Med 42: 428-437.

Nielsen R (1976). Pleuropneumonia of swine caused by Haemophilus parahaemolyticus. Studies on the protection obtained by vaccination. Nord Vet Med 28: 337-348.

Nicolet J, Sholl E (1981). Hemophilus infections. In Leman AD (ed): "Diseases of Swine" 5th Ed. The Iowa State University Press, p 368-377.

Sanford SE, Josephson GKA (1981). Porcine Haemophilus pleuropneumonia epizootic in southwestern Ontario: clinical, microbiological, pathological and some epidemiological findings. Can J Comp Med 45: 2-7.

Germfree Research: Microflora Control and Its Application to the Biomedical Sciences, pages 415–419

INTESTINAL MICROECOLOGY AND IMMUNE RESPONSE OF GERMFREE AND CONVENTIONAL MICE IN MIXED TRICHINELLOSIS

Zdzislow W. Przyjalkowski, D. Vet. Med.

Institute of Parasitology
Polish Academy of Sciences
Warszawa, 00-973, Poland

Germfree (GF) and conventional (CV) C3H inbred male mice, 6 months old, were infected simultaneously with: Trichinella spiralis and Trichinella pseudospiralis, 500 larvae each. The GF techniques for preparation of larval inoculum, infection and animal isolation in Trexler's plastic isolator system were described earlier (Przyjalkowski 1968). A virulent strain of Trichinella spiralis (Owen 1835) was supplied by the Trichinosis Unit of the Agricultural Research Center, Beltsville, MD., USA. The Trichinella pseudospiralis (Garkavi 1972), was obtained originally from the Soviet Union.

Autopsies were performed every 5 days from the 10th to the 30th day post infection (dpi). Intestinal trichinellae of both species were counted in situ by the trichinoscopic method. Fragments of the small intestine as well as the masseter and diaphragm muscles were fixed and sections were stained routinely with hematoxylin-eosin for histopathological studies.

In GF mice ten dpi intestinal trichinellae averaged 40.5 per mouse. In CV mice, 75 trichinellae per mouse were recovered. Trichinella pseudospiralis females outnumbered Trichinella spiralis females in the small intestine. Males of these species could not be differentiated. On the 15th dpi, no intestinal trichinellae were recovered from GF mice, while in CV animals the mean number of trichinellae were amounted to 9.5 per mouse. Trichinella pseudospiralis females again outnumbered Trichinella spiralis females. The last intestinal trichinellae were observed in CV mice 20 dpi, then they were expelled. Thus GF mice expelled their intestinal trichinellae in much shorter time than comparable CV animals.

The leukocyte count was higher in CV than in GF mice. The eosinophil count was similar in peripheral blood, whereas neutrophilia, characteristic for individual Trichinella pseudospiralis infection and high eosinophilia for the Trichinella spiralis infection were not observed in the mixed infection of both trichinellae.

In the histopathological examinations of CV mice (Figures 1-3), the first changes in the mucous membrane of the small intestine appeared 10 dpi as numerous monocyte infiltrations. Single trichinella larvae in the masseter and diaphragm were also observed with characteristic basophilic transformation (degeneration) of the muscle fibers. By the 15th dpi, inflammatory cellular reactions in the intestinal mucosa decreased and by the 25th dpi they were minimal, while the muscles contained abundant trichinella larvae and numerous foci of basophilic transformation. Cellular infiltrations around some muscular larvae (probably Trichinella spiralis) were slight, while around other larvae (probably Trichinella pseudospiralis) infiltrations were absent.

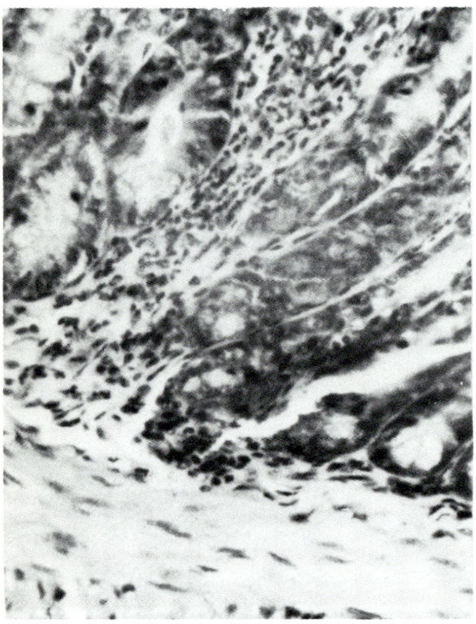

Fig. 1

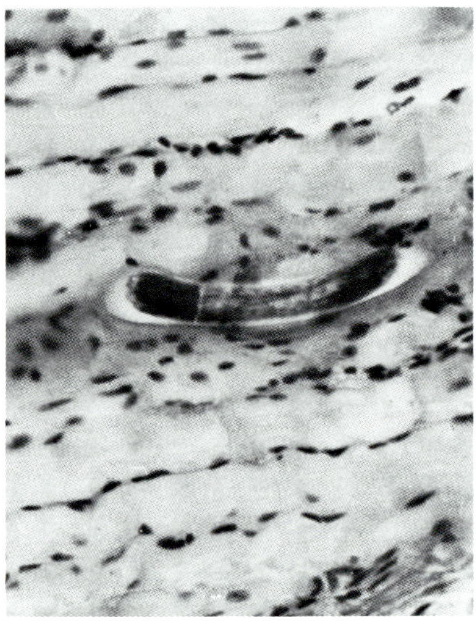

Fig. 2

Fig. 1. Profuse neutrophilic and monocytic infiltrations in the small intestine of conventional mouse 4 dpi. (HE x 160).
Fig. 2. Muscular larva (most probably of T. pseudospiralis) without any inflammatory cell reaction. CV mouse 10 dpi, x 160).

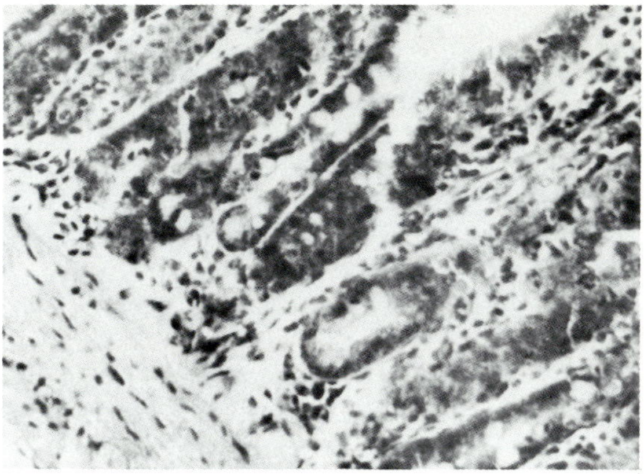

Fig. 3. Minimal inflammatory cell reactions in the intestinal mucous membrane. Conventional mouse 20 dpi, (HEx160).

In GF mice (Figures 4-6), inflammatory changes were weaker than in CV animals. In the intestines they were observed only on the 15th dpi. First foci of basophilic transformation of muscle fibers appeared 10 dpi, althouth in the muscles no larvae were observed 15 dpi. In general, inflammatory cellular reactions in the intestines and muscles were more pronounced in CV than in GF mice. These reactions appeared relatively early, but underwent a rapid reduction in the course of mixed trichinellosis. This may indicate an antagonism between *Trichinella spiralis* and *Trichinella pseudospiralis* which was expressed in a form of suppression of the host cellular response.

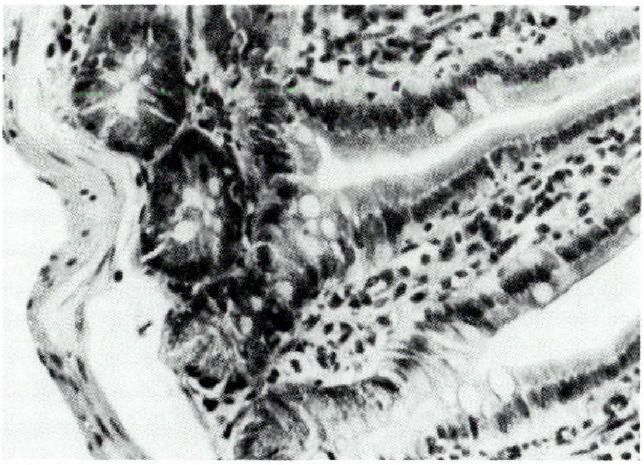

Fig. 4. Slight infiltrations of monocytes and neutrophils in the intestinal mucosa. GF mouse, 10 dpi, (HEx160).

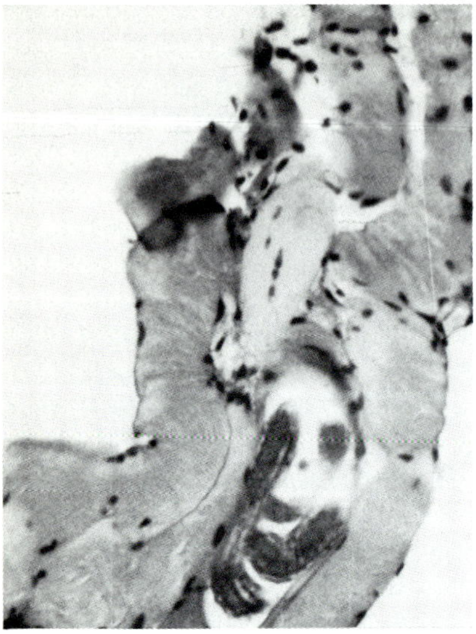

Fig. 5

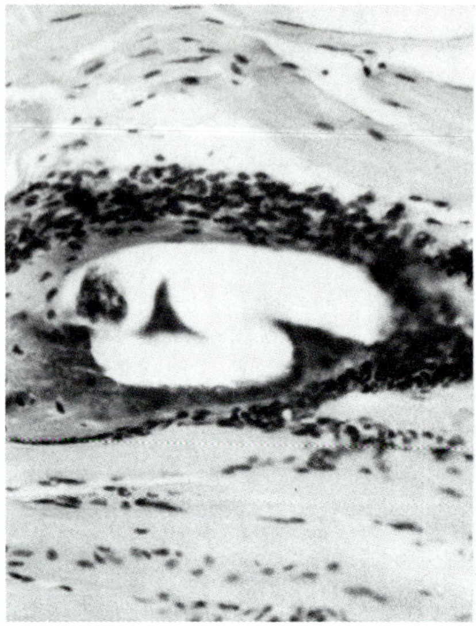

Fig. 6

Fig. 5. Parasite muscular larva without any inflammatory reaction around it, 15 dpi, GF mouse, (HEx160).
Fig. 6. Neutrophilic infiltration around muscular trichinella larva (probably T. spiralis). GF mouse, 15 dpi, (HEx160).

Macrophage migration tests carried out in the course of mixed infections with both trichinellae (Przyjalkowski, Schollengerger, Frymus 1981) showed a clear stimulation of macrophate migration instead of inhibition. According to Fox, MacSween and Rajeraman (unpublished), we dealt in that case, not with appearance of cell-mediated immunity but cell mediated immuno-suppression, caused most probably by Trichinella pseudospiralis.

By indirect hemagglutination tests in GF and CV mice, antibody production was found only in mice with single Trichinella spiralis or Trichinella pseudospiralis infections. Mice infected with mixed Trichinella spiralis and Trichinella pseudospiralis larvae did not produce any antibody. Thus this reaction was also suppressed in the host (Przyjalkowski, Cabaj 1981), suggesting a complete blockade of humoral immunity.

Therefore, Trichinella pseudospiralis appears to be a mysterious parasite to parasitologists, epidemiologists and clinicians. Further investigation in this direction is necessary, since experimental studies in monkeys (Teppema,

Blomjouse, Elgersma and Ruitenberg 1981) suggest that *Trichinella pseudospiralis* should be considered as a possible human pathogen, both in single infections and in mixed infections with *Trichinella spiralis*.

Fox RJ, MacSween M, Rajeraman K (personal communcation). Macrophage migration stimulation factor.

Przyjalkowski ZW (1968). Effect of intestinal flora and of a monoculture of *E. coli* on the development of intestinal and muscular *Trichinella spiralis* in gnotobiotic mice. Bull. Acad. Polon. ci., Ser. Sci. Biol. 29:51.

Przyjalkowski ZW, Schollenberger A, Frymus T (1981). Macrophage migration inhibition test in germfree and conventional mice infected with *T. spiralis* and *T. pseudospiralis* in single of mixed doses of both parasites. Ibidem, 29:72.

Przyjalkowski ZW, Cabaj W (1981). Host response of germfree and conventional mice to single or mixed *T. spiralis* and *T. pseudospiralis* infections. II. Serological investigations. Ibidem 29:61.

Teppema JS, Blomjouse F, Elgersma A, Ruitenberg R (1981). *T. pseudospiralis* and *T. spiralis* infections in monkeys. III. Pathological aspects. "Trichinellosis". Chertsey: Kim, p 209.

VIII TRANSPLANTATION

Germfree Research: Microflora Control and Its Application to the Biomedical Sciences, pages 423–427

GNOTOBIOTICS AND HUMAN MARROW TRANSPLANTATION

C. Dean Buckner, M.D.

Fred Hutchinson Cancer Research Center

1124 Columbia Street, Seattle, WA 98104

Allogeneic marrow transplantation is being increasingly utilized for the treatment of non-malignant and malignant diseases of the hematopoetic system (6). Over the past 15 years there has been considerable improvement of survival following allogeneic marrow transplantation (6). Much of this improvement comes from better patient selection, improved supportive care, increased effectiveness of antibiotics and more experience. Despite these improvements marrow transplantation is still associated with a significant mortality from the procedure limiting the application to patients with a fatal prognosis in a short period of time. Two of the major problems contributing to transplant failure are graft-versus-host disease (GVHD) and immunodeficiency (6).

A large body of data exists in experimental animal systems demonstrating very clearly that acute GVHD can be abolished or ameliorated by performing the allogeneic marrow transplant in germ free or decontaminated animals. Attempts have been made in a variety of institutions to evaluate or to utilize uncritically the principles of isolation and decontamination established in animal research. The purpose of this paper is to review the experience with isolation and decontamination procedures in marrow transplant patients in Seattle. It is beyond the scope of this report to evaluate and discuss work performed at other institutions.

The isolation facilities at the Fred Hutchinson Cancer Research Center consist of Laminar Air Flow (LAF) rooms suitable for single patient occupancy (Sci-Med, Minneapolis, Minn.). Patients were placed in LAF rooms shortly before beginning the preparative regimen used to facilitate engraftment (either cyclophosphamide alone or a combination of cyclophosphamide and total body irradiation) and patients generally remained in the rooms until they were well enough to be managed in a outpatient setting. Patients received oral non-absorbable antibiotics and sterile food. All materials that entered the rooms were also sterile.

In 1978 we concluded a randomized trial comparing the results of transplanting patients in LAF rooms versus conventional rooms (1). Control patients did not receive non-absorbable antibiotics. We were able to confirm the observations of others that patients acquired fewer infections in a protective environment than in a conventional room. However, as in other situations, there was no survival advantage. This was not surprising considering that the major causes of death in both LAF and non-LAF patients were recurrent leukemia and interstitial pneumonitis often due to reactivation of or transfusion induced cytomegalovirus infection.

A decrease in infections was observed despite the fact that the majority of patients did not take the prescribed oral non-absorbable antibiotics due to mucositis and overall debility. We also failed to demonstrate an effect on the incidence of acute GVHD. However, patients with aplastic anemia appeared to have a delayed onset of GVHD from a median of 24 days in the control group to 48 days in the LAF group.

In a parallel study we also demonstrated that the administration of daily prophylactic granulocyte (PG) transfusions from HLA compatible donors significantly decreased the frequency of severe infections in transplant recipients as compared to patients transplanted in conventional rooms (3). We then conducted a randomized trial of LAF versus PG transfusions without the inclusion of a control group in patients with acute leukemia (2). In this study we demonstrated that the 2 modalities were equally effective in preventing infections during the period of granulocytopenia following marrow transplantation. However, since LAF could be extended for a longer period of time than could PG transfusions there were less infections in the LAF group after recovery of granulocyte function. Again, no survival difference could be detected. There was a lower incidence of acute GVHD in the LAF group but this was not statistically significant.

We then analyzed separately patients who were transplanted for aplastic anemia (4,5). These patients receive cyclophosphamide alone in preparation for the transplant as contrast to patients with leukemia who received cyclophosphamide as well as total body irradiation. The latter regimen is considerably more toxic resulting in interference with the administration of oral non-absorbable antibiotics. In patients with aplastic anemia there was a significant increase in survival and a concommitant decrease in the incidence of acute GVHD for patients transplanted in LAF rooms as compared to patients in conventional rooms. In addition we were able to confirm our previous observations that patients with aplastic anemia who developed acute GVHD developed it later than control patients and had a better survival.

In our most recent study, which has yet to be completely analyzed, we randomized patients with acute leukemia between LAF, PG, and prophylactic systemic antibiotics (PSA). PSA consisted of an aminoglycoside, a synthetic penicillin and vancomycin administered during the period of granulocytopenia (< 500 granulocytes/mm3). Patients in LAF were further randomized to receive or not to receive PSA. A preliminary analysis of this data suggests that the most effective method of preventing infection acquisition is the combination of LAF and PSA. However, there were no differences in survival between the 3 groups and no significant differences in the incidence and severity of acute GVHD. LAF patients had less infections after engraftment as compared to either the PG or PSA groups.

These studies consistently demonstrated superior infection prevention for patients with leukemia and aplastic anemia transplanted in LAF rooms as compared to control patents. Patients with aplastic anemia had a significant decrease and a delay in the onset of acute GVHD with an improved long term survival. Since this was not observed in patients with acute leukemia we looked for an explanation. One finding of note was improved decontamination as measured by survellence cultures from various parts of the body. The following table shows the percent of patients with leukemia or aplastic anemia who on monitoring cultures had consistently negative cultures or transiently negative cultures. Patients with aplastic anemia were more likely to have consistently negative monitoring cultures than patients with acute leukemia, especially from the rectum which is probably the most important site. Patients with acute leukemia often had early clearing of bacteria with reappearance in the 3rd and 4th weeks after transplantation. The efficacy of decontamination correlated fairly well with the quantity of oral non-absorbable antibiotics taken by the patients.

SURVEILLANCE CULTURES - LAF PATIENTS

	Aplastic Anemia	Leukemia
No. Patients	90	36
Neg. Cultures (%)		
Nose	66	90
Mouth	21	31
Rectal	48	74
Skin	91	87
Transient Clearing (%)		
Nose	17	11
Mouth	37	33
Rectal	60	13
Skin	5	3

The results of isolation and decontamination in terms of preventing or delaying acute GVHD and improving survival in patients with acute leukemia have been disappointing. However, the overwhelming evidence of efficacy in animal systems plus the results in patients with aplastic anemia encourage us to continue to explore isolation and decontamination procedures as a method of improving overall transplant survival. We need to improve the efficacy of the decontamination procedures. A reduction in the toxicity of the preparative regimens and/or an increase the palatibility and acceptance of the non-absorbable antibiotics would be of great benefit. In addition serious consideration needs to be given to performing a carefully designed and well controlled clinical trial of extended time in LAF before and after the transplant.

REFERENCES

1. Buckner CD, Clift RA, Sanders JE, et al (1978). Protective environment for marrow transplant recipients. A prospective study. Ann Intern Med 89:893.
2. Buckner CD, Clift RA, Thomas ED et al. (1983). Early infectious complications in allogeneic marrow transplant recipients with acute leukemia: Effects of prophylactic measures. Infection 11:243.
3. Clift RA, Sanders JE, Thomas ED et al (1978). Granulocyte

transfusions for the prevention of infection in patients receiving bone-marrow transplants. N Engl J Med 298:1052.
4. Navari RM, Buckner CD, Clift RA et al (1984). Prophylaxis of infection in patients with aplastic anemia receiving allogeneic marrow transplants. Am J Med 76:564.
5. Storb R, Prentice RL, Buckner CD et al (1983). Graft-versus-host disease and survival in patients with aplastic anemia treated by marrow grafts from HLA-identical siblings. Beneficial effect of a protective environment. N Engl J Med 308:302.
6. Thomas ED, Clift RA, Storb R (1984). Indications for marrow transplantation. Ann Rev Med 35:1.

ACKNOWLEDGEMENTS: These investigations were supported by Grant Numbers Ca 18029, Ca 15704, Ca 30924, and Ca 9319 awarded by the National Cancer Institute, DHHS.

Germfree Research: Microflora Control and Its Application to the Biomedical Sciences, pages 429–432

SELECTIVE OR COMPLETE ELIMINATION OF THE ENDOGENOUS FLORA DURING ALLOGENEIC BONE MARROW TRANSPLANTATION?

H.F.L. Guiot, J.W.M. van der Meer, F.E. Zwaan, C.J.J. Mulder, J. Biemond, and E.C. Klasen.
Depts. of Infectious Diseases and Haematology, Gastroenterology, Human Genetics, University Hospital Leiden, J.A. Cohen Institute, 2333 AA Leiden, The Netherlands.

INTRODUCTION

In our hospital infection prevention in adult granulocytopenic patients is performed by selective elimination of the aerobic potentially pathogenic micro-organisms from the digestive tract. This procedure has been called selective antimicrobial modulation (SAM) and the efficacy has been shown in a double-blind placebo-controlled study in patients undergoing chemotherapy for acute myelogenous leukaemia (Guiot 1983). Also in allogeneic bone marrow transplantation (BMT) SAM has shown to be effective (Van der Meer 1984).

An important question is whether the large numbers of bacteria (anaerobes and facultative streptococci), persisting in the gut during SAM increase the risk of graft-versus-host disease (GVHD) compared with patients given nonabsorbable antibiotics to eliminate all bacteria (total antibiotic decontamination - TAD). It has been suggested that bacterial cell-wall antigens pass through the epithelium of the gut and activate donor lymphocytes. These activated lymphocytes might crossreact with antigens on the target tissue, resulting in the signs of GVHD. Although there is some evidence that this occurs in animals, it is not clear which bacteria are involved and whether it may be extrapolated to the human situation.

With the aim to elucidate this question, we started a controlled study to compare SAM with TAD with respect to the effect on GVHD. The present interim report shows that the differences between both groups are small, but the lack of objective parameters to measure the severity of GVHD may be a source of bias in favour of the patients on SAM.

METHODS

Patients on SAM: From day 14 before BMT until the first day of the conditioning phase, the patients were placed on the standard regimen, consisting of a combination of oral polymyxin B, neomycin, nalidixic acid, and amphotericin B. During both the conditioning phase and the first weeks after BMT when the patients may have difficulties swallowing SAM, cotrimoxazole is administered intravenously to maintain the decontaminated state in these patients (Van der Meer 1984). When they were able to swallow again, the oral SAM was reinstated until the risk of gastrointestinal GVHD was ceased (around day 60). Antimicrobial prophylaxis was supported by local antimicrobial treatment in the oral cavity. The patients were nursed in laminar flow rooms, and a low-bacteria diet was prescribed.

Patients on TAD: These patients were given SAM and cotrimoxazole in the same way as the "patients on SAM", but in addition they were given cephaloridin 2 x 500 mg dd orally, or if they were not able to swallow the drug, clindamycin 3 x 600 mg dd intravenously.

Classification of GVHD: according to Thomas (1975)

α_1-antitrypsin: In two patients with GVHD of the gut α_1-antitrypsin was measured quantitatively in 24-hour portions of faeces and in the serum, once weekly by radial immunodiffusion (Weisdorf 1983).

RESULTS

At the time of this interim report the results obtained in 14 patients could be evaluated. The patients characteristics of both groups were similar and they were given the same conditioning regimen. There are no differences in the incidence and severity of GVHD, i.e.: in both groups a mean grade of 1.6 occurred (range: GVHD grade 0 (no signs of GVHD) in one patient, grade III in another patient). There is a slight suggestion that the onset of GVHD, that is serious enough to be treated, is delayed in the TAD group (Fig. 1).

The only difference seems to be the cumulative percentage of patients that develop diarrhoea most probably caused by GVHD (Fig. 1); although ultimately at day 60 the differences are not significant, there is a strong suggestion that diarrhoea starts earlier and occurs more often in patients given TAD compared with patients given SAM only. To find out whether these differences were related with the administration of TAD perse (disturbed intestinal osmoregulation), or by differences in severity of GVHD of the gut, the daily plasma loss in the gut during the period of GVHD was measured, with the aim to get information on the quantity of mucosal damage. This was done by measuring the faecal α_1-antitrypsin clearance in one patient on TAD and another on SAM (Weisdorf 1983). The results showed similar α_1-antitrypsin clearance suggesting similar mucosal damage. This presumption was confirmed by X-ray photographs. In both patients equally large areas of the distal part of the small gut had lost their normal structure. The patient on TAD had a higher loss of water via the stools during 15 weeks, i.e. 430 ml per day, against a mean of 215 ml per day in the patient on SAM during 10 weeks.

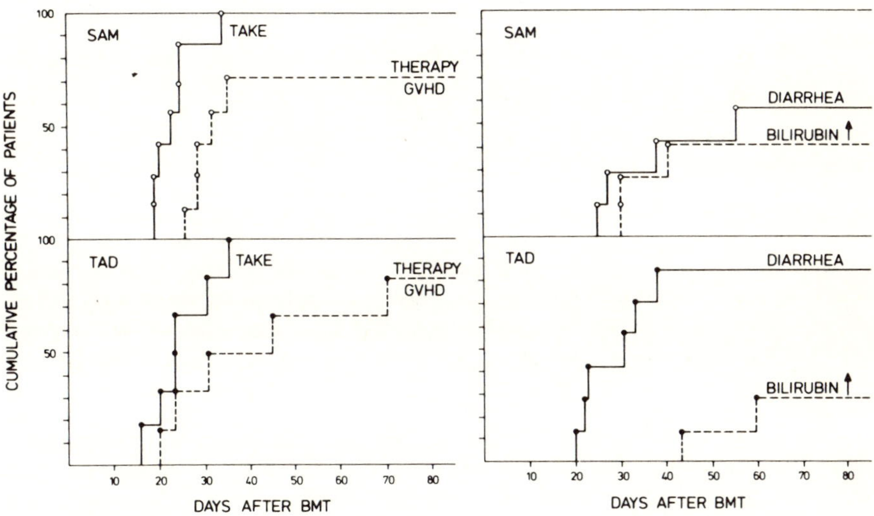

Fig. 1. Cumulative percentage of patients with a take of the BMT (granulocytes > 500/mm^3), with therapy for GVHD (high dose methylprednisolone or monoclonal anti-T-cell antibodies), with diarrhoea (watery stools > 3 days), and with bilirubin serum concentration > 60 µg/ml).

CONCLUSION

The main conclusion drawn from these data is that patients on TAD have more diarrhoea in case of GVHD of the gut than patients on SAM. This may be the result of disturbed osmoregulation of the gut which is germfree or contains extremely low numbers of bacteria. If this is true indeed, than it might be that the relative "absence" of diarrhoea in patients on

SAM masks severity of GVHD, whereas the relative "presence" of diarrhoea in patients on TAD discloses GVHD of the gut. In other words, it might be that the classification of the grade of GVHD was biassed in favour of the group of patients on SAM, i.e. the grade of GVHD is underestimated. With this in mind, it might be that the differences between patients on TAD and patients on SAM with respect to the onset of GVHD are more important than expected on the basis of the results given in this interim report.

REFERENCES

Guiot HFL, Broek PJ van den, Meer JWM van der, Furth R van (1983). Selective antimicrobial modulation of the intestinal flora of patients with acute nonlymphoblastic leukemia: A double-blind, placebo-controlled study. J Infect Dis 147: 615-623.

Meer JWM van der, Guiot HFL, Broek PJ van den, Furth R van (1984). Infections in bone marrow transplant recipients. Semin Hematol 21:123-140.

Thomas ED, Storb R, Clift RA, et al. 1975). Bone-marrow transplantation. New Engl J Med 292:832-843 and 895-902.

Weisdorf SA, Salati LM, Longsdorf JA, Ramsay NKC, Sharp HL (1983). Graft-versus-host disease of the intestine: a protein losing enteropathy characterized by fecal α_1-antitrypsin. Gastroenterology 85:1076-1081.

Germfree Research: Microflora Control and Its Application to the Biomedical Sciences, pages 433–435

THE INFLUENCE OF THE MICROBIAL FLORA ON INCIDENCE OF GRAFT-VERSUS-HOST DISEASE IN HUMAN ALLOGENEIC BONE MARROW TRANSPLANTATION

Th. Schmeiser, E. Kurrle, R. Arnold, W. Heit,
D. Krieger, B. Kubanek, H. Heimpel
Zentrum Innere Medizin, Universität Ulm
Steinhövelstr. 9, 7900 Ulm, FRG

Infections and graft-versus-host disease (GvHD) are severe complications after allogeneic bone marrow transplantation (BMT). Animal studies and results from human BMT gave evidence that allogeneic BMT under germfree conditions established by total decontamination in strict reverse isolation (ITD) reduces significantly the incidence of infections and might mitigate the incidence and severity of GvHD (Heit 1977, Vossen 1981, Buckner 1983, Schmeiser 1983, Storb 1983). Selective decontamination (SD) - another method of antimicrobial prophylaxis which aims to eliminate potentially pathogenic microorganisms - was also able to prevent GvHD in animals (Heidt 1981). In patients with acute leukemia SD was similar effective as ITD in the prevention of infections. In the study presented here the efficacy of both prophylactic regimens (ITD and SD) was investigated in allogeneic BMT. Both treatment groups were analysed for the incidence and severity of GvHD and the influence of distinct species of microorganisms on the occurrence of GvHD.

MATERIAL AND METHODS

Twenty-eight patients treated with a HLA-identical allogeneic BMT for aplastic anemia (n=6), acute leukemia (n=18) or chronic granulocytic leukemia (n=4) received ITD or SD as antimicrobial prophylaxis. Seventeen patients got total decontamination in strict reverse isolation (group ITD), whereas 11 patients received selective decontamination of the gastrointestinal tract in single rooms (group SD). Patients with aplastic anemia were given cyclophosphamide 200 mg/kg body weight, patients with acute leukemia or chronic granulocytic leukemia received total body irradiation (10 Gy to 12 Gy) and cyclophosphamide 120 mg/kg body weight. In group ITD an oral combination of vancomycin (2 g/day), neomycin (2 g/day), polymyxin B (0.4 g/day) and amphotericin B (2 g/day) was used for decontamination of the gastrointestinal tract. The regimen for SD consisted of polymyxin B (0.4 g/day), cotrimoxazole (2.88 g/day) and amphotericin B (2 g/day). Patients in both groups received sterile drinks and food, daily disinfection of the skin and a local antimicrobial treatment of the body orifices. Twice weekly surveillance cultures from faeces and oral washings were routinely done for the isolation and identification of aerobic microorganisms. Infections were diagnosed bacteriologically, clinically or radiologically. Diagnosis and grading of GvHD was based on histological investigations.

RESULTS AND DISCUSSION

Both regimens for decontamination (ITD and SD) were effective in the prevention of severe infections caused by gram-negative microorganisms. Four septicaemias occurred in group ITD as compared to 3 in group SD, no septicaemia was caused by gram-negative bacilli. Four out of 17 patients in group ITD did not develop any infections, whereas all 11 selectively decontaminated patients had fever episodes and infections. The ITD group had a lower incidence of fever days during severe granulocytopenia (38% in group ITD versus 65% in group SD, $p < 0.01$). The incidence of GvHD of grade II-IV was 5/17 (29%) in group ITD and 7/11 (64%) in group SD. It was analysed whether the colonization of patients with distinct species of microorganisms was associated with an increased risk of GvHD. In order to investigate this question, "risk indices" of individual species of aerobic bacteria and yeasts isolated from faeces and oral washings were calculated as described previously (Kurrle 1981). In both groups significantly increased "risk indices" were found only for species of Enterobacteriaceae, e.g. E. coli and Proteus spp. ($p < 0.01$). This is reflected by the observation that 8/11 (73%) of the patients who were colonized with Enterobacteriaceae before the onset of GvHD developed GvHD of grade II-IV as compared to 6/17 (35%) patients who were free of Enterobacteriaceae ($p < 0.01$).

These results indicate that the colonization of patients with Enterobacteriaceae of the digestive tract during the early posttransplantation period may increase the risk for acute GvHD. These observations are in agreement with animal studies and preliminary clinical results (Pollard 1976; Vossen 1981). By which mechanism Enterobacteriaceae might influence GvHD is yet unclear.

Buckner CD, Clift RA, Thomas ED, et al. (1983) Early infectious complications in allogeneic marrow transplant recipients with acute leukemia: effects of prophylactic measures. Infection 11: 243.

Heit H, Heit W, Kohne E, et al. (1977). Allogeneic bone marrow transplantation in conventional mice: I. Effect of antibiotic therapy on long-term survival of allogeneic chimeras. Blut 35: 143.

Heidt PJ, Vriesendorp HM, Zurcher C, et al. (1981). Selective gastrointestinal decontamination in dogs undergoing allogeneic bone marrow transplantation. In Sasaki S et al. (eds): "Recent Advances in Germfree Research," Tokyo: Tokai University Press, p 579.

Kurrle E, Bhaduri S, Krieger D, et al. (1981). Risk factors for infections of the oropharynx and the respiratory tract in patients with acute leukemia. J Infect Dis 144: 128.

Pollard M, Chang CF, Srivastava KK (1976). The role of microflora in development of graft-versus-host disease. Transplant Proc 8: 533.

Schmeiser T, Arnold R, Kurrle E, et al. (1983). Infection prophylaxis in bone marrow transplantation with two different methods of gnotobiotic care. Exp Hematol 10 Suppl. 13: 107.

Storb R, Prentice RL, Buckner CD, et al. (1983). Graft-versus-host disease and survival in patients with aplastic anemia treated by marrow grafts from HLA-identical siblings. N Engl J Med 308: 302.

Vossen JM, Heidt, PJ, Guiot HFL, et al. (1981). Prevention of acute graft versus host disease in clinical bone marrow transplantation: complete versus selective intestinal decontamination. In Sasaki S et al. (eds): "Recent Advances in Germfree Research," Tokyo: Tokai University Press, p 573.

Germfree Research: Microflora Control and Its Application to the Biomedical Sciences, pages 437–441

THE INFLUENCE OF CALCIUM LEUCOVORIN ON THE CO-TRIMOXAZOLE INDUCED INHIBITION OF LEUKOCYTE REGENERATION AFTER BONE MARROW TRANSPLANTATION IN MICE

P.J. Heidt, M.J. Kieskamp and C.P.J. Timmermans

Radiobiological Institute TNO
151 Lange Kleiweg, 2288 GJ Rijswijk
The Netherlands

INTRODUCTION

One of the most life-threatening complications in patients with a severe long-lasting granulocytopenia is the occurrence of bacterial and fungal infections (Bodey et al., 1966). One method to prevent these infections is the selective decontamination (SD) of the gastrointestinal (GI) tract (Van der Waaij, Berghuis-de Vries, 1974), which results in the selective elimination of the potential pathogenic microorganisms from the GI tract.

One of the drugs frequently used for SD is co-trimoxazole (SxT), a combination of trimethoprim and sulfamethoxazole (Heidt, 1979; Sleijfer et al., 1980; Dekker et al., 1981). However, these drugs can have toxic side effects since after profound bone marrow suppression sometimes a slower regeneration of the granulocytes is observed in patients receiving this drug combination (Dekker et al., 1981; Wade et al., 1983; Schey and Kay, 1984), while also SxT-attributed leukopenia, agranulocytosis and thrombocytopenia have been described under therapy with this drug combination (Lawson, 1983). As was also suggested by Golde et al. (1978) and Rivard et al. (1981), these side-effects can be explained by a possible inhibition of the conversion of p-aminobenzoic acid to dihydrofolate by sulfamethoxazole and the blocking of dihydrofolate-reductase by trimethoprim (Otten, Plempel, 1975).

Therefore, we investigated whether mice which were treated with SxT show a slower regeneration of the leukocytes after isologous bone marrow transplantation (BMT) and whether this slower regeneration of the leukocytes can be counteracted by simultaneous treatment with calcium leucovorin (d.1-5-formyltetrahydrofolate; calcium salt).

MATERIALS AND METHODS

The design of the study is given in Fig. 1. Three groups of 30 male C3H/LwRij mice were completely deconta-

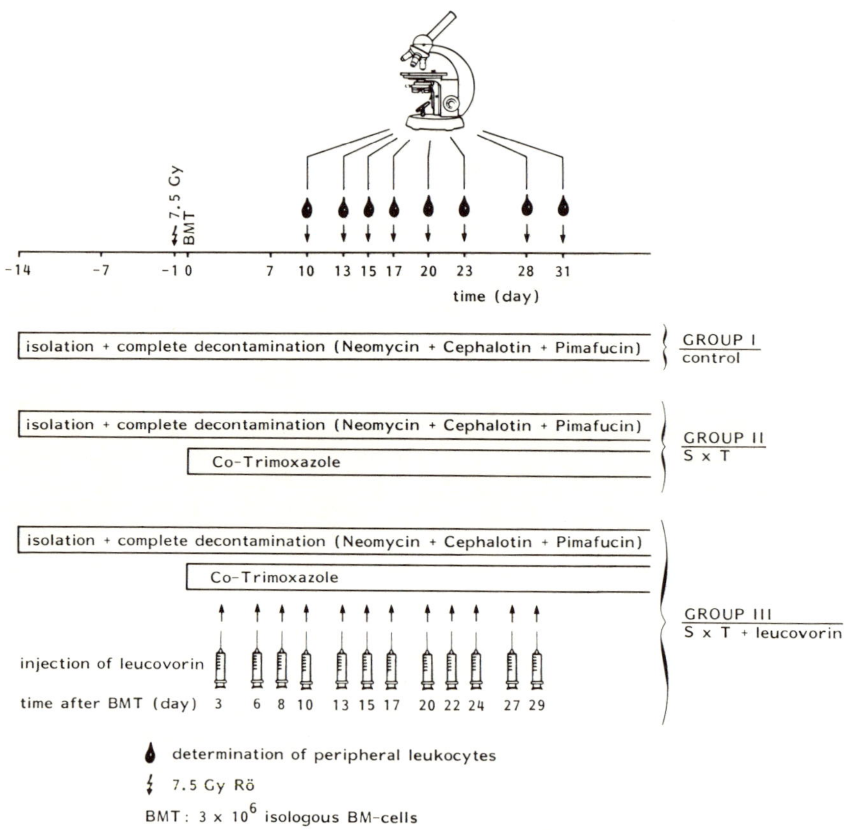

Fig. 1: Design of the study.

minated to rule out possible stimulation of the regenerating bone marrow by components from the microflora in the control group. All animals were housed in macrolon cages (5 animals per cage) in a laminar flow isolator and were fed sterile food and drinking water ad libitum. For complete decontamination, Neomycin (2.5 g per l), Cephalotin (2.5 g per l) and Pimafucin (0.1 g per l) were added to the drinking water, as was sugar (50 g per l) to improve the taste.

The control group was completely decontaminated only while the second group received SxT (1920 mg per l) in the drinking water in addition to the antimicrobial drugs for decontamination. The third group of completely decontaminated animals also received SxT in the drinking water and was treated i.m. with leucovorin (9 µg per mouse, 3 times per week). After all animals were successfully decontaminated, they were irradiated (7.5 Gy) and transplanted with 3×10^6 isologous bone marrow cells. Injections of leucovorin, which were started on day 3 after BMT, were performed with a Hamilton microliter syringe. Starting on day 10 after BMT, we sacrificed 4 animals from each group three times per week and leucocyte counts of the peripheral blood were determined.

RESULTS

The results are given in Fig. 2, which shows the mean data obtained from 4 animals per group each time. The time period after which a certain number of leukocytes in the peripheral blood is reached after bone marrow transplantation is delayed by treatment with SxT. This delay can be reversed by treatment of the animals with leucovorin. The difference in the time period after BMT that normal values of peripheral leukocytes ($> 10 \times 10^6$ per ml blood) are reached between both the control and the SxT plus leucovorin treated groups of animals and the group which was treated with SxT alone is statistically significant ($p < 0.05$).

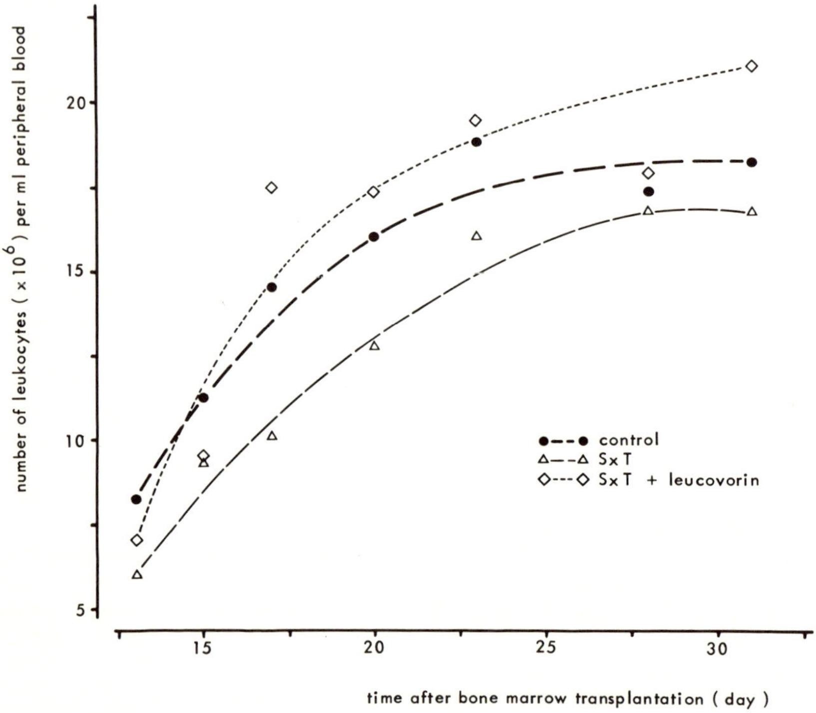

Fig. 2: Mean number of leukocytes (x 10^6) per ml of peripheral blood in decontaminated control, SxT and SxT plus leucovorin treated C3H/LwRij mice after total body irradiation and isologous bone marrow transplantation.

CONCLUSIONS

Although the number of determinations is limited, these findings suggest that the delay in the regeneration of the leukocytes after BMT or after remission induction therapy for lymphoproliferative diseases as a result of SxT given for infection prevention in patients can be prevented by simultaneous administration of calcium leucovorin.

REFERENCES

Bodey GP, Buckley M, Sathe YS, Freireich EJ (1966). Quantitative relationship between circulating leukocytes and infection in patients with acute leukemia. Ann. Intern. Med. 64: 328.

Dekker AW, Rozenberg-Arska M, Sixma JJ, Verhoef J (1981). Prevention of infection by trimethoprim-sulfamethoxazole plus amphotericin B in patients with acute non-lymphocytic leukaemia. Ann. Intern. Med. 95: 555.

Golde DW, Bersch N, Quan SG (1978). Trimethoprim and sulfamethoxazole inhibition of haematopoiesis in vitro. Brit. J. Hematol. 40: 363.

Heidt P (1979). Selective decontamination of the digestive tract in various animal species. In: Van der Waaij D, Verhoef J (eds.): "New criteria for antimicrobial therapy: Maintenance of digestive tract colonization resistance", Amsterdam: Exerpta Medica, p. 54.

Lawson DH (1983). Adverse effects of co-trimoxazole. In: Hitchings GH (ed.): "Handbook of experimental pharmacology, vol. 64: Inhibition of folate metabolism in chemotherapy". Berlin: Springer, p. 207.

Otten H, Plempel M (1975). Antibiotika und Chemotherapeutika in Einzeldarstellungen: Chemotherapeutika mit breitem Wirkungsspektrum. In: Otten H, Plempel M, Siegenthaler W (eds.): "Antibiotika-Fibel", Stuttgart: Georg Thieme, p. 109.

Rivard GE, Momparler LF, Momparler RL (1981). Effects of trimethoprim on leukaemic cells in vitro. Brit. J. Haematol. 47: 221.

Schey SA, Kay HEM (1984). Myelosuppression complicating cotrimoxazole prophylaxis after bone marrow transplantation. Brit. J. Haematol. 56: 179.

Sleijfer DT, Mulder NH, de Vries-Hospers HG, Fidler V, Nieweg HD, van der Waaij D, van Saene HKF (1980). Infection prevention in granulocytopenic patients by selective decontamination of the digestive tract. Eur. J. Cancer 16: 859.

Van der Waaij D, Berghuis-de Vries JM (1974). Selective elimination of Enterobacteriaceae species from the digestive tract in mice and monkeys. J. Hyg. 72: 205.

Wade JC, de Jongh CA, Newman KA, Crowley J, Wiernik PH, Schimpff SC (1983). Selective antimicrobial modulation as prophylaxis against infection during granulocytopenia: Trimethoprim-sulfamethoxazole vs. Nalidixic acid. J. Infect. Dis. 147: 624.

Germfree Research: Microflora Control and Its Application to the Biomedical Sciences, pages 443–446

"Comparative Analysis: 4% Chlorhexidine Gluconate versus 7.5% Povidone-Iodine Surgical Scrubs in Bone Marrow Transplant Recipients"

Karen A. Baker, R.N., B.S.N., Carol McGarigle, R.N., Hollis Bodman, M.S., Joel M. Rappeport, M.D.
Divisions of Nursing, Hematology and Microbiology, Brigham and Women's Hospital
Boston, MA 02115

Bone Marrow Transplantation (BMT) has been utilized for 12 years at the Brigham and Women's Hospital (BWH) and Children's Hospital Medical Center (CHMC) in Boston. From 1972 to 1983, 133 allogeneic and syngeneic bone marrow transplants were performed on patients with the diagnosis of aplastic anemia, acute and chronic leukemia, severe combined immunodeficiency and other congenital hematological disorders. The morbidity and mortality rates of infectious complications in leukemia patients have been significant following the administration of chemo-radiation preparative therapy or as a result of the disease itself. Protected environments have been shown to contribute to a decreased incidence of serious nosocomial infections when used concomitantly with oral, non-absorbable antibiotics (Schimpff 1975). Patients at BWH and CHMC are hospitalized in Laminar Air Flow (LAF) or sterile protective isolation rooms to decrease infectious complications associated with BMT. The LAF unit provides a horizontal flow of air which filters air-borne microorganisms out of the room, thus rendering the air 99% sterile. One important aspect in the care of the patient is a daily bath using a select cleansing agent which begins upon entrance to the LAF room and continues until recontamination, approximately 2 weeks prior to discharge.

Prior to the study period, patients bathed with 7.5% Povidone-Iodine (Betadine[R]) Surgical Scrub and then applied a layer of Betadine[R] solution to provide further antimicrobial coverage to the skin. Betadine is a rapid acting complex of polyvinylpyrollidone and iodine, an effective topical antiseptic. Some patients developed erythema and dryness of their skin after daily use of Betadine and also complained about the brown film remaining on their skin from the Betadine solution. Some patients were begun on 4% Chlorhexidine Gluconate surgical scrub (Hibiclens[R]), a bisdiguanide antiseptic. Some patients noted dryness of their skin with Hibiclens after the bath. These patients were not required to "paint" a layer of Hibiclens on their skin because of the reported sustained antimicrobial coverage with the cleanser (Peterson 1978).

In 1981 we began to compare the effects of the cleansers, Betadine and Hibiclens, on the daily condition of the patient's skin and the patient's weekly skin surveillance cultures. We randomized 20 patients to Betadine (10) or Hibiclens (10) baths. A non-sterile isolation room was used for 3 BMT recipients at that time, while the remaining 17 patients were housed in LAF rooms. Age range was 16 to 46 years and sex distribution was male (9), female (1) in the Betadine group and male (8), female (2) in the Hibiclens group. Patients were cultured weekly and results were measured in mean colony counts per surface area, 25 cm^2. Skin decontamination was begun 3 to 11 days prior to BMT with concurrent placement in LAF or non-LAF isolation. Gut decontamination was achieved through the administration of non-absorbable antibiotics 4 times/day. The cleanser was discontinued if the patient complained of skin irritation or experienced signs of sensitivity and the other agent was initiated.

The skin condition sheets of all 20 patients were reviewed during the first 2 weeks post-BMT to evaluate the incidence of early toxicity of the patient's skin. Note in Table 1 that 4/10 patients in the Betadine group had symptoms or complaints severe enough to warrant switching to Hibiclens 2-42 days after beginning the study due to severe itching (4), pain (1), dryness (1), and erythema (2). No evidence of GVHD was present at this time.

The skin condition sheets of all 20 patients were reviewed during the 3rd to 5th weeks post-BMT to evaluate late toxicity as noted in Table 2. Note that only 6 patients were evaluable on Betadine at this time because 4 patients were switched to Hibiclens. Some dryness and itching was common among the patients in the Betadine group and 7/14 patients in the Hibiclens group had no erythema. Persistent erythema was seen in 3/6 patients in the Betadine group and 4/14 in the Hibiclens group.

EARLY TOXICITY

	BETADINE[R] N=10	HIBICLENS[R] N=10
DRYNESS	5	8
ITCHING	2	2
ERYTHEMA	5	5
CROSS-OVER DUE TO TOXICITY	4	0

Table 1. Comparison of Skin Irritation - 2 Weeks Post-BMT.

LATE TOXICITY
COMPARISON OF SKIN IRRITATION

	BETADINE[R] N=6		HIBICLENS[R] N=14	
	DRYNESS/ ITCHING	ERYTHEMA	DRYNESS/ ITCHING	ERYTHEMA
ABSENT	1/6	1/6	3/14	7/14
TRANSIENT	2/6	2/6	4/14	3/14
PERSISTENT	3/6	3/6	7/14	4/14

Table 2. Comparison of Skin Irritation 3-5 Weeks Post-BMT.

We compared the skin cultures of both groups and found there was no significant difference in overall growth in either group except for one patient in the Hibiclens group whose cultures revealed 4+ growth of gram-rods. The patient later succumbed to Klebsiella pneumonia. There appeared to be no difference between culture results in patients in LAF or non-LAF isolation.

Among the many complications after BMT are dermatologic problems secondary to the preparative therapy or the BMT itself. The clinical disorder known as acute graft-versus host disease (GVHD) may occur when the donor's bone marrow produces lymphocytes that attack the immunoincompetent host (Parkman 1980). The skin, liver and gastrointestinal tract may be affected and the skin manifestation usually seen is a maculopapular rash beginning on the face, anterior thorax and behind the ears which may progress to generalized erythema or total exfoliation similar to a third degree burn. In this population of 20 patients studied we noted 6/10 patients in the Betadine group and 2/10 patients in the Hibiclens group had evidence of acute skin GVHD with histologic confirmation on skin biopsy.

In conclusion, we have studied a small sample of patients randomized to Betadine or Hibiclens baths for the purposes of decontamination. We noted the incidence and severity of toxicity to be greater in the Betadine group. There was no apparent difference in the skin surveillance cultures between the two agents. Our findings suggest a correlation between the severity of skin GVHD and the use of Betadine but they need further substantiation in a larger study.

Parkman R, Rappeport JM, Rosen F (1980). Human graft versus host disease. J Invest Derm 74:276.

Peterson AF, Rosenberg A, Alatary SD (1978). Comparative evaluation of surgical scrub preparations. Surg Gynec Obs 146:55.

Schimpff SC, Greene WH, Young VM, Fortner CL, Jepsen L, Cusack N, Block JB, Wiernik PK (1975). Infection prevention in acute non-lymphocytic leukemia. Laminar room reverse isolation with oral, non-absorbable antibiotic prophylaxis. Ann Int Med 82:357.

Germfree Research: Microflora Control and Its Application
to the Biomedical Sciences, pages 447–450

XENOGENEIC BONE MARROW CHIMERISM IN GERMFREE RATS

Morris Pollard, Phyllis H. Luckert and A. Meshorer

Lobund Laboratory, University of Notre Dame

Notre Dame, Indiana 46556 U.S.A.

Bone marrow (BM) transplantation is considered one of the most encouraging of procedures for treatments of hematologic dyscrasias, and more recently for treatments of genetic diseases (Reviews - Parkman et al, 1983; Bayever and Feig, 1983).

One of the serious limitations in BM therapies is the development of graft vs host (GVH) disease in patients who receive BM cells from histo-incompatible donors. This is characterized by inflammatory lesions involving skin, lung, intestine, liver and kidney. GVH disease develops in conventional (CV) mice and rats following lethal total body irradiation (TBI) and IV inoculations of BM cells from histo-incompatible donors. The spectrum of lesions observed in them is similar to that noted above in humans. Animals that receive histo-compatible BM cells are spared GVH disease.

Thirteen years ago, Jones et al (1971) demonstrated that gnotobiotic (GF) mice, following lethal doses of x-rays (TBI) and then IV-inoculated H^2 histo-incompatible BM cells, rarely developed even marginal evidence of GVH disease. The same experimental protocol in CV mice resulted in lethal GVH disease. Allogeneic BM chimerism resulted in incomplete reconstitution of immunological competence (Loughman and Nordin, 1973).

Young GF AKR mice were treated with lethal TBI and H^2 histo-incompatible BM cells, and they did not develop the anticipated lymphatic leukemia (Pollard and Truitt, 1973). Leukemic mice were subjected to the same experimental protocol, and significant numbers were cured (Truitt, Pollard and Srivastava, 1974). Syngeneic BM cells reversed the acute damaging effects of TBI; however, the treated mice were not spared the eventual development of leukemia. Similar results were demonstrated with SJL/J mice, a strain with very high incidence of lethal reticulum cell sarcoma (RCS): this disease was prevented and reversed following TBI and transfusion of allogeneic BM cells (Pollard and Truitt, 1974; Truitt and Pollard, 1976). Syngeneic BM cells did not protect against RCS. GVH disease developed in irradiated CV SJL/J mice which were inoculated with allogeneic BM cells.

While GF mice and rats with allogeneic BM chimerism were spared the pathogenic effects of GVH disease and survived, mice and rats with xenogeneic BM cells usually died after a relatively brief period of recovery from the effects of irradiation (Jones et al, 1971). Early reports on xenogeneic chimerism indicated that CV chimeric mice with rat BM cells produced blood with characteristics of rats; however, long-term survivals were unusual (Nowell et al, 1956).

During the past 10 years, we have investigated xenogeneic BM chimerism in GF rats. The basic experimental protocol was similar to that noted above with GF mice. In 7 trials, GF rats of 2 strains (L-W and Sprague-Dawley) were given 10^3R x rays (TBI) and 24 hours later each was given BM cells washed from the femurs of GF mice (3 strains: C3H, SJL, CFW). Survival rates were related to the numbers of BM cells inoculated into each irradiated rat: as cell numbers were increased, survival periods were extended; and surviving individuals were recorded as long as 241 and 486 days. In two recent trials, each irradiated Sprague-Dawley (S-D) rat was given BM cells from 24 femurs from GF CFW mice. After 100 days, 2 were killed for examinations (Table 1) which included blood counts, hematocrit levels, hemagglutination, leukocyte alkaline phosphatase, electrophoresis pattern of hemoglobin, and examinations of spleen imprints for immunofluorescence with fluorescein-tagged antiserum to mouse or rat gamma globulin. Based on levels of white blood cell counts (12,200 + 11,600/mm^3) and of hematocrits (45% + 44%) and on specific tests (Table 1), the rats were actively producing mouse blood. Spleens, Peyer's patches, and lymph nodes developed small primary follicles without germinal zones. Bone marrows were repopulated with multiple cell types. The thymuses were small, with defined cortical and medullary areas. The functional status of these immunoresponsive tissues was not determined. The intestinal mucosa was intact and active with mitotic figures. The livers and lungs showed no evidence of necrosis, nor of inflammation.

At present, two GF rats with mouse BM cells are surviving without evidence of disease at 170 days, and two others are surviving at 150 days. All control rats died within 3 weeks after TBI.

Table 1

Examinations of Xenogeneic BM Chimeric Sprague-Dawley Rats

	Rat	Mouse
Hemagglutination	-	+
Leukocyte Alkaline Phosphatase	-	+
Hemoglobin Electrophoresis	-	+
Immunofluorescence	-	+

DISCUSSION

Engraftments of mouse BM cell transplants in irradiated GF rats were successful on the basis of long-term survivals and production of mouse blood cells. Production of and maintenance of physiological levels of blood cells were regulated by the host. Since erythrocytes survive 120 days in the normal host, we can assume that the erythrocytes being produced by the rats were of mouse origin, which is confirmed by the specific tests noted in Table 1.

The xenogeneic BM chimeric GF rats show no gross, nor microscopic evidence of GVH disease.

Bayever E, Feig SA (1983). Bone marrow transplantation for inherited metabolic disorders. In "Recent Advances in Bone Marrow Transplantation," New York: Alan R. Liss, p. 185.

Jones JM, Wilson R, Bealmear PM (1971). Mortality and gross pathology of secondary disease in germfree mouse radiation chimeras. Radiation Research 43(3):577.

Loughman BE, Nordin AA (1973). Studies of the immunological capacity of germ-free mouse radiation chimeras. II. Thy-

mus cell function in a humoral immune response to sheep erythrocytes. Cellular Immunology 9:118.
Nowell PC, Cole LJ, Habermeyer JG, Roan PL (1956). Growth and continued function of rat marrow cells in x-radiated mice. Cancer Res 16:258.
Parkman R, Rappeport JM, Geha R, Rosen FS, Nathan DG (1983). Bone marrow transplantation lymphoid and hematopoietic immunodeficiencies. In "Recent Advances in Bone Marrow Transplantation," New York: Alan R. Liss, p. 173.
Pollard M, Truitt RL (1973). Allogeneic bone marrow chimerism in germfree mice. I. Prevention of spontaneous leukemia in AKR mice. Proc Soc Exp Biol and Med 144:659.
Pollard M, Truitt RL (1974). Allogeneic bone marrow chimerism in germfree mice. II. Prevention of reticulum cell sarcomas in SJL/J mice. Proc Soc Exp Biol and Med 145:488.
Truitt RL, Pollard M (1976). Allogeneic bone marrow chimerism in germfree mice. IV. Therapy of "Hodgkin's-like" reticulum cell sarcoma in SJL mice. Transplantation 21:12.
Truitt RL, Pollard M, Srivastava KK (1974). Allogeneic bone marrow chimerism in germfree mice. III. Therapy of leukemic AKR mice. Proc Soc Exp Biol and Med 146:153.

Germfree Research: Microflora Control and Its Application to the Biomedical Sciences, pages 451–454

ISOLATION OF SUPPRESSOR-LIKE T CELLS FROM BACTERIA-FREE ALLOGENEIC BONE MARROW CHIMERAS

Robert L. Truitt and Charles C-Y. Shih

Winter Research Laboratory, University of Wisconsin Medical School- Milwaukee Clinical Campus, Milwaukee, WI 53233

Over the past few years we have developed and used a murine model of acute T lymphoblastic leukemia to evaluate the principles of adoptive immunotherapy in a clinically relevant experimental system (Truitt 1978; Truitt et al. 1979; Truitt, Shih 1982). The unique feature of the experimental model was the application of germ-free techniques to the treatment of AKR leukemia by allogeneic bone marrow transplantation. Specifically, antibiotic decontamination (DC) and isolation in a laminar airflow protective environment (PE) were used to avoid or mitigate the "secondary disease" syndrome which is responsible for a major portion of deaths attributed to graft-vs-host (GVH) disease.

A significant survival advantage was observed in leukemic AKR (H-2^k) mice (AKR-L) undergoing an acute and/or chronic GVH reaction following allogeneic (SJL; H-2^s) bone marrow transplantation. The increased survival rate was dependent upon several factors including (a) induction of a remission with chemotherapy, (b) use of both DC and PE, (c) the interval of pretransplant DC, (d) histocompatibility relationship between donor and host, (e) dose of donor T cells transplanted, and (f) timing of reconventionalization.

The immunologic status of the allogeneic "leukemia-cured" chimeras in this experimental model was studied (Truitt, Shih 1981). Cells from allogeneic bone marrow chimeras (ABMC) responded to both T and B cell mitogens. Short-term chimeras failed to produce any antibodies to sheep RBC after primary or secondary sensitization, wheras long-term chimeras responded with both IgM and IgG production, albeit in a delayed fashion. Most ABMC responded to unrelated H-2 alloantigens in MLC. The MLC response to allo-H-2 was not influenced significantly by time post-transplant. ABMC exhibited a diminished MLC response to host AKR alloantigens in comparison to normal SJL mice or donor-syngeneic (SJL→SJL) chimeras (SBMC) regardless of the number of days post-transplant (unpublished). The ability of ABMC to generate a cytolytic (CML) response to unrelated allo-H-2 was depressed (Truitt, Shih 1981); against host alloantigens, the response was negligible (unpublished).

In contrast to cells from normal SJL mice or SBMC, stimulation of cells from ABMC with host AKR antigens in a 5-day MLC reaction resulted in a decrease rather than an increase in the frequency of anti-host CTL (unpublished data). Limiting dilution microcytotoxicity assays (LDA) were used to quantify the frequency of anti-host reactive CTL (technique described in Truitt et al. 1983). The median frequency of anti-AKR reactive CTL in the spleens of AKR hosts given SJL BM and LN cells peaked at 7 days post-transplant (approx. one per 2,500 cells) then declined (less than 1/10,000 beyond 100 days). Some ABMC were found to have cells which "down-regulated" CTL activity when added to normal SJL splenocytes. These animals could be identified by a characteristic "kink" in their LDA regression curves using AKR or AKR-L targets.

Transplantation tolerance is evidenced by immunologic unreactivity of donor cells towards host but not third-party alloantigens. To determine whether ABMC had cells that specifically suppressed anti-host and not anti-third party CTL, we co-cultured ABMC spleen cells in LDA with stimulator cells that expressed both host ($H-2^k$) and third-party ($H-2^d$) antigens (i.e., with (AKR x DBA/2)F_1cells). The contents of each individual microculture well were split after 8 days and tested for cytolytic activity against host (AKR-L) targets and unrelated allo-$H-2^d$ (P815 tumor cells). Cells which down-regulated generation of anti-host CTL were present in these cultures as evidenced by the characteristic kinked LDA regression curve against AKR-L targets. However, these cells had no effect on development of CTL against allo-$H-2^d$. That is, they were specific in their effect. This observation led us to speculate that suppressor cells may be involved in the diminished ability of ABMC to respond to host alloantigens. Specifically, they may play a role in maintaining donor-host tolerance.

Suppressor-like cells were cultured from microwells of LDA in which spleen cells from ABMC were used as responder. Wells that contained rapidly proliferating, noncytotoxic cells, but which should have contained a significant number of CTL based on LDA frequency estimates, were subcultured in the presence of irradiated AKR stimulator cells and secondary MLC supernatant (2^oSN) as a source of T cell growth factors (e.g., IL-2). These cells have been designated SAC for SJL→AKR chimera. Of several cell lines derived in this way, the SAC-9 line has been maintained the longest (i.e., over two years in IL-2 conditioned medium without antigen stimulation). Data from flow cytometric studies indicated that SAC-9 expressed Thy-1.2 and $H\text{-}2^s$ antigens of SJL origin, but not the Thy-1.1 and $H\text{-}2^k$ antigens of AKR. Preliminary flow cytometry analysis indicated that the line was $Lyt\text{-}1^+2^-$.

SAC-9 cells proliferated in response to H-2-encoded antigens of the k haplotype, but not b, d, f, p, q, r, and s in the absence of exogenous IL-2. The SAC-9 cells and clones derived from them suppressed generation of CTL in MLC/CML assays in a dose and time dependent manner, but without apparent antigen specificity. In contrast, in the presence of exogenous IL-2 (2^oSN), SAC-9 cells suppressed CTL generation in an antigen specific fashion. Further, in preliminary studies we found that SAC-9 cells injected together with SJL BM + LN cells into lethally irradiated AKR hosts completely suppressed generation of antihost reactive CTL as measured by LDA in vitro. Suppression was antigen specific.

It is possible (if not probable) that the use of DC/PE facilitated detection of suppressor cell populations in our chimeras. That is, when confounding causes of death such as infection and secondary disease were eliminated, suppressor maintained donor-host tolerance was detectable. Failure of some investigators to detect specific suppressor cells also may be due to their use of T cell depleted BM. By using a DC/PE protocol we could keep alive a significant proportion of chimeras given BM plus post-thymic T cells.

Little is known about donor-host tolerance in adult allogeneic BM chimeras. Evidence from certain rodent models has suggested that donor-host tolerance is an active process, whereby cells promoting tolerance in vivo have suppressor activities in vitro (e.g., Tutschka 1983). An important role for serum blocking factors has not been demonstrated in man or in rodents. Clonal deletion remains a viable explanation for tolerance in BM transplantation; such a theory does not preclude a suppression mechanism that works within tolerant animals to produce an apparent clonal deletion (i.e., functional clonal deletion). The availability of cloned suppressor cells that maintain donor-host tolerance in vivo could help resolve fundamental questions concerning the mechanism of tolerance in bone marrow transplantation.

Truitt RL (1978). Application of germfree techniques to the treatment of leukemia in AKR mice by allogeneic bone marrow transplantation. In Waters H (ed): "Handbook of Cancer Immunology", Volume 5, New York: Garland STPM Press, p. 431.

Truitt RL, Bortin MM, Rimm AA (1979). Successful adoptive immunotherapy of AKR spontaneous T cell leukemia following antibiotic decontamination and protective isolation. In Fliedner T, Heit H, Niethammer D, Pfieger H (eds): "Clinical and Experimental Gnotobiotics", Stuttgart: Gustav Fischer Verlag, p. 215.

Truitt RL, Shih C-Y (1981). Recovery of the immune system following adoptive immunotherapy of bacteria-free leukemic AKR mice. In Sasaki S, Ozawa A, Hashimoto K (eds): "Recent Advances in Germfree Research", Tokyo: Tokai University Press, p. 645.

Truitt RL, Shih C-Y (1982). Modification of host microflora during adoptive immunotherapy of spontaneous AKR leukemia. In Vitetta E, Fox CF (eds): "B and T Cell Tumors", New York: Academic Press, p. 487.

Truitt RL, Shih C-Y, LeFever AV, Tempelis LD, Andreani M, Bortin MM (1983). Characterization of alloimmunization-induced T-lymphocytes reactive against AKR leukemia in vitro and correlation with graft-vs-leukemia activity in vivo. J Immunology 131:2050.

Tutschka PJ (1983). The rat model as a preclinical system for bone marrow transplantation. In Gale RP (ed): "Recent Advances in Bone Marrow Transplantation", New York: Alan R. Liss, Inc., p. 511.

Acknowledgements

This work was supported by Grants CA-18440, CA-26245 and CA-20484 from the National Cancer Institute, Department of Health and Human Services. The authors thank Cabrina Truitt for typing the manuscript.

Germfree Research: Microflora Control and Its Application to the Biomedical Sciences, pages 455–459

HISTOLOGICAL AND IMMUNOLOGICAL EVALUATION OF GNOTOBIOTIC C3H/He MICE FOLLOWING I.V. ADMINISTRATION OF 8-AMINOGUANOSINE AND 2'DEOXYGUANOSINE

Michael J. Herrmann, M.S.,* Patricia M. Bealmear, Ph.D., and Joshua Epstein, D.Sc.
Roswell Park Memorial Institute and SUNY @ Buffalo School of Medicine*
Buffalo, NY 14263, USA

INTRODUCTION

Successful allogeneic bone marrow transplantation depends on the removal of competent T cells responsible for graft-vs-host disease (GVHD) from the bone marrow graft and prevention and/or limitation of the complications of infection (van Bekkum, Knaan 1977; Winston et al. 1984). Current methods employed to pre-treat bone marrow prior to transplantation, e.g., lectin (Reisner et al. 1983) or monoclonal antibody treatment (Reinherz et al. 1982) have met with only limited success.

The search for a safe, economical, and reproducible technique led to the investigation of several drugs which exhibit inhibitory activity on enzymes of the purine metabolic pathway (Fox et al. 1981), a pathway which has been shown to be vital to proper immune function. A very potent inhibitor of purine nucleoside phosphorylase, 8-aminoguanosine (8-AGuo), when used in combination with its analog 2'-deoxyguanosine (2'dGuo), exhibited selective T-cell cytotoxicity in preliminary in vitro studies (Bealmear et al. 1983a; Kennedy 1983). It was also demonstrated that pre-treatment of a histoincompatible bone marrow/spleen inocula from the mouse (a cell source that contains the same cell distribution as bone marrow alone in the human) with 8-AGuo and 2'dGuo prevents fatal GVHD (Kennedy and Bealmear 1982; Kennedy et al. 1983a) and permits survival for a normal life span without any evidence of histopathology (Kennedy 1983).

The purpose of this study was to evaluate any toxic or immune effects the treatment with 8AGuo and 2'dGuo might have following in vivo administration. If any practical application is to be made of this study, the drug would have to survive clinical trials with the FDA.

MATERIALS AND METHODS

C3H/He mice mono-associated with Staphylococcus epidermidis were injected i.v. daily for 1 wk with 100 μM 8-AGuo + 100 μM 2'dGuo in suspension media. Control mice were injected with media alone. At the end of the treatment period, the mice were sacrificed and histological specimens were prepared. In addition, spleen cells were prepared from both groups and used for in vitro evaluation of immune function employing mitogens and the mixed lymphocyte reaction (MLR). Protocols for these blastogenesis assays were described in detail by Kennedy and Bealmear (1981). Comparisons of the histology and immunological function between the treatment and control groups were then made.

Human peripheral blood lymphocytes were isolated by ficoll-hypaque centrifugation from buffey coat cell collected by plasmapheresis; they were washed and treated with 100 μM 8AGuo and 100 μM 2'dGuo for use in blastogenesis assays, i.e., PHA, Con A, PWM, and MLR. Comparisons were made between the stimulation indices (S.I.) of the treated and untreated groups (S.I. = CPM stimulated/CPM unstimulated). A percent inhibition of the responses due to the drug was calculated by: %Inhibition = 100 - CPM with drug/CPM without drug.

RESULTS

Histological examination of lymphoid and nonlymphoid tissues showed no evidence of histopathology for the drug treated or control group (Figure 1).

There were no significant differences in the blastogenesis responses of the experimental and control animals after i.v. treatment with 8 AGuo and 2'dGuo (Table 1).

Inhibition of mitogen responses by pretreatment of

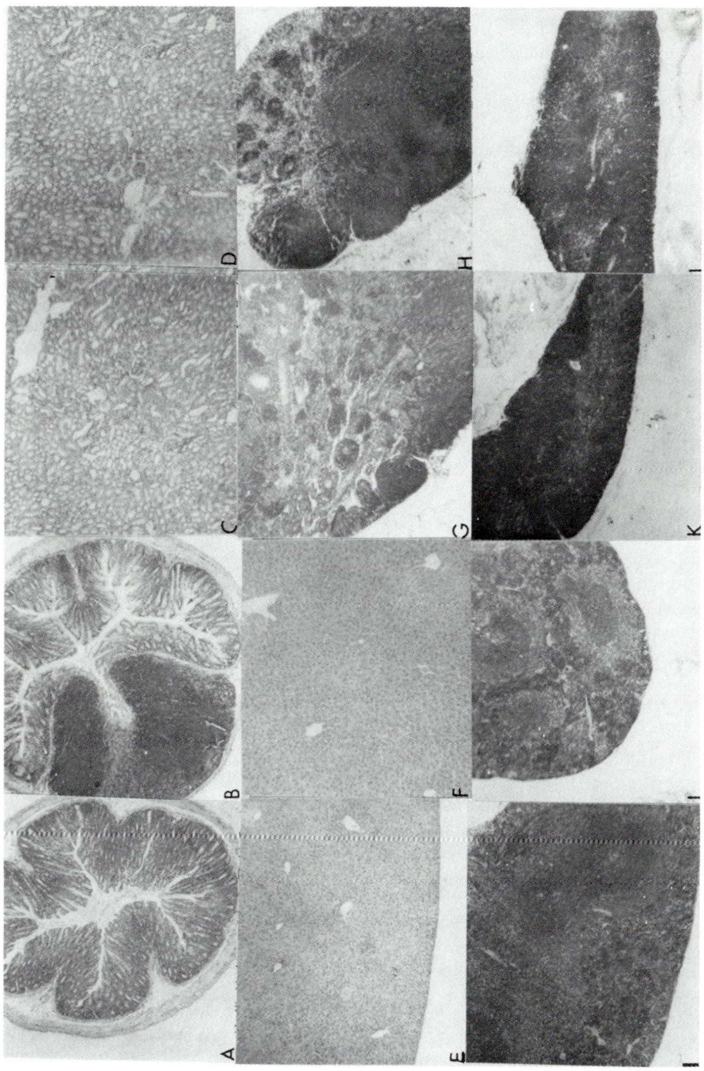

Fig. 1. Tissues of mice treated with drug daily for 1 wk; untreated tissues (C) are on the left and the treated (E) are on the right for each group. A. C-ileum, B. E-ileum; C. C-kidney, D. E-kidney; E. C-liver, F. E-liver; G. C-lymph node, H. E-lymph node; I. C-spleen, J. E-spleen; K. C-thymus, L. E-thymus. X350.

Table 1

STIMULATION INDICES FOR MONO-ASSOCIATED MICE AFTER I.V. INJECTIONS DAILY FOR 1WK

	N	PHA	Con A	PWM	LPS	MLR
EXPERIMENTAL 100 μM 8 AGuo + 100 μM 2'dGuo	5	2.82 ± 1.09	4.00 ± 1.12	1.83 ± 0.51	1.68 ± 0.73	2.42 ± 0.4
CONTROL Media Alone	5	3.32 ± 1.47	5.05 ± 2.33	1.78 ± 0.48	1.67 ± 0.55	2.27 ± 0.5

human peripheral blood lymphocytes and bone marrow cells with 8 AGuo and 2'dGuo at various concentrations exhibit the abiltiy of this drug combination to inhibit T-cell dependent immune responses. Survival studies for x-irradiated germfree C3H/He mice who received allogeneic spleen and bone marrow cells treated with 8 AGuo and 2'dGuo showed 85% survival when compared to 0% in the untreated group.

CONCLUSIONS

Daily in vivo administration of a specific concentration of 8 AGuo + 2'dGuo in C3H/He mice for 1 wk produced no evidence of histopathology or overall alteration in the immune response as measured by blastogenesis assays with mitogens and H-2 antigens. In vitro addition of the drug combination was capable of inhibiting T-cell dependent responses as observed in the mitogen and MLR assays. Allogeneic spleen and bone marrow cells pre-treated with 8 AGuo + 2'dGuo before transplantation reduced the occurrence of fatal GVHD to 85%. Further evaluation of this drug combination may provide a satisfactory method for preparing allogeneic bone marrow for transplantation and elimination of the complications of GVHD.

ACKNOWLEDGEMENTS

We gratefully acknowledge the technical assistance of Ms. Joyce Jividen for the germfree animal part of the work and of Ms. Frances Robinson for the photomicrography.

Michael J. Herrmann was supported by Training Grant # 1 T35-AGGM-00089, National Institute on Aging, Department of Health and Human Services.

REFERENCES

Bealmear PM, (1981). Host defense mechanisms in gnotobiotic animals. In Gershwin ME, Merchant B (eds.): "Immune defects in laboratory animals", Vol. 2, Part IX, Chapter 11, New York: Plenum Press, p 261.

van Bekkum DW, Knaan S (1977). Role of bacterial microflora in development of intestinal lesions from graft-versus-host reaction. J Nat Cancer Inst 58:787.

Fox RM, Kellford RF, Tripp EH, Taylor IW, (1981). G_1-phase arrest of cultured human leukemic T cells induced by deoxyadenosine. Cancer Res 41:5141.

Jones JM, Wilson R, Bealmear PM. (1971). Mortality and gross pathology of secondary disease in germfree mouse radiation chimeras. Radiat Res 45:577.

Kennedy DW, (1983). "Prevention of lethal graft-vs-host disease following bone marrow transplantation." Ph.D. Dissertation, Buffalo, NY: SUNY @ Buffalo. 152 p.

Kennedy DW, Bealmear PM, (1981). Possible co-dominance of the nude and wildtype genes in SPF C3H/He heterozygous nu/+ mice. In Sasaki S, Ozawa A, Hashimoto K (ed.): "Recent Advances in Germfree Research," Tokyo: Tokai University Press, p 611.

Reinherz EL, Geha R, Rappeport JM, Wilson M, Penta AC, Hussey RE, Fitzgerald KA, Daley JF, Levine H, Rosen FS, Schlossman SF, (1982). Reconstitution after transplantation with T-lymphocyte-depleted HLA haplotype-mismatched bone marrow for severe combined immunodeficiency. Proc Natl Acad Sci USA 79:6047.

Reisner Y, Kapoor N, Kirkpatrick D, Pollack MS, Cunningham-Rundles S, Dupont B, Hodes MZ, Good RA, O'Reilly RJ, (1983). Transplantation for severe combined immunodeficiency with HLA-A, B, D, DR incompatible parental marrow cells fractionated by soybean agglutinin and sheep red blood cells. Blood 61:341.

Winston DJ, Ho WG, Champlin RE, Gale RP (1984). Infectious complications of bone marrow transplantation. Exp Hematol 12:205.

Germfree Research: Microflora Control and Its Application
to the Biomedical Sciences, pages 461–466

INHIBITION OF ADENOSINE DEAMINASE AND PURINE NUCLEOSIDE PHOSPHORYLASE AND T-CELL FUNCTION IN GERMFREE MICE AND HUMAN PERIPHERAL BLOOD

Sheila C. Wiedl, Ph.D., Patricia M. Bealmear, Ph.D., and Joshua Epstein, D.Sc.
Departments of Dermatology, Medical Oncology, and Viral Leukemia, Roswell Park Memorial Institute, Buffalo, NY 14263, USA

INTRODUCTION

Bone marrow transplantation (BMT) is currently employed in the therapy of a variety of diseases including leukemia and immune deficiencies. Two major problems that limit the practical application of BMT are graft-vs-host disease (GVHD) and infection (van Bekkum, Knaan 1977; Winston et al. 1984). Germfree animals are useful for the study of GVHD in transplantation because the complications of infection are eliminated (Jones, Wilson, Bealmear 1971).

GVHD may be prevented by the elimination of competent T lymphocytes from the BM graft (Bealmear 1975; Bealmear 1981)). Current methods for this elimination include E-rosetting and monoclonal antibody (Reinherz et al. 1982) or lectin (a protein that binds to carbohydrates) treatment (Reisner et al. 1981; Reisner et al. 1983); however, these methods are often time consuming, present a risk of bacterial or viral contamination, and are not 100% effective. A correlation between lymphocyte function and enzymes of the purine metabolic pathways has been demonstrated. Abnormal levels of these enzymes--adenosine deaminase (ADA) and purine nucleoside phosphorylase (PNP)--have been associated with defective T-cell functions (Fox et al. 1981; Giblett et al. 1975), therefore, selective inhibition of one or both of these enzymes may eliminate T lymphocyte function from the BM graft, and thus prevent GVHD. To test this hypothesis: 1) human peripheral blood was pretreated with inhibitors of ADA or PNP before it was set up in vitro in blastogenesis assays, and 2) mouse BM and spleen cells were pretreated with the same inhibitors before transplantation to allogeneic recipients; a 1:1 ratio of BM to spleen cells in the mouse provides the same cell distribution that BM alone provides in the human.

MATERIALS AND METHODS

Germfree C3H/He mice were exposed to 1000 Rads of x-ray, and transplanted i.v. with equal numbers of bone marrow (5 x 106) and spleen (5 x 106) cells from allogeneic DBA/2 mice. The inoculum was pre-treated with 10 μM 2'deoxycoformycin (dCF), an inhibitor of ADA, and 100 μM deoxyadenosine (dAdo), or 100 μM 8-aminoguanosine (8AGuo), an inhibitor of PNP, and 100 μM 2' deoxyguanosine (2'dGuo). Untreated allogeneic and syngeneic control groups were also transplanted. Transplanted animals were observed for survival and hematological and immunological reconstitution.

The levels of ADA and PNP were measured in the cells to be transplanted before treatment and after 1 hr treatment. In vitro mitogen and mixed lymphocyte blastogenesis assays were also conducted using cell populations comparable to those used in the in vivo experiments. The procedures for the blastogenesis assays were described previously by Kennedy and Bealmear (1981). ADA levels in spleen cells from the chimeras were determined at 8, 14, and 21 days post-transplant for the in vivo experiments and at 0, 1, and 84 hours for the in vitro experiments.

In the human experiments, human peripheral blood lymphocytes were prepared by ficoll-hypaque centrifugation, washed, counted and pre-treated for 1 hr with either 10 μM dCF and 100 μM dAdo or 100 μM 8 AGuo and 100 μM 2'dGuo, then plated in mitogen assays (phytohemagglutinin [PHA], concanavalin A [Con A], and pokeweed mitogen [PWM]. The enzyme levels in the cultured cells were measured at 0, 1, and 84 hrs.

For the enzyme assays, the cells were washed 2 times with PBS and resuspended in sucrose buffer, then they were disrupted by freeze-thawing and clarified by centrifugation @ 50,000 x g for 30 min. The supernatants were used for the assays. Enzyme determinations were made with a spectrophotometer by a modification of Kalckar's technique (Kalckar 1947); measurement of ADA activity was based on absorption change at 265 nm for the conversion of adenosine to inosine; the PNP assay was based on the formation of uric acid from inosine and measured at 293 nm . Data were expressed as μ mole of substrate turned over/hr/10^6 cells.

RESULTS

All syngeneic chimeras (Syn.) survived and all allogeneic chimeras (Allo.) died at 8 days post-transplant. In both allogeneic groups that received either dCF- or 8 AGuo-treated cells (Allo.-dCF and Allo.-8AGuo, respectively), 80% survived for the 15-month-period of observation with no gross symptoms of GVHD (Kennedy, 1983). Spleen cells taken from the Allo.-dCF-treated group had no ADA activity at 1 hr (time of the transplant-Table 2), had regained 55.5% of normal ADA activity by 8 days and 72.6% by 21 days post-transplant (Table 1). ADA activity was 6.6 X

Table 1

ADENOSINE DEAMINASE LEVELS* IN ALLO.-dCF GERMFREE MICE

Source of Cells 5×10^6 Bone Marrow Cells $+ 5 \times 10^6$ Spleen Cells	Days Post-Transplant 8	 14	 21
Syngeneic			
(C3H/He to C3H/He)	0.062	0.031	0.098
Syngeneic + 10 μM dCF + 100 μM dAdo			
(C3H/He to C3H/He)	0.357		
Allogeneic			
(DBA/2 to C3H/He)	0.850	all dead	
Allogeneic + 10 μM dCF + 100 μM dAdo			
(DBA/2 to C3H/He)	0.071	0.058	0.093

* measured as μMoles of adenosine substrate converted to inosine @ 37 C @ 265 nm per hr per 10^6 cells.

normal levels in the Allo. at day 8 when they were dying of acute GVHD. In the in vitro experiments (Table 2), ADA activity was absent for the entire 84 hr-culture period in

Table 2

ADENOSINE DEAMINASE LEVELS* OF DRUG-TREATED SPLEEN AND BONE MARROW CELLS FROM GERMFREE DBA/2** MICE DURING THE MLR

Source of Cells	Time In Vitro (hrs)		
5 x 10^6 Bone Marrow Cells + 5 x 10^6 Spleen Cells	0	1	84
DBA/2	0.128	0.128	
DBA/2 + 10 μM dCF + 100 μM dAdo	0.128	0.000	0.000
DBA/2 + 100 uM 8AGuo + 100 uM 2'dGuo	0.128	0.094	0.005

* measured as μMoles of adenosine substrate converted to inosine @ 37 C @ 265 nm per hr per 10^6 cells.

** DBA/2 mice are the donors for all allogeneic bone marrow-spleen transplants.

the dCF-treated cells, whereas, ADA activity was reduced at 1 hr, but not eliminated in the 8AGuo-treated cells. Results indicated a 92% inhibition of the MLR with dCF treatment. In the mitogen assays, the PHA-, Con A-, and PWM-responding cells were inhibited, but not the LPS (lipopolysaccharide)-responding cells; T helper and T suppressor cells plus thymic-dependent B cells were sensitive to the drug, but not the thymic-independent B cells. When human peripheral blood lymphocytes were also pre-treated with dCF and dAdo for 1 hr then plated in mitogen assays, both the PHA- and Con A-responding cells were inhibited (Table 3). ADA activity was completely inhibited during the entire 84-hr assay period.

DISCUSSION

Pretreatment of cells with dCF and dAdo for 1 hr inhibits ADA activity, and inhibition is maintained for the duration of the mitogen and mixed lymphocyte assays, even if the cells are washed following drug treatment. Pretreatment with 8AGuo and 2'dGuo diminishes, but does not eliminate, ADA activity; this inhibition is lost if the cells are washed. Both drug combinations, i.e., dCF and dAdo, and 8AGuo and 2'dGuo, have significant potential in the clinical application to ameliorate GVHD in bone marrow transplant patients who must receive allogeneic grafts.

Table 3

ADENOSINE DEAMINASE LEVELS* OF DRUG-TREATED HUMAN PERIPHERAL BLOOD CELLS DURING THE MITOGEN RESPONSES

Source of Cells		Time In Vitro (hrs)	
Normal Blood Donors (2×10^7 PB Cells)		0	84
UNTREATED CONTROLS			
Alf (2/18)	Control	0.072	0.007
	PHA	0.072	0.006
	Con A	0.072	0.006
	PWM	0.072	0.003
New (2/17)	Control	0.088	0.013
	PHA	0.088	0.011
	Con A	0.088	0.012
	PWM	0.088	0.010
Pat (5/10)	Control	0.068	0.032
	PHA	0.068	0.030
	Con A	0.068	0.035
	PWM	0.068	0.036

		Time In Vitro (hrs)		
		0	1	84
DRUG TREATED				
Pet (5/10)	Control	0.068	0.071	
	PHA + dCF & dAdo	0.068	0.000	0.000
	PHA + 8AGuo & 2'dGuo	0.068	0.075	0.032
	Con A + dCF & dAdo	0.068	0.000	0.000
	Con A + 8AGuo & 2'dGuo	0.068	0.075	0.021
	PWM + dCF & dAdo	0.068	0.000	0.000
	PWM + 8AGuo & 2'dGuo	0.068	0.075	0.035
Sch (5/2)	Control	0.050	0.050	0.033
	PHA + dCF & dAdo	0.050	0.000	0.000
	PHA + 8AGuo & 2'dGuo	0.050	0.043	0.024
	Con A + dCF & dAdo	0.050	0.000	0.000
	Con A + 8AGuo & 2'dGuo	0.050	0.043	0.020
	PWM + dCF & dAdo	0.050	0.000	0.000
	PWM + 8AGuo & 2'dGuo	0.050	0.043	0.027

* measured as μMoles of adenosine substrate converted to inosine @ 37 C @ 265 nm per hr per 10^6 cells.

REFERENCES

Bealmear PM, (1975). Amelioration of graft-vs-host disease in germfree allogeneic chimeras by pre-treatment of bone marrow with anti 0 serum. Vth International Symp. on Gnotobiology, Stockholm, Sweden, p 18.

Bealmear PM, (1981). Host defense mechanisms in gnotobiotic animals. In Gershwin ME, Merchant B (eds.): "Immune defects in laboratory animals", Vol. 2, Part IX, Chapter 11, New York: Plenum Press, p 261.

van Bekkum DW, Knaan S (1977). Role of bacterial microflora in development of intestinal lesions from graft-versus-host reaction. J Nat Cancer Inst 58:787.

Fox RM, Kellford RF, Tripp EH, Taylor IW, (1981). G_1-phase arrest of cultured human leukemic T cells induced by deoxyadenosine. Cancer Res 41:5141.

Giblett ER, Ammann AJ, Sandman R, Wara DW, Diamond LK, (1975). Nucleoside phosphorylase deficiency in a child with severely defective T-cell immunity and normal B-cell immunity. Lancet 2:1010.

Jones JM, Wilson R, Bealmear PM. (1971). Mortality and gross pathology of secondary disease in germfree mouse radiation chimeras. Radiat Res 45:577.

Kalcker HM, (1947). Differential spectrophotometry of purine compounds by means of specific enzymes. J Biol Chem 167:461.

Kennedy DW, (1983). "Prevention of lethal graft-vs-host disease following bone marrow transplantation." Ph.D. Dissertation, Buffalo, NY: SUNY @ Buffalo. 152 p.

Kennedy DW, Bealmear PM, (1981). Possible co-dominance of the nude and wildtype genes in SPF C3H/He heterozygous nu/+ mice. In Sasaki S, Ozawa A, Hashimoto K (ed.): "Recent Advances in Germfree Research," Tokyo: Tokai University Press, p 611.

Reinherz EL, Geha R, Rappeport JM, Wilson M, Penta AC, Hussey RE, Fitzgerald KA, Daley JF, Levine H, Rosen FS, Schlossman SF, (1982). Reconstitution after transplantation with T-lymphocyte-depleted HLA haplotype-mismatched bone marrow for severe combined immunodeficiency. Proc Natl Acad Sci USA 79:6047.

Reisner Y, Kapoor N, Kirkpatrick D, Pollack MS, Cunningham-Rundles S, Dupont B, Hodes MZ, Good RA, O'Reilly RJ, (1983). Transplantation for severe combined immunodeficiency with HLA-A, B, D, DR incompatible parental marrow cells fractionated by soybean agglutinin and sheep red blood cells. Blood 61:341.

Reisner Y, Kapoor N, Kirkpatrick D, Pollack MS, Dupont B, Good RA, O'Reilly RJ, (1981). Transplantation for acute leukemia with HLA-A and B nonidentical parental marrow cells fractionated by soybean agglutinin and sheep red blood cells. Lancet 2:327.

Winston DJ, Ho WG, Champlin RE, Gale RP (1984). Infectious complications of bone marrow transplantation. Exp Hematol 12:205.

Germfree Research: Microflora Control and Its Application to the Biomedical Sciences, pages 467–471

GROWTH OF DOG INTESTINAL MUCOSA XENOGRAFTS IN NUDE MICE

Suzanne Harbin, M.S.*
James B. Heneghan, Ph.D.
Department of Surgery, Louisiana State University School of Medicine
New Orleans, Louisiana 70112, U.S.A.

Introduction:

Several types of non-malignant human tissue, including skin (Rygaard, 1974) and bone marrow (Thorling and Pedersen, 1974) have been successfully xenografted onto nude mice. Valerio et al. (1981) were able to maintain longterm (89 days) xenografts of normal adult human colonic mucosa in nude mice; however, the architectural integrity of the xenografted mucosal epithelium was severely compromised. The purpose of these experiments was to develop an in vivo organ culture system for gastrointestinal mucosa by using nude mice as the recipient of the xenograft and to maintain the normal mucosal architecture of the implanted tissue for an extended period of time.

Materials and Methods:

Hypothymic nude mice Hap: (nu/nu)BR were reared in flexible film isolators and fed Purina 5010-C lab chow and sterile tap water ad-libitum. Colon tissue specimens were obtained from conventional or germfree dogs. Adult dog tissue was dissected free from underlying muscle and connective tissue layers, whereas fetal dog tissue was left intact before transplantation, which was performed in a laminar air-flow hood (Germ Free Laboratories, Miami, Florida).

In the first experiment conventional adult dog tissue and fetal dog colon were rinsed several times with normal saline (22°C), cut into discs (1 cm in diameter) and implanted subcutaneously in the interscapular area of the mice under a 1.2 cm disc of Gelman Acropore filter media. In experiment 2, conventional adult dog tissue was rinsed in normal saline then incubated in CMRL-1066 tissue culture media L-glutamine (GIBCO) containing 50 U penicillin, 50ug streptomycin/ ml, GIBCO for 30 minutes at either 4°C or 22°C. The tissue was then implanted as in experiment 1. In the 3rd experiment, conventional adult dog tissue was rinsed with normal saline (22°C) then incubated in CMRL containing 100 U penicillin, 100 ug streptomycin/ml and 1.5% dimethyl sulfoxide (DMSO, Sigma Chemical Co.) at 4°C for either 30 minutes or 60 minutes or at 22°C for 30 minutes; the tissue was again implanted as in experiment 1. In experiment 4, tissue from conventional and germfree dogs was incubated for either 2, 5, 7 or 14 days in CMRL-1066 supplemented with DMSO (1.5%), glucose (15 mM, GIBCO), methionine (7uM, GIBCO), hydrocortisone succinate (1.5 uM, GIBCO), penicillin (100 U/ml), streptomycin (100ug/ml), and gentamicin (50ug/ml, GIBCO). The conventional dog tissue was preincubated for 60 minutes in the CMRL media containing 300 U penicillin. 300 ug streptomycin/ml. The tissue was implanted directly under the skin in the flank area of the mice.

Results:

As with many other tissues, when removed from its natural environment and cultured in a foreign system, the gastrointestinal mucosa underwent a degenerative process. However, in each experiment, the tissue began to recover and restore its original morphology. Three general stages of tissue recovery were observed. In the first stage (Figure 1), the viable tissue resembled ringlets of cuboidal and columnar glandular epithelial cells, the center of which was filled with mucus. This has also been observed by other investigators (Senior et al., 1982) in in vitro organ culture and has been described as remnants of crypt cells which will eventually proliferate and re-establish the continuity of the mucosal surface. In the second stage (Figure 2) the individual crypt-like cell units either expanded or coalesced to form a simple lumen, the interior of which was filled with mucus. In the 3rd stage the tissue strongly resembled mature gastrointestinal mucosal villi (Figures 3 and 4). A summary of the experimental work can be found in Table 1.

In experiment 1, viable implants from conventional dog mucosa were recovered from 4 to 9 weeks after implantation.

The implants recovered at 4 to 7 weeks resembled stage 1 of regeneration while those from weeks 8 and 9 resembled the second stage. Viable implants from fetal tissue were recovered at 4 weeks (first stage), 7 and 17 weeks (both in 3rd stage of recovery).

Table 1

Experiment	No. Implants	Recovery (%)	weeks (implants)	Stage of Recovery
1.				
a) conventional dog tissue	59	6(10)	4(1)	1
			5(2)	1
			6(1)	2
			8(1)	2
			9(1)	3
b) fetal tissue	21	3(14)	2(1)	1
			7(1)	3
			17(1)	3
2. MRL				
a) 30 min, 22°C	23	2(9)	1(1)	1
			4(1)	1
b) 30 min, 4°C	15	1(7)	1(1)	1
3. CMRL (1.5%DMSO)				
a) 30 min, 4°C	6	1(17)	1(1)	1
b) 60 min, 4°C	3	0		
c) 60 min, 22°C	3	1(33)	1(1)	1
4. Days *in vitro*				
2 (germfree dog)	8	5(63)	1(2)	1
			2(3)	2
5 (germfree dog)	8	0		
7 (germfree and conventional dog)	12	0		
14 (germfree and conventional dog)	12	0		

Discussion and Conclusions:

The gastrointestinal mucosa of conventional, germfree and fetal dogs can be cultured in vivo in the nude mouse host. The length of time required to regenerate a "normal" mucosal architecture appeared to be a function of the pre-treatment of the tissue before implantation. Fetal tissue had the best potential for growth; samples of this tissue which received no special pretreatment revealed the re-generation of a mature gastrointestinal mucosal surface after only seven weeks in vivo. Adult conventional dog tissue with no pretreatment required a longer time (9 weeks) before this regeneration was observed. Low temperature appeared to be detrimental to gastrointestinal mucosa; whereas a maximum of 2 days of in vitro culture (36°C) in enriched media significantly improved the viability of the mucosal implants. These experiments to develop an in vivo organ culture system have been very successful. Not only have mucosal xenografts been grown in nude mice for a longer period of time (up to 119 days) than previously reported but also the xenografts have demonstrated a mature "normal" gastrointestinal mucosal morphology. Once these techniques have been adapted to human gastrointestinal mucosa, the effects of carcinogens, hormones, and trophic factors can be studied directly on human tissue in an environment that is physiologically more normal than that of artificial in vitro culture systems. This work was supported by the American Cancer Society Grant PDT-223. *Recipient of Association for Gnotobiotics Travel Grant Award.

Bibliography:

Rygaard J (1974). Skin grafts in nude mice. 3. Fate of grafts from man and donors of other taxonomic classes. Acta Pathol Microbiol Scand !A1 82: 105.

Senior PV, Pritchett CJ, Sunter JP, Appleton DR, Watson AJ (1982). Crypt regeneration in adult human colonic mucosa during prolonged organ culture. J Anat 134(3): 459.

Thorling EB, Pedersen B (1974). Proliferation of human bone marrow in NMRI nu/nu Bom mice. Acta Pathol Microbiol Scand !A1 82: 345.

Valerio MG, Fineman EL, Bowman RL, Harris CC, Stoner GD, Autrup H, Trump BF, McDowell EM, Jones RT (1981). Long-term survival of normal adult human tissues as xenografts in congenitally athymic mice. JNCI 66(5): 849.

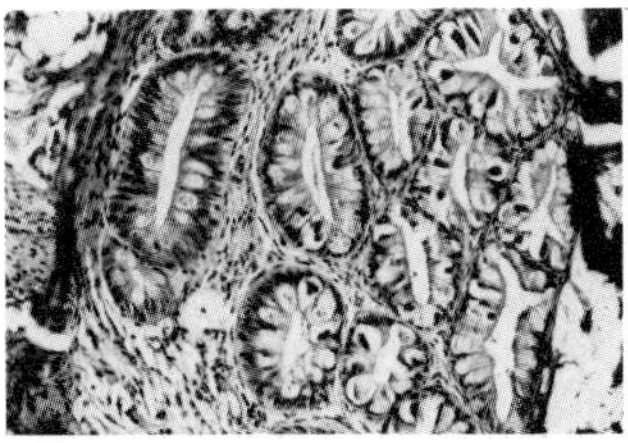

Figure 1. Stage 1 of gastrointestinal mucosa recovery (200X)

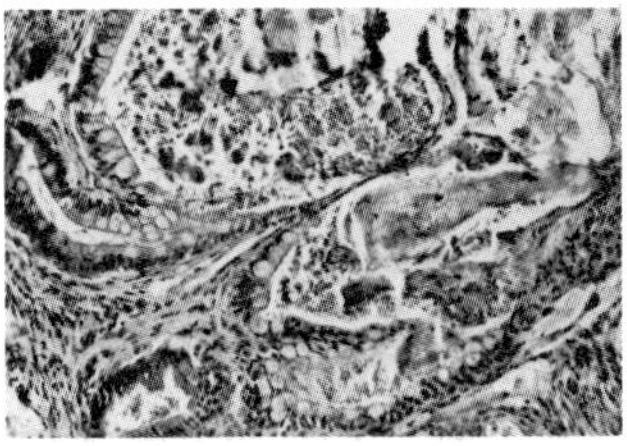

Figure 2. Stage 2 of gastrointestinal mucosa recovery (200X)

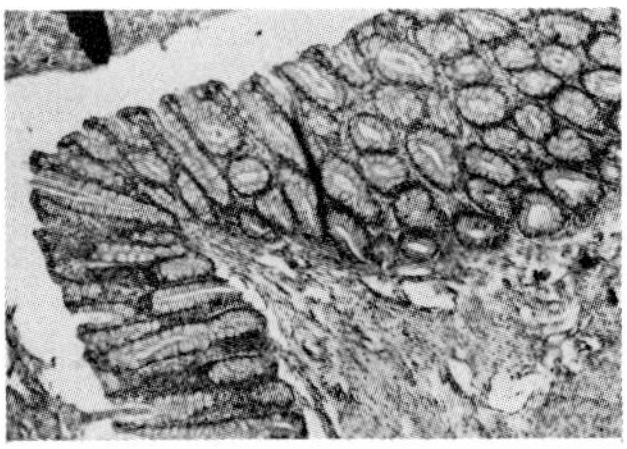

Figure 3. Stage 3 of gastrointestinal mucosa recovery (100X)

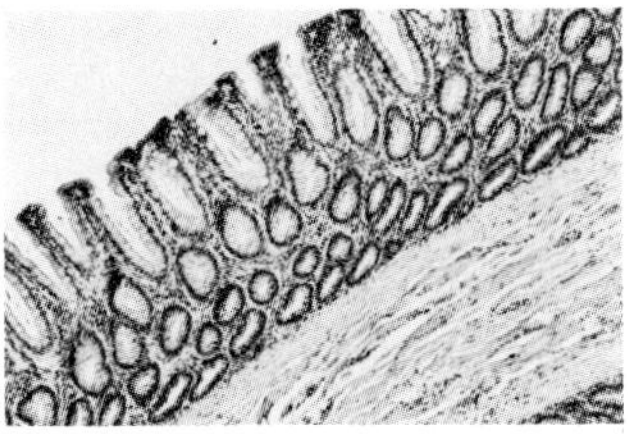

Figure 4. Colonic mucosal villi: normal control (100X)

IX DAVID – THE BOY IN THE BUBBLE

Germfree Research: Microflora Control and Its Application to the Biomedical Sciences, pages 475–489

DAVID'S STORY: THE GIFT OF 12 YEARS, 5 MONTHS, AND 1 DAY

Patricia M. Bealmear, Ph.D., Mary Ann South, M.D.
Raphael Wilson, Ph.D.
Roswell Park Memorial Institute, Buffalo, NY 14263; National Institute for Neurological and Communicative Disorders and Stroke, Bethesda, MD 20205; King's College, Wilkes Barre, PA 18711

This story was presented as part of a 3-hr evening symposium. Dr. Patricia M. Bealmear was chairwoman; and Drs. Mary Ann South and Raphael Wilson made presentations. Dr. John R. Montgomery and Drs. South and Wilson were the three original Principal Investigators on the case when David was born; they were on the staff of Texas Children's Hospital, the Department of Pediatrics, and the Division of Experimental Biology, respectively, at Baylor College of Medicine. A formal discussion followed the presentations with Drs. Edward Balish, Dean Buckner, Manfred Dietrich, Peter Heidt, James Heneghan, Bernard Saint Martin, Cyrille Tancrede, Philip Trexler, and Dirk vander Waaij serving as discussants.

In the course of David's life and death many misrepresentations and misinterpretations of his story crept into the popular press. This session was held to set the record straight, to put things into perspective, and to look at the positive aspects of David's case. David's life was a success story. He, his family, and his friends had 12+ eventful years filled with much happiness that would not have been possible had he not been the 'Bubble Boy.' The chronological story of his life is portrayed in Figs. 2-51.

IN RETROSPECT

David's story did not begin on September 21, 1971 with his birth, but rather in 1967 when the first of many patients was admitted to Texas Children's hospital and ultimately was diagnosed as having severe combined immune deficiency disease (SCID). The treatment available at the time was not successful, and the child died of *Pneumocystis carinii* pneumonia. In the subsequent three years, eight additional children were admitted with SCID and died of severe infections without responding to treatment, or of GVHD following misdirected efforts to reconstitute the immune system. By this time, Dr. Robert A. Good of the University of Minnesota had attempted bone marrow transplants (BMT) on SCID patients, with complete success when the donor and recipient were histocompatible. When the last of these children, named David Joseph (DJ), was admitted, he and his sister were matched at the HL-A complex, and she became the donor for a BMT; however, he was already seriously ill and died of *Pneumocystis carinii* pneumonia before he could be reconstituted.

When David's mother became pregnant again, an amniocentesis was performed to determine the baby's sex since the form of SCID in DJ, if genetic, was an x-linked trait. When the parents were told that they could expect a boy with a 50-50 chance that he might have the disease, they agreed to a "super-sterile" cesarian delivery. Then the child was placed into a plastic isolator within seconds of his birth (Figs. 2-3). Dozens of Baylor specialists were called in to assist in every aspect of David's delivery and subsequent care and development. David's first isolators were modifications of the LOBUND animal isolators and one designed by Dr. Wilson for delivery of a child in a similar situation in Birmingham, AL. At the beginning, it was felt that David needed to live in protective isolation until he either outgrew his immune defect, a matched bone marrow donor was found, or other treatment procedures became available. By the time of his death at age 12, David's living quarters had expanded from a "crib" isolator (Figs. 2, 13, 25) to a 4-chambered unit that consisted of a transporter (Fig. 48), sleep isolator (Fig. 9), supply isolator, and a sealed laminar flow playroom (Figs. 26-29).

Extensive testing was initiated immediately after David's birth (South et al., 1977). No thymic shadow was shown on the x-rays. The range of his absolute lymphocyte count was 399-440/mm^3, and the lymphocytes showed no proliferative response to phytohemagglutinin (PHA). Both the cell mediated and humoral arms of the immune system were severely defective. No HLA match was found among members of the immediate family. Four HLA matches were later found in searches of national and international tissue-typing registries, but David's lymphocytes elicited a positive mixed lymphocyte reaction from each donor.

Phagocytic functions, adenosine deaminase (ADA) levels, and serum complement levels were normal except that Clq was 30% of normal (Table 1). To 47 months of age serum IgM levels were generally low, and IgG was undetectable (Table 2). At 2 months of age the IgM present had an abnormally fast electrophoretic mobility, was of relatively small molecular size (4s), and contained μ heavy chain determinants but no detectable light chains. This paraprotein was still present at 25 months along with three other "immunoglobulin" fractions. One showed abnormally high electrophoretic mobility; the other two were within the normal range. None of these reacted with monospecific antisera for IgM, IgG, or IgA. This μ-chain paraprotein was also found in the serum of all the other SCID patients, both x-linked and autosomal, at the CRC. No serum IgA was detected until 39 months, when the assays indicated IgA at the lowest range of the test's sensitivity. Ultracentrifugal analysis of serum revealed no 19s material at 24 months, but at 36 months both 7s and 19s materials were present. At 44 months these fractions were still present, the abnormal 4s component had disappeared, an IgM component of normal electrophoretic mobility was detected for the first time, and IgD was present.

Thymosin assays of serum obtained at birth, 5.5 months, and 8.5 months of age showed values of 0 when compared to 1/4 in an adult control subject and 1/128 in a 10-yr-old control subject. This lack of serum thymosin was compatible with the lack of normal functioning thymus tissue. At the age of 2.5 years, studies were made of the effect of thymosin on PHA stimulation of David's cells and those of an age-matched control (Table 3). The results suggested that some potentiation of the PHA responsiveness of the mononuclear cells may have occurred with high thymosin concentrations (300-600 μg/ml), however, the changes may have been related to reduced unstimulated activity.

Before injection of keyhole limpet hemocyanin (KLH) at 1 month, antibody titer was 0-0 (the first number represents total antibody titer ($\log_2$), and the second number,

Table 1. *Serum complement levels at 20 months of age*

	Complement fraction, μg/ml			
	C3	C4	C5	C1q
SCID[1] patient	1,467	845	61	48
Normal serum pool	1,847	945	101	160

[1] Severe combined immune deficiency.

Table 2. *Serum immunoglobulins (Ig)*

	Immunoglobulins, mg/100 ml		
Age	IgM	IgG	IgA
Birth	16	560	0
1 wk	22	560	0
3 wk	30	ND[1]	ND
7 wk	10	305	0
3 mo	32	130	0
4 mo	29	135	0
5 mo	13	130	0
6 mo	22	170	0
7 mo	17	105	0
10 mo	11	37	0
11 mo	8	27	0
12 mo	11	28	ND
13 mo	7	0	0
28 mo	6	0	0
39 mo	15	0	1.2
42 mo	17	0	3.7
43 mo	12	0	6.7
46 mo	10	0	<5.8

Not done.

Table 3. *Effect of thymosin on phytohemagglutinin (PHA) stimulation of mononuclear cells from severe combined immune deficiency (SCID) patient and normal donor at 30 months*

Thymosin,[1] μg/well	PHA, μg/well	SCID patient, cpm ± SE[2]	SI[3]	Normal control, cpm ± SE	SI
0	0	92 ± 3	1.0	100 ± 14	1.0
	0.5	97 ± 5	1.1	3,322 ± 131	33.2
	1.0	93 ± 2	1.0	1,788 ± 98	17.9
	2.5	60 ± 6	0.7	277 ± 29	2.8
100	0	84 ± 4	1.0	99 ± 21	1.0
	0.5	69 ± 10	0.8	1,974 ± 291	19.7
	1.0	85 ± 8	1.0	2,238 ± 178	22.4
	2.5	85 ± 3	1.0	75 ± 11	0.8
200	0	123 ± 16	1.0	98 ± 9	1.0
	0.5	114 ± 12	0.9	3,580 ± 341	36.0
	1.0	83 ± 9	0.7	2,284 ± 105	23.0
	2.5	99 ± 3	0.8	102 ± 16	1.1
300	0	93 ± 7	1.0	69 ± 29	1.0
	0.5	98 ± 4	1.1	3,787 ± 213	54.9
	1.0	108 ± 5	1.2	2,071 ± 46	30.0
	2.5	86 ± 8	0.9	89 ± 5	1.3
600	0	144 ± 11	1.0	90 ± 10	1.0
	0.5	109 ± 7	0.8	3,300 ± 347	36.7
	1.0	88 ± 3	0.6	4,610 ± 396	51.2
	2.5	80 ± 3	0.6	235 ± 42	2.6

[1] Reference 65.
[2] Mean counts per min of triplicate cultures ± SE.
[3] Stimulation index (stimulated/unstimulated).

(Published with permission, South et al., 1977.)

the titer after mercaptoethanol digestion). A skin test was negative to KLH at 7 days post-injection. Serum antibody titers tested 7, 10, 21, and 56 days after the first injection were insignificant. David was again injected with KLH several times with negative results until at 11 months of age, on the 5th rechallenge, he had an erythematous reaction, but on subsequent tests the reactions were less pronounced and antibody responses were still insignificant. Transfer factor (TF) was administered at 1 year of age, and the day after the first dose of TF the old KLH skin test site again showed the erythematous reaction; antibody responses remained negative. A skin test at 42 months was negative.

At 1.5 months of age, an injection of typhoid antigen was given intradermally in six places on the arms and legs. Febrile agglutinins tested 7-9 days later were negative. No antibody response was observed when typhoid antigen was given for the second time 17 days after the first injection.

Isohemagglutinin titers, both anti-A and anti-B, were negative when the patient was tested at 3 weeks, 1 month, and 40 months of age. Blood type was ABO group A. Tissue typing showed phenotype A1 B7/Aw31 B14.

David's lymphocytes showed very minimal or no blastogenesis in response to PHA, PWM, or in mixed lymphocyte culture (MLC) in tests using the isolated leukocyte technique (Tables 4 and 5). In tests using the whole blood

Table 4. *Mitogenic response of patient's lymphocytes in tests using isolated leukocyte technique (birth to 28 months)*[1]

Age, months (+ days)	Stimulation index:[2] patient (control) MLC	PHA	PWM
(Birth)		<1.0 (8.0)	1.5 (12.0)
0 (1)		0.6 (8.6)	1.5 (1.5)
1		1.0 (5.0)	1.3 (15.0)
2 (22)	<1.0 (25.9)		
	(5.9)		
3 (16)	<1.0 (10.0)		
4 (5)		1.4 (30.0)	1.3 (9.0)
5 (3)		1.2 (5.0)	0.9 (3.5)
5 (20)	1.5 (15.8)		
	1.2 (18.5)		
6 (30)			
			<1.0 (83.0)
11 (23)		<1.0 (453.0)	1.0 (77.0)
13 (20)		0.7 (483.0)	1.1 (188.0)
15 (20)		0.9 (127.0)	
15 (23)		0.9 (101.0)	
16 (12)	2.6 (31.0)	0.6 (130.0)	0.4 (114.0)
	2.4 (22.7)		
18 (8)		0.8 (250.0)	1.2 (122.0)
18 (23)	1.8 (11.0)	0.7 (24.0)	1.0 (27.0)
27 (26)		0.7 (15.0)	2.0 (15.0)

[1] MLC: mixed lymphocyte culture; PHA: phytohemagglutin; PWM: pokeweed mitogen.
[2] The stimulation index is ratio of counts per min in stimulated lymphocytes to counts per min in unstimulated lymphocytes in radioactive assay. Values for control subject are within parentheses.

Table 5. *Phytohemagglutinin (PHA)-induced blastogenesis of purified T cells from patient*[1]

Age, months (+ days)	Mitogen	Mean cpm ± SE	SI
35 (13)	Media control	54 ± 7	1.0
	PHA 2.5 μg/well	468 ± 33	8.7
	PHA 2.0 μg/well	940 ± 112	17.4
	PHA 1.0 μg/well	421 ± 21	7.8
	PHA 0.5 μg/well	ND	
	PHA 0.25 μg/well	104 ± 22	1.9
36 (13)	Media control	45 ± 4	1.0
	PHA 2.0 μg/well	91 ± 8	2.0
	PHA 1.0 μg/well	112 ± 14	2.4
	PHA 0.5 μg/well	156 ± 11	3.5
37 (26)	Media control	134 ± 5	1.0
	PHA 2.5 μg/well	487 ± 32	3.6
	PHA 1.0 μg/well	418 ± 31	3.1
	PHA 0.5 μg/well	332 ± 64	2.5
	PHA 0.25 μg/well	343 ± 39	2.6
37 (8)	Media control	92 ± 4	1.0
	PHA 2.5 μg/well	60 ± 10	0.7
	PHA 1.0 μg/well	97 ± 8	1.0
	PHA 0.5 μg/well	97 ± 3	1.0

[1] SI: stimulation index (ratio of counts per min in stimulated lymphocytes to counts per min in unstimulated lymphocytes in radioactive assay); PHA: phytohemagglutinin ND: not done; cells per culture = 2.5×10^5.

(Published with permission, South et al., 1977.)

culture technique, transient, low responses to PHA were observed but were not maintained consistently. No increases in lymphocyte responsiveness to nonspecific mitogens were observed after TF was administered at 11 months of age.

Cytotoxicity and lymphokine responses induced by complement (C3) membrane receptors were investigated at 30 months of age. T cells forming rosettes (E-RFC) with fresh sheep red blood cells (SRBC) were first separated from non-E-rosetting cells (non-E-RFC), then the non-E-RFC were activated to form EAC rosettes, a B-cell characteristic. The cytotoxic activities and lymphotoxin production of lymphoid subpopulations were quantitated using [^{51}Cr]-labelled human melanoma (SH-1) and lung fibroblast (WI-38) target cells (Table 6). E-RFC and non-E-RFC of both David and the normal controls gave negligible direct cytolytic responses; activated EAC-RFC from normal donors gave cytotoxic responses independent of antibody or EAC constituents. Activated EAC-

RFC from David failed to express cytotoxicity significantly greater than that obtained with non E-RFC incubated with either individual EAC constituents, autologous sera, or EAC alone.

Soluble mediator production by complement receptor-activated lymphocytes (CRL) was measured at 30 months of age (Table 6). The preliminary findings suggested that EAC-rosetting of CRL from David had induced LT-like activities. Consistent with the LT responses, CRL from David also produced MIF activity greater than the background.

Studies were done at 24-36 months using specific markers to distinguish B-cell types and T-cell types (Fig. 1). Seven studies done at intervals during the first and second years of life, indicated that surface membrane immunoglobulin-bearing cells (SmIg) ranged from 55-100% of the total lymphocytes (mean was 80%) (South et al., 1977). Peripheral blood counts during the first 2 years showed that the average white blood cell count was 5.3 x $10^3/mm^3$ with a mean of 35% lymphocytes (Fernbach, Starling, and Falletta, 1977). Incubation with calf thymosin in vitro failed to increase the number of lymphocytes showing T-cell markers.

Between 2.5 and 4 years of age, striking changes were observed (Fig. 1); at specific times the percentage of B cells decreased and T cells increased. By 3 years of age 20-40% of the lymphocytes bore SmIg and 10.5-39% bore EAC-RFC receptors; E-RFC increased to 19-60%. This represented a dramatic shift toward a normal distribution.

Table 6. *Soluble mediator production by complement receptor-activated lymphocytes (CRL) from normal donors and a severe combined immune deficiency (SCID) patient at 30 months*[1]

Lymphoid subpopulations	Supernatant dilution	Normal donors (*n*)	Normal donors *Mean % ± SE*	SCID patient *(DV)* (*n*)	SCID patient *(DV)* Mean % ± SE
Lymphotoxin					
E-RFC	1:3	(7)	1.3 ± 0.2[2]	(5)	1.1 ± 0.1
	1:6	(10)	1.2 ± 0.1		ND[3]
E-RFC + PHA	1:3	(3)	2.6 ± 0.2		ND
(1 μg/ml)	1:6	(6)	4.4 ± 0.9		ND
EAC-RFC	1:3	(6)	8.6 ± 0.6	(5)	4.0 ± 0.2
	1:6	(12)	7.5 ± 0.3	(3)	3.7 ± 0.1
Leukocyte migration	1:2	(5)	60 ± 7[4]	(5)	40 ± 16
inhibition factor (EAC-RFC)	1:3	(3)	38 ± 3	(3)	10 ± 3

[1] *n*: Number of experiments (for normal donors each experiment represents a separate donor); E-RFC: lymphocytes capable of forming rosettes with sheep erythrocytes; PHA: phytohemagglutinin; EAC-RFC: lymphocytes capable of forming rosettes with the erythrocyte-antibody-complement complex.

[2] Mean percentage of ^{51}Cr release ± SE.

[3] Not done.

[4] Mean percentage of migration inhibition ± SE.

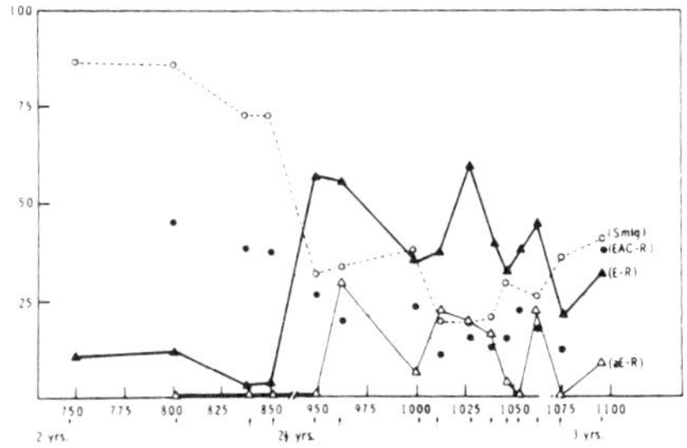

Fig. 1. Changes in percentages of T- and B-cell types in the peripheral blood lymphocytes of severe combined immune deficiency patient between the ages of 24 and 36 months (39). SmIg; surface membrane immunoglobulins; EAC-R: erythrocyte, antibody, complement-rosette-forming cells; E-R: erythrocyte-rosette-forming cells. aE-R: active erythrocyte rosettes. Arrows denote ages when studies were done.

(Published with permission, South et al., 1977.)

Studies of the ultrastructure of the lymphocytes at 3-10 months of age showed that they were atypical (Criswell et al., 1975). There was an increased ratio of cytoplasmic to nuclear area, but the cytoplasm appeared empty with abundant smooth but no rough cytoplasmic reticulum. The nucleus showed no aggregated ribosomes or nuclear pores, and the membrane appeared to pull away from the nuclear space. A great majority of the cells had many villous projections when viewed with scanning electron microscopy. At 15 months of age, 4 months after the administration of the first dose of TF, a second population of smaller lymphocytes was present. The new cell type had more characteristics of normal lymphocytes with more dense cytoplasm, detectable amounts of rough endoplasmic reticulum, and a more intact nuclear membrane. These new cells increased in number from 10% at 15 months to 93% at 4 years of age.

By the time David was 4 years of age, approximately 35 species of microorganisms, mostly transient contaminants, had been isolated (Wilson et al., 1977). Those isolated frequently and in sufficiently high concentration to indicate colonization were speciated as follows: anaerobes--*Propionibacterium acnes*, *Lactobacillus catenaforme* (disappeared spontaneously), *Bacteroides oralis* ss. *elongatus*, *Clostridium* (*perenne*, *hastiforme*, *bifermentans*), *Bacteroides clostridiiformis* ss. *clostridiiformis*; aerobes--*Alcaligenes faecalis* (eradicated by antibiotics), *Staphylococcus epidermidis*, *Enterobacter agglomerans*, *Micrococcus* sp. subgroup 1, *Bacillus pulvifaciens* (disappeared spontaneously); yeasts--*Candida* (*tropicalis*, *parapsilosis*). No viruses or protozoa were isolated to date. Results of tests for viruses at the time of his death have not yet been reported. At 3 years of age, the mean quantitation of anaerobic cells was 7.9 x 10^7 viable cells/g feces (much less than the mean anaerobic load from normal children). The mean aerobic concentration was 1.2 x 10^8 viable cells/g feces (similar to normal children). David had no evidence of infection, was always in excellent health until January 1984, and even tolerated contamination with potential pathogens (e.g., *Pseudomonas aeruginosa*) without any serious consequences.

At 7 years, 8 months of age, David's mitogen responses were still low and continued to be low through 11 years, 9 months of age (Paschall et al., 1984). However, the MLC response was 3.0 (S.I.) at 10 years, 9 months.

At 11 years, 3 months of age, natural killer (NK) cell function was characterized by normal target cell binding by NK cells (9.5% vs 8.1% for the control) but severely depressed NK cell cytotoxicity (1.2% vs 21.3% for the control) (Paschall et al., 1984).

With the passage of time, David did not develop any natural immunity on his own. He did develop physically, mentally, psychologically, and spiritually through the combined efforts of his family, physicians, scientists, and countless other dedicated individuals who worked with him.

In the areas of motor development and locomotor activity David showed early indications of precocious development (Freedman et al., 1976). At less than 4 months he could squirm to and retrieve objects, sit with slight support, and get up on his hands and knees (6.5-month achievements). At 4 months he coordinated hand and eye (5-month achievements). At 6 months he crawled on all fours (10-month norm), recovered objects he dropped, used both hands in coordination, transferred objects from hand to hand, and handled two objects independently (all 8-9-month achievements). At 8 months he sat alone for long periods (10-month norm), stood without support (13-14-month norm), and crawled skillfully (12-15-month norm), could walk alone, and he could also squat or stoop (17-month norm). By 11 months he walked easily and could turn around and sit in a child's chair (18-month norm). A campaign of intensified stimulation, in relation to caretaking activities, social interplay, and language learning was instituted when he was 9-10 months of age because he seemed to disregard his environment, demonstrated a delay in prelanguage development, and at times was involved in rhythmical rocking behavior. Improvements were observed in a few weeks. He spoke his first complete sentence at 22 months. By the age of 24 months, he was a happy, outgoing mischievous little boy; he played actively with others and displayed a capacity for involvement well within the normal range for 2-year olds; he could feed himself, sing his ABC's, count to 50, and identify a trapezoid, parallelogram, hexagon, octagon, etc. At 31 months of age when he went into his laminar flow playroom for the first time, he was asked what he saw when he looked up. He spontaneously identified the filter system as "two rectangles". By the age of three, he had developed an interest in geography and could identify all the states by their shapes and tell where they belonged in a jigsaw puzzle of the United States. By the time that the International Germfree Research Symposium was held in Stockholm in 1975, he had also mastered the map of Europe (not quite 4 years of age).

David began formal schooling with a homebound teacher in the fall of 1977, and he started first-grade schoolwork the next September. By the fall of 1979 he had a telephone hookup linking him to his second grade classroom in the community's public school system. His academic progress was rapid. By the next year he was spending more time at home and less at the hospital. Perhaps because of his confinement and use of the "spacesuit", made for him by NASA, David developed a passion for outer space items (Fig. 47). He learned to recite poetry, and even played a musical instrument. By the age of 10, he was reported well ahead of his peers. During 1982, he acquired a videocassette recorder to see movies such as "Star Wars," and he intensified his use of the telephone. As reported in the Houston Post, he was at least three grade levels above normal. On his 12th birthday, David was reportedly delighted with a new encyclopedia set, and he was learning to play a miniature piano. In his transporter isolator he went to a "real" theater for a private showing of "Return of the Jedi."

On October 21, 1983, David received the bone marrow transplant that his physician, Dr. William Shearer, and his family hoped would cure his immune deficiency and permit him to move into the world outside of his bubble. The decision to proceed with the transplant was made after much consideration by both David and his family. David's sister was the donor for the mismatched bone marrow, which was treated with monoclonal antibodies (anti-T_{12}) to eliminate mature T cells.

On January 2, 1984, the first symptom of David's final illness appeared in the form of elevated body temperature. Later in the month he developed diarrhea, and the fever increased. On February 7, he was removed from his bubble and placed in "reverse isolation" in a special hospital room in order to treat his "flu-like" symptoms. On February 21, he was placed on the critical list after fluid collected around his heart and in his lungs. On February 22, less than 12 hours after he was placed on a breathing device, death came from heart failure.

David died from a massive proliferation of his own B cells, which had invaded the intestine, lungs, and spleen (Simmons, 1984). "Multiple ulcerations" were found throughout his gastrointestinal tract, appearing as "craters" amidst the proliferation. Shearer said that the inciting factor for David's heart failure was the accumulation of fluid in the pericardial sac, arising from inflammation and vasculitis. There were never any overt signs of infection, but his symptoms pointed to a possible viral infection. David did not have GVHD, and it appeared that none of his sister's grafted cells ever established themselves in the host.

According to Dr. Michael Blaese, Chief of the Cellular Immunology Section, Metabolism Branch, National Cancer Institute, David is not the first patient with SCID to develop B-cell proliferation (Simmons, 1984). "There have been now more than a dozen children with SCID...who have been treated by one transplant procedure or another, who have developed the same kind of [abnormal] growth," Blaese told JAMA Medical News. "In each case, this cancer has been associated with the Epstein-Barr virus," a common, herpes-type virus that causes infectious mononucleosis and whose main target is B lymphocytes, he said. "This virus has the ability of transforming a B cell--essentially making it become malignant--in tissue culture," Blaese said. In normal persons, however, the T-cell system (specifically, suppressor T cells) restrains the B cells infected with the virus from dividing abnormally. Whether or not David's cells were infected with the Epstein-Barr or any other virus, Blaese said that "trying to find ways of controlling these B-cell outgrowths is going to be a very important consideration in the future for doing these transplants. Maybe we'll have to try to select donors who have not yet been infected with the virus."

From a scientific point of view also, David's was a very full life. Until his bone marrow transplant, David had been the longest living untreated survivor of SCID. Much information about the basic disorder was learned from him, e.g., the immunologic, hematologic, and nutritional patterns in the SCID patient. Such an opportunity to study the basic pathology and evolution of this immune disorder had previously never been available. In addition, much was learned in this unique experience about the role of the environment on mental, psychomotor, and psychosocial devel-

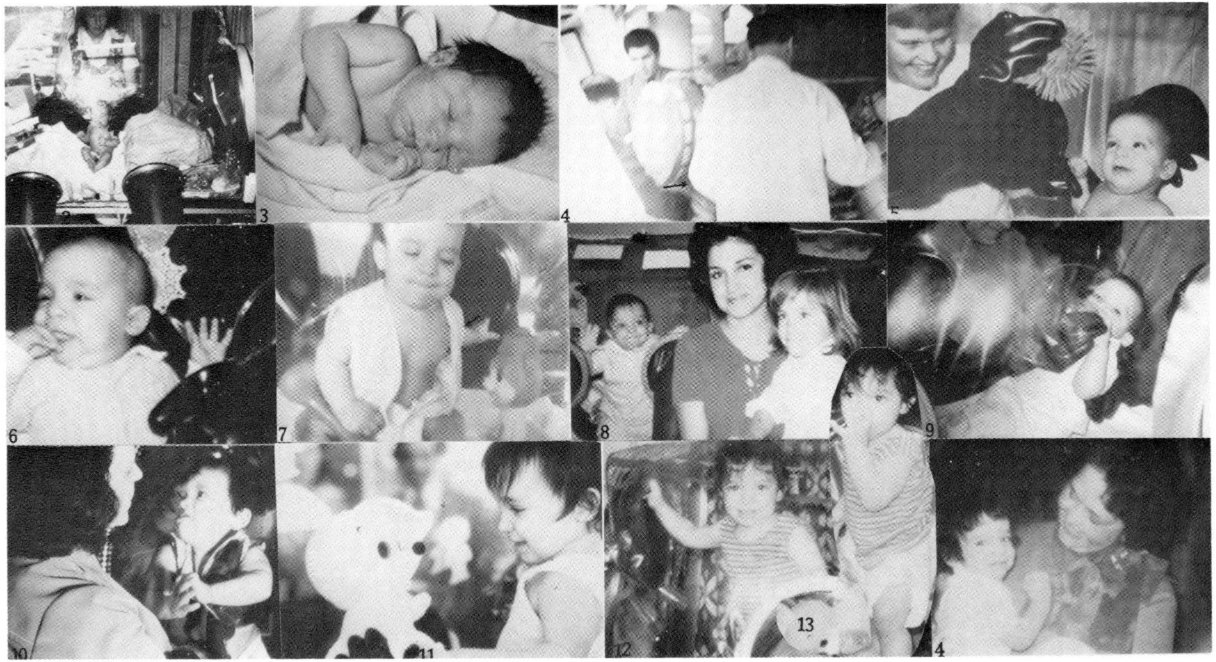

Figs. 2-3. newborn; 4. David's godfather, Dr. Wilson assists in transporting him home for the 1st time @ Thanksgiving (2 mon); 5. testing his eye coordination after his trip home @ Christmas (4 mon); 6. Valentine's Day trip home (5 mon); 7. the day after his skin graft (6 mon); 8-9. Easter @ home with his mother and sister (7 mon); 10. 11 mon; 12. first birthday (BD); 12-13. he outgrew his isolator and his head reached the top @ 13 mon (thumb sucking and "baggy britches"); 14. David & his godmother, Dr. Bealmear (16 mon).

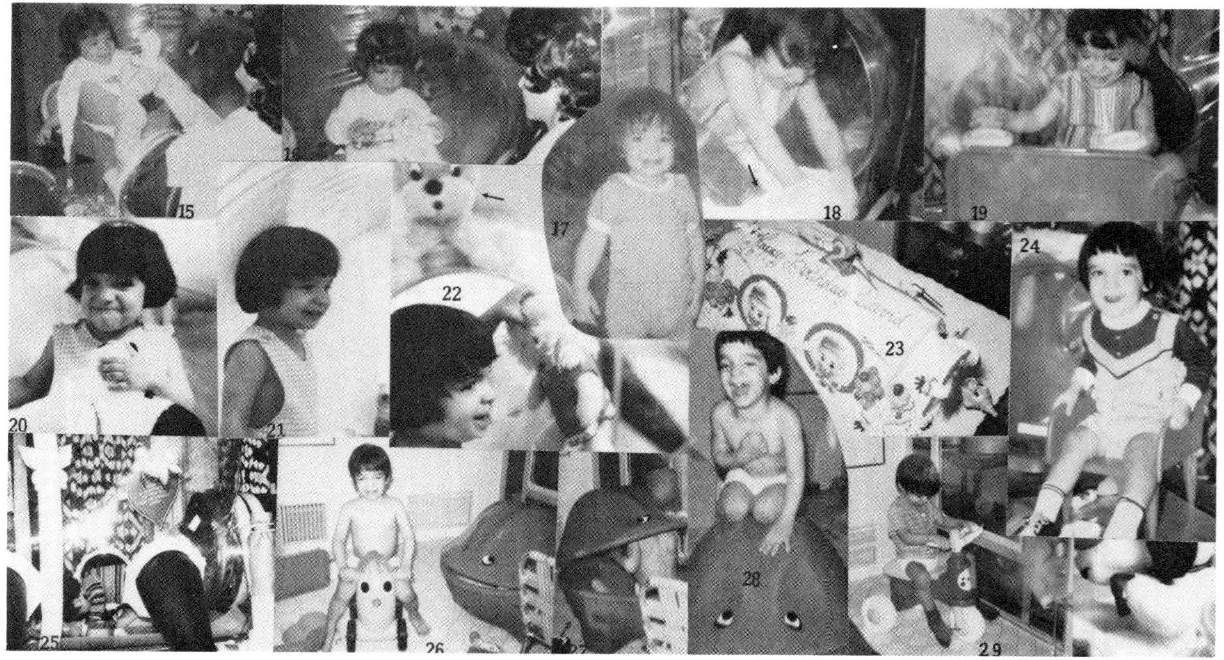

Figs. 15-16. David's mother dresses him with his ND shirt (18 mon); 17. exercise @ home at Easter "fogs up" the isolator requiring the addition of a 2nd set of filters (19 mon); 18-19. the object of his first sentence, "What is this?" (22 mon); 20. his first Snoopy dog (23 mon); 21. playing hide and seek (23 mon); 22. the object of his first words, "Bow Wow" (19 mon); 23. isolator BD cake for 2nd BD; 24. 2nd BD, 1st hard sole shoes; 25. Valentine's day in isolator made by a waterbed manufacturer (29 mon); 26-28. 1st time in laminar air flow playroom on a solid floor, 1st riding toy, and 1st "private room" (31 mon); 29. 3rd BD, 1st tricycle with pedals.

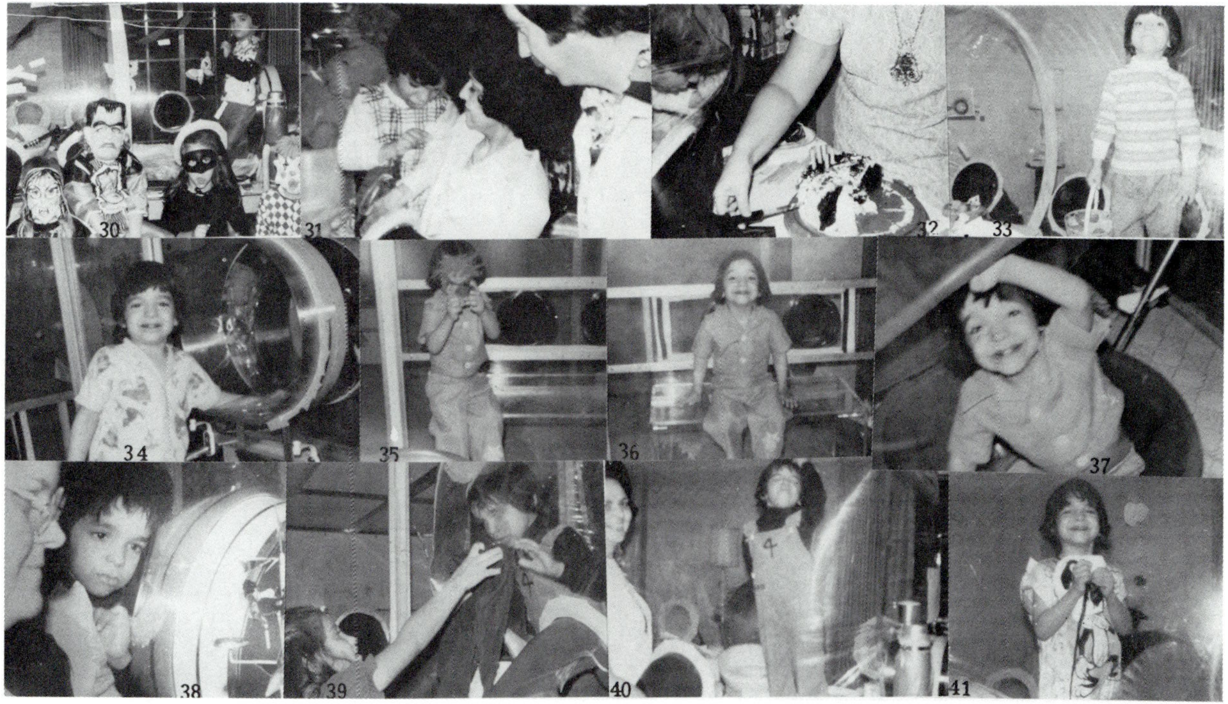

Figs. 30. Halloween @ home with the neighborhood children (37 mon); 31-32. celebrating his mother's BD at the hospital, cutting her cake (41 mon); 33. Easter @ home (43 mon); 34. climbing stairs in playroom (44 mon); 35-37. playing with his Troll from Norway (45 mon); 38. lesson on energy--the candle (4 yrs); 39-40. 4th BD, his sister opens his gifts and he shows off his ND blue and gold outfit; 41. Halloween at home with his friends (49 mon).

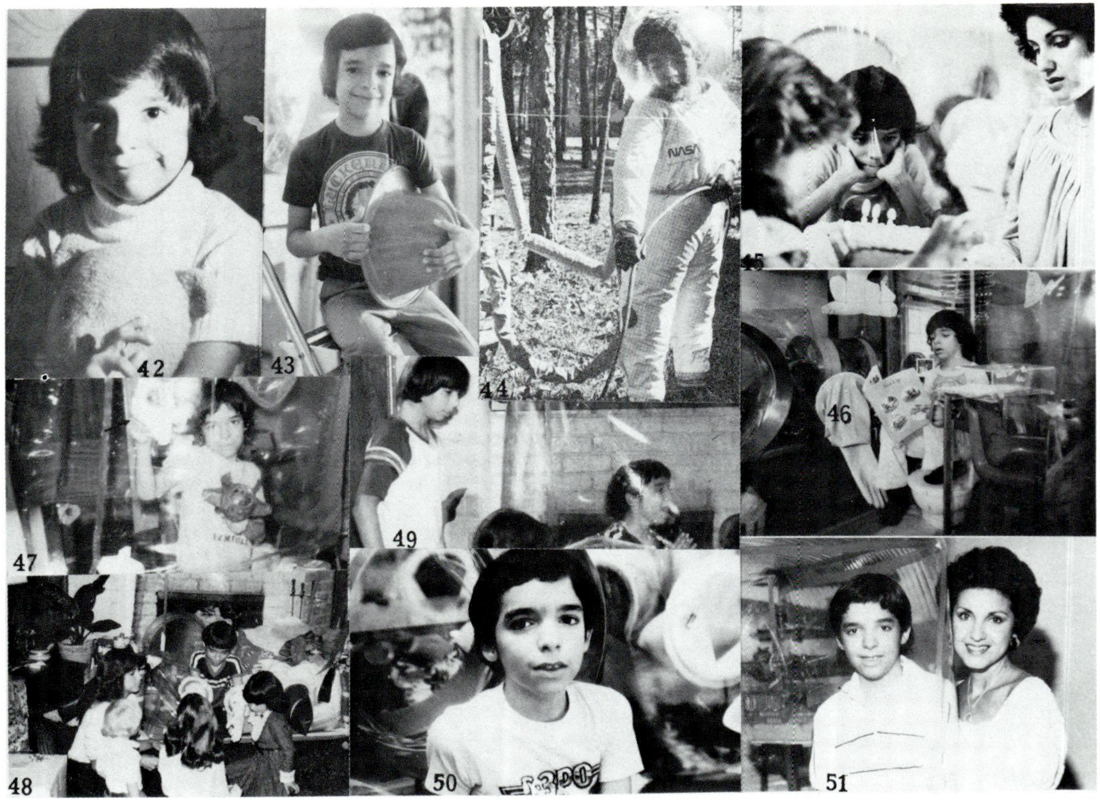

Figs. 42. 55 mon; 43. 6th BD with his fireman's hat (he was an honorary member of the Conroe Fire Dept.); 44. 6 1/2 yrs in space suit with garden hose; 45. 7th BD with mother and sister; 46. 7 yrs; 47. 9th BD with space toys; 48-49. 1Cth BD in transport isolator with relatives and friends in the family room; 50. 10 yrs in star wars shirt; 51. 11th BD.

opment. The isolation systems had worked better than anticipated, proving to be adaptable to his changing needs as he grew from a baby to a young man. He was kept free of infection throughout his entire life. The problems arose from another source, not the isolation system. Finally, last but not least, David's story brought modern immunology to the attention of the general public.

REFERENCES

Criswell BS, South MA, Jordan HW, Kimzey SL, Montgomery JA, Heim L, (1975). Fine structure of lymphocytes from an immune deficient child before and after administration of transfer factor. Exp Hematol 3:327.

Fernbach DJ, Starling KA, Falletta JM, (1977). Hematology. In Williamson A (ed.): "A special report: four-year study of a boy with combined immune deficiency maintained in strict reverse isolation from birth,": Pediatric Res 11, p 78.

Freedman DA, Bealmear PM, Montgomery JR, South MA, Wilson R, (1976). Further observations on the effect of reverse isolation from birth on affective development. J Amer Acad Child Psychiat 15:593.

Paschall VL, Brown LA, Lawrence EC, Karol RA, Lotzova E, Brown BS, Shearer WT, (1984). Immunoregulation in an isolated 12-year-old boy with congenital severe combined immunodeficiency. Pediatric Res 18:723.

Simmons K, (1984). Physicians continue to study cause(s) of 'bubble' boy's death. J Amer Med Assoc 251:1929.

South MA, Montgomery JR, Richie E, Mukhopadhyay N, Criswell BS, Mackler BF, DeFazio SR, Bealmear P, Heim LR, Trentin JJ, Dressman GR, O'Neill P, (1977). Immunologic studies. In Williamson A (ed.): "A special report: four-year study of a boy with combined immune deficiency maintained in strict reverse isolation from birth,": Pediatric Res 11, p 71.

Wilson R, Taylor GR, Kropp K, Montgomery JR, South MA, Trentin JJ, Hilton EN, Mastromarino A. Gnotobiotic care and infectious disease prevention. In Williamson A (ed.): "A special report: four-year study of a boy with combined immune deficiency maintained in strict reverse isolation from birth,": Pediatric Res 11, p 67.

Index